Functionally Graded Structures

Modelling and computation of static and dynamical problems

Online at: https://doi.org/10.1088/978-0-7503-5301-4

Functionally Graded Structures

Modelling and computation of static and dynamical problems

Edited by
Snehashish Chakraverty
National Institute of Technology Rourkela, Rourkela, Odisha, India

Subrat Kumar Jena
Indian Institute of Technology Delhi, New Delhi, India

Ömer Civalek
Civil Engineering Department, Engineering Faculty, Akdeniz University, Antalya, Turkey

IOP Publishing, Bristol, UK

ISBN 978-0-7503-5301-4 (ebook)
ISBN 978-0-7503-5299-4 (print)
ISBN 978-0-7503-5302-1 (myPrint)
ISBN 978-0-7503-5300-7 (mobi)

DOI 10.1088/978-0-7503-5301-4

Version: 20230801

IOP ebooks

British Library Cataloguing-in-Publication Data: A catalogue record for this book is available from the British Library.

Published by IOP Publishing, wholly owned by The Institute of Physics, London

IOP Publishing, No.2 The Distillery, Glassfields, Avon Street, Bristol, BS2 0GR, UK

US Office: IOP Publishing, Inc., 190 North Independence Mall West, Suite 601, Philadelphia, PA 19106, USA

Dedicated to the pursuit of knowledge and the advancement of human understanding.

Contents

Preface

Functionally graded materials (FGMs) were initially developed for use in a thermal barrier material capable of withstanding large temperature variations over a very thin cross section for a space plane project. Since then, FGMs have gained popularity as advanced structural materials for heat-shield applications in various industrial sectors, including aerospace, nuclear power, automobiles, aviation, space vehicles, biomedicine, and steel. FGMs are inhomogeneous materials made of ceramic–metal composites with continuously varying compositions or volumes of elements in one or more specific directions. The ceramic component can withstand high temperatures due to its low thermal conductivity, making it suitable for use in high-temperature environments such as chemical plants, nuclear reactors, and the production of high-speed vehicles. Meanwhile, the ductile metal component inhibits fractures caused by strains induced by large temperature gradients; the use of such components may increase mechanical strength while reducing the weight of the structure, resulting in substantial cost savings.

With a focus on providing a timely contribution to the challenging field of FGMs, this book offers a comprehensive examination of their static and dynamic behaviors along with wave propagation in functionally graded structures, including various complicating effects. In this regard, different analytical, semi-analytical, and numerical methods are also presented. Serving as a valuable resource for researchers, undergraduate and graduate students, and industry professionals, this book provides a one-stop resource for exploring the static and dynamic behaviors of functionally graded structures. In addition, it serves as a useful tool for teaching classical and advanced functionally graded structures to students and researchers from diverse fields.

This work comprises a total of 13 chapters, which offer comprehensive coverage of research related to the analysis of the static, dynamic, and wave propagation characteristics of functionally graded structures. The initial four chapters, viz. chapters 1–4, are devoted to static behavior; they specifically focus on beams, sandwich beams, and plates composed of functionally graded materials. On the other hand, chapters 5–9 delve into the topic of dynamic analysis, including the buckling and vibration of functionally graded beams, functionally graded nanobeams, functionally graded sandwich beams, and functionally graded shells. The subsequent two chapters, i.e. chapters 10 and 11, investigate the effect of functionally graded magneto-electro-elastic (MEE) facings on the damped nonlinear transient response of sandwich plates and concentrate on the study of the free vibration response of thin functionally graded plates, respectively. Finally, chapters 12 and 13, respectively, examine Lamb-type waves in functionally graded orthotropic piezoelectric plates and wave propagation in generalized thermoelastic functionally graded plates.

Chapter 1 investigates the nonlinear static characteristics of beam-type structures made of functionally graded composites at the micro- or nanoscale using unconventional continuum theories of elasticity, including the nonlocal strain gradient and nonlocal couple stress theories. The study focuses on identifying unconventional load–deflection curves associated with the nonlinear stability of beams under axial compressive loads and the nonlinear flexural behavior of beams subjected to uniformly distributed loads. An isogeometric approach using proper nonuniform rational B-spline basis functions is employed to ensure that accurate geometrical descriptions are used when solving the developed size-dependent beam models. The focus of **Chapter 2** is the bending analysis of functionally graded sandwich beams subjected to uniformly distributed loads on elastic foundations. Two types of sandwich beam are studied: Type 1, which has homogeneous cores and functionally graded skins, and Type 2, which has functionally graded cores and homogeneous skins. The material properties are assumed to change gradually through the thickness, and the governing differential equations are derived based on the Bernoulli–Euler beam theory and the principle of minimum total potential energy. Navier's solution method is employed to obtain the solution of simply supported sandwich beams, while the Winkler foundation model is used to consider the interaction between the functionally graded sandwich beam and the elastic foundation.

Chapter 3 examines the static deformations of different FGM structures subjected to symmetric and asymmetric static loads, considering various boundary conditions and environmental factors. The study employs the first-order shear deformation theory (FSDT) to derive the governing equations and solves them using the semi-analytical polynomial method (SAPM). The study analyzes the impact of various factors, including loading conditions, the type of FGM, boundary conditions, and environmental factors, on large deformation strain fields. The results are compared to those from various references to validate the approach. The chapter provides useful insights for designing FGM structures under different loadings and environmental conditions. **Chapter 4** presents a static bending analysis of porous functionally graded sandwich beams made of two porous face sheets constructed of FGMs and a homogeneous isotropic ceramic core. The material properties of the FG sandwich beam vary continuously in the thickness direction based on the volume fraction of constituents as defined by the modified rule of mixtures, which includes the porosity volume fraction and distinct types of porosity distribution pattern across the cross section. The governing differential equations are derived using Hamilton's principle, and a Navier-type analytical solution is presented for the static bending problem of porous/nonporous FG sandwich beams that have simply supported ends. This formulation is validated through detailed comparisons with available published data, and parametric studies are performed to examine the influences of various factors on the axial and vertical displacements as well as the axial and shear stresses of FG sandwich beams.

Chapter 5 discusses an investigation of the dynamic behavior of doubly-curved shell structures made of FGMs using the equivalent single layer (ESL) approach with higher-order theories. The shell geometry is described using differential geometry principles, and a mapping is adopted for arbitrarily shaped domains. A smooth variation of the constituent materials is characterized using a generalized five-parameter expression, and the elastic constants are determined using a homogenization procedure. The governing equations for doubly-curved laminated structures with FGM layers are derived from Hamilton's principle, and the generalized differential quadrature (GDQ) method is used for numerical implementation. The computational efficiency of the model is demonstrated by a preliminary validation analysis, and the influence of the FGM and porosity distribution parameters on the structural response is investigated through a parametric study. The impact of porosity on the buckling behavior of functionally graded beams is investigated in **Chapter 6**, in which the carrier's mechanical properties are assumed to vary in the thickness direction. The system of differential equations is obtained using Hamilton's principle, and a state-space approach is employed to solve the problem. The accuracy of the solution is validated by comparing it with other solutions available in the literature. A thorough parametric study is conducted to examine the effect of porosity on the buckling behavior of FG beams.

Chapter 7 focuses on the vibration and buckling problems of FGM beams and plates and explores various mathematical models and solutions, both analytical and numerical. The effects of material property variation, structure size, boundary conditions, and loads on FGM structures are extensively discussed. FGM structures have great potential in engineering applications and are attracting global attention. **Chapter 8** explores the free vibration analysis of porous functionally graded sandwich beams using two porous FGM face sheets and a homogeneous isotropic ceramic core. The modified rule of mixtures is used to characterize the material properties of the face sheets, taking into account the porosity volume fraction and different porosity distribution patterns. A higher-order shear deformation theory is applied without the need for shear correction factors. Hamilton's principle is used to derive the equations of motion, and Navier's method is used to obtain analytical solutions for free vibration under simply supported boundary conditions. The formulation is validated through comparisons with published results, and numerical examples are presented to investigate the effects of the skin–core–skin thickness ratio, power-law index, span-to-depth ratio, porosity distribution pattern, and porosity volume fraction on the natural frequencies of FG sandwich beams.

In **Chapter 9**, the vibrational behavior of functionally graded nanobeams under magnetic fields is investigated using Eringen's nonlocal elasticity theory. The torque effects of the axial magnetic field on the FG nanobeam are determined using Maxwell's equations. The material properties are assumed to vary according to the power law in the radial direction. The Adomian decomposition method is used to solve the governing differential equations. Three classical boundary conditions, namely simply supported–simply supported (S–S), clamped–simply supported (C–S), and clamped–clamped (C–C), are considered. The effects of the magnetic field and elastic foundation parameters on the vibration frequencies are analyzed for different boundary conditions. **Chapter 10** focuses on the free vibrational behavior of thin FGM plates incorporating the rotary inertia effect using the dynamic stiffness method (DSM). The power law is used to model the property variations of the FGM plate. The equation of motion is formulated by employing Kirchhoff–Love plate theory and the physical neutral surface concept, and a Levy-type closed-form solution is used to solve the governing differential equation of motion. The dynamic stiffness matrix is developed through boundary conditions, and the Wittrick–Williams algorithm is used as a solution technique to obtain accurate natural frequencies and mode shapes. The computed DSM frequencies are compared with existing published results, and new results are presented for the design and development of FGM plates.

Chapter 11 employs the finite element (FE) method to study the controlled nonlinear transient response of sandwich plates consisting of functionally graded MEE face sheets and an agglomerated carbon nanotube (ACNT) core with active constrained layer damping (ACLD) treatment. The ACLD patch includes a piezoelectric (PE) layer for control feedback and a viscoelastic (VE) layer for passive damping. The equations of motion are derived using layer-wise shear deformation theory (LSDT) and Hamilton's principle, incorporating the synergy of three-field coupling, agglomeration, and electromagnetic (EM) circuit effects. Numerical examples are provided to analyze the impact of carbon nanotube (CNT) distributions, the MEE composite's functional gradation, EM circuits, and coupling. This research serves as a valuable benchmark for further investigation of the active control of agglomerated CNT structures and the design of smart structures and systems. **Chapter 12** investigates the propagation behavior of Lamb waves in functionally graded orthotropic piezoelectric material plates with spatial variation in the material constants. The power series technique is used to solve the dynamic equations, and the dispersion equation is derived using appropriate boundary conditions. The effects of functional grading and material parameters are illustrated through numerical analysis and graphical representation. The results of this study can aid in the design and improvement of transducers using functionally graded materials and in nondestructive ultrasonic evaluation. Finally, in **Chapter 13**, an analytical investigation is presented of the dynamic analysis of Lamb wave propagation in a functionally graded plate using the dual-phase-lag (DPL) thermoelastic model. The analytical solution is obtained by applying the Legendre orthogonal polynomial series expansion and solving the secular equation. The effects of thermal modes are demonstrated by comparing the dispersion curves of an inhomogeneous thermoelastic plate with those of an equivalent pure elastic plate.

We trust that the consolidated, challenging, and comprehensive approaches used in all the chapters will benefit readers in their future studies and research. The Editors greatly appreciate all the authors for their commendable efforts and support in contributing and submitting the chapters on time. Finally, we would also like to thank the IOP team for their assistance and support throughout this project.

The Editors
S Chakraverty, Subrat Kumar Jena, and Ömer Civalek

Acknowledgements

The first editor, Professor S Chakraverty would like to thank his beloved parents, the late Sh. Birendra K Chakraborty and the late Parul Chakraborty, for their blessings. He would also like to thank his wife, Shewli, and daughters, Shreyati and Susprihaa, for their support and for being a source of inspiration during this project. The support of the NIT Rourkela administration is also gratefully acknowledged.

Dr Subrat Kumar Jena, the second editor, is immensely grateful to his family members, specifically Sh. Ullash Chandra Jena, Sh. Durga Prasad Jena, Sh. Laxmidhara Jena, Smt. Urbasi Jena, Smt. Renu Bala Jena, Smt. Arati Jena and his sisters Jyotrimayee, Truptimayee, and Nirupama, for their unwavering love, constant motivation, unrelenting support, and blessings. Dr Subrat Kumar Jena will always be indebted to his younger brother, Dr Rajarama Mohan Jena, and sister-in-law, Dr Sujata Swain, for their encouragement, love, and support. Moreover, Dr Subrat Kumar Jena would also like to appreciate his girlfriend, Miss Bhubaneswari Mishra from the Department of Mathematics, National Institute of Technology Rourkela, for her unconditional love, moral support, faith, and encouragement, which reinvigorated his appetite for research and motivated him to achieve beyond his expectations. Furthermore, the second editor would like to express his deep appreciation to his PhD supervisor, Professor S Chakraverty, for his invaluable guidance and support throughout the early stages of his research. Finally, he would like to thank the administration of the Indian Institute of Technology, Delhi, and his postdoctoral mentor, Professor S Pradyumn, for their assistance and support.

We, the three editors, extend our sincere gratitude and acknowledgment to the reviewers for their invaluable feedback and appreciation provided during the development of the book proposal. We express our utmost appreciation to all the chapter contributors for their timely submissions and excellent efforts. We are also immensely grateful to the entire team at IOP for their unwavering support, cooperation, and assistance, which have enabled us to publish this book according to schedule. Finally, we are deeply indebted to the authors and researchers cited in the bibliography/reference sections at the end of each chapter.

Editor biographies

Professor S Chakraverty

Professor S Chakraverty has 30 years of experience as a researcher and teacher. He currently works in the Department of Mathematics (Applied Mathematics Group), National Institute of Technology, Rourkela, Odisha, as a senior (higher administrative grade) professor. Prior to this, he was with CSIR-Central Building Research Institute, Roorkee, India. After graduation from St. Columba's College (Ranchi University), his career started at the University of Roorkee (now the Indian Institute of Technology, Roorkee); he completed MSc (Mathematics) and MPhil. (Computer Applications) at the said institution and secured his first positions at the University. Dr Chakraverty received his PhD from IIT-Roorkee in 1993. Thereafter, he did his postdoctoral research at the Institute of Sound and Vibration Research (ISVR), the University of Southampton, UK, and at the Faculty of Engineering and Computer Science, Concordia University, Canada. He was also a visiting professor at Concordia and McGill universities, Canada, during 1997–1999 and visiting professor at the University of Johannesburg, Johannesburg, South Africa, during 2011–2014. He has authored/coauthored/edited 31 books and published 430 research papers (to date) in journals and conferences, and two books are in production. He is on the editorial boards of various international journals, book series, and conferences. Professor Chakraverty is the chief editor of the *International Journal of Fuzzy Computation and Modelling* (IJFCM), Inderscience Publishers, Switzerland (http://www.inderscience.com/ijfcm), an associate editor of *Computational Methods in Structural Engineering*, *Frontiers in Built Environment*, and *Curved and Layered Structures* (De Gruyter) and is also an editorial board member of *Springer Nature Applied Sciences*, *IGI Research Insights Books*, *the Springer Book Series of Modeling and Optimization in Science and Technologies*, *Coupled Systems Mechanics* (Techno Press), *the Journal of Composites Science* (MDPI), *Engineering Research Express* (IOP), and *Applications and Applied Mathematics: An International Journal.* He is also a reviewer for around 50 national and international journals of repute. He was the president of the mathematical sciences section (including Statistics) of the Indian Science Congress (2015–2016) and vice president of the Orissa Mathematical Society (2011–2013). Professor Chakraverty is a recipient of prestigious awards, viz. an Indian National Science Academy (INSA) nomination under the International Collaboration/Bilateral Exchange Program (with the Czech Republic), the Platinum Jubilee ISCA Lecture Award (2014), the CSIR Young Scientist Award (1997), BOYSCAST Fellowship (DST), the UCOST Young Scientist Award (2007, 2008), the Golden Jubilee Director's (CBRI) Award (2001), the INSA International Bilateral Exchange Award ([2010–2011] selected but could not undertake [2015] selected), and Roorkee University Gold Medals (1987, 1988) for first positions in MSc and MPhil (Computer Application). He has recently been listed in the 2% of world scientists (2020 and 2021) in the Artificial Intelligence and Image Processing category based on an independent study by Stanford University scientists. Professor Chakraverty received the IOP Publishing Top-Cited Paper Award for one of the most-cited articles over three years (2018–2020) from India published across the entire IOP Publishing journal portfolio. The same article also featured in the top 1% of most-cited papers in the materials subject category. This data is from the citations recorded in Web of Science. He has already guided 22 PhD students and 12 are ongoing. Professor Chakraverty has undertaken around 16 research projects as a principal investigator funded by international and national agencies totalling about Rs. 1.5 crores. He has hosted around eight international students with different international/national fellowships to work in his group as PDF and PhD visiting researchers for different periods. He has also organized a good number of international and national conferences, workshops, and training programs. His current research areas include differential equations (ordinary, partial, and fractional), numerical analysis and computational methods, structural dynamics (FGM, nano) and fluid dynamics, mathematical and uncertainty modelling, soft computing and machine intelligence (artificial neural network, fuzzy, interval, and affine computations).

Dr Subrat Kumar Jena

Dr Subrat Kumar Jena currently holds the position of postdoctoral fellow at the Department of Applied Mechanics in the Indian Institute of Technology Delhi, New Delhi, India. He has recently concluded an honorary postdoctoral fellowship in 'Computational Solid Mechanics via Numerical Solution' at the Nonlinear Multifunctional Composites—Analysis and Design (NMCAD) Laboratory, Department of Aerospace Engineering, Indian Institute of Science (IISc), Bengaluru, India. Dr Jena completed his PhD at the Department of Mathematics, National Institute of Technology Rourkela, Odisha, India. His research expertise encompasses structural dynamics, computational solid mechanics, applied mathematics, mathematical modelling, and uncertainty modelling. Dr Jena earned his MSc in Mathematics and Scientific Computing from the Motilal Nehru National Institute of Technology Allahabad (NIT Allahabad), Prayagraj, India. His research contributions are widely recognized; he has 32 research papers in peer-reviewed international journals, two international conference papers, nine book chapters, and three books (one book is in press, two other books are in production) to his credit. Recently, Dr. Jena has been honoured with the "Mathematics 2022 Best PhD Thesis Award" by the Mathematics Journal, Basel, Switzerland, for his outstanding PhD thesis. In addition, Dr Jena received IOP Publishing Top-Cited Paper Awards in 2021 and 2022 for India, published across the entire IOP Publishing journal portfolio in the preceding three years i.e. 2018–2020 and 2019–2021, respectively, for his paper that ranked among the top 1% of the most-cited papers in the materials subject category. In 2020–1, one of his papers published in ZAMM, the *Journal of Applied*

Mathematics and Mechanics (Wiley) was among the most cited papers. Furthermore, three of his papers have been recognized as the most-cited papers published in *Curved and Layered Structures* (De Gruyter). His contributions to the field of shell buckling, particularly in the area of static or dynamic analysis of general structures, have been acknowledged, as he is included in the Shell Buckling website under 'Shell Buckling People.' He was included in Stanford University's Top 2% Most Influential Scientists list in 2021 based on an independent study by Stanford University scientists. Dr Jena's eminence as a reviewer and guest editor for many prestigious international journals is underscored by his review of over 70 manuscripts.

Professor Ömer Civalek

Professor Ömer Civalek has been serving as a professor at the Faculty of Engineering in the University of Akdeniz since 2004. He has received two PhD degrees in Structural Engineering and Applied Mechanics from Dokuz Eylül University and the University of Firat, respectively. He has made substantial contributions to the field of engineering in the form of over 350 refereed journal papers, more than 50 conference papers, a few book chapters, and 70 papers published in various national journals. His research interests are primarily focused on continuum mechanics, solid mechanics, nanoscale mechanics, the vibration and buckling of plates and shells, computational mechanics, laminated structures, structural dynamics, the modeling of nanostructures, and biomechanics. Professor Civalek has been appointed as a visiting professor for a term of three years (2020–2023) at the China Medical University in Taichung, Taiwan.

List of contributors

Bekir Akgöz
Civil Engineering Department, Engineering Faculty, Akdeniz University, Antalya, Turkey

Mehmet Avcar
Department of Civil Engineering, Faculty of Engineering, Suleyman Demirel University, Cunur, Isparta, Turkey

Rabia Benferhat
Laboratory of Geomatics and Sustainable Development, University of Tiaret, Tiaret, Algeria

S Chakraverty
National Institute of Technology Rourkela, Rourkela, Odisha, India

Prashant Choudhary
Department of Mechanical Engineering, GMR Institute of technology, Razam, Andhra Pradesh, India

Ömer Civalek
Civil Engineering Department, Engineering Faculty, Akdeniz University, Antalya, Turkey

Shahriar Dastjerdi
Civil Engineering Department, Engineering Faculty, Akdeniz University, Antalya, Turkey

Sudarshan Dhua
NIT Andhra Pradesh, India

Rossana Dimitri
Department of Innovation Engineering, University of Salento, Lecce, Italy

Kada Draiche
Department of Civil Engineering, University of Tiaret, Tiaret, Algeria

Ankit Gupta
Department of Mechanical Engineering, School of Engineering, Shiv Nadar University, Noida, Uttar Pradesh, India

Lazreg Hajdi
Department of Civil Engineering, University of Tiaret, Tiaret, Algeria

Dineshkumar Harursampath
Department of Aerospace Engineering, Indian Institute of Science, Bangalore, India

Somnath Karmakar
National Institute of Technology Rourkela, Rourkela, Odisha, India

Subodh Kumar
Department of Mechanical Engineering, SR University, Ananthasagar, Hasanparthy, Warangal, Telegana, India

Vinyas Mahesh
NIT Silchar, Assam, India

Vishwas Mahesh
Indian Institute of Science (IISc), Sit Tumkur, Bangalore, India
Arpita Maji
NIT Andhra Pradesh, India

Arjun Siddharth Mangalasseri
RWTH, Aachen, Aachen, Germany

Arindam Nath
NIT Andhra Pradesh, India

Sathiskumar Anusuya Ponnusami
City University, London, UK

Sahmani Saeid
School of Science and Technology, The University of Georgia, Tbilisi, Georgia

Babak Safaei
Department of Mechanical Engineering, Eastern Mediterranean University, Famagusta, North Cyprus, Turkey

V R Shanmukhasundaram
Department of Mechanical Engineering, Madanapalle Institute of Technology and Science, Andhra Pradesh, India

Jay Prakash Srivastava
Department of Mechanical Engineering, Gandhi Institute of Technology and Management, Hyderabad, Telangana, India

Hassaine Daouadji Tahar
Laboratory of Geomatics and Sustainable Development, University of Tiaret, Tiaret, Algeria

Youcef Tlidji
Materials and Structures Laboratory, University of Tiaret, Tiaret, Algeria

Francesco Tornabene
Department of Innovation Engineering, University of Salento, Lecce, Italy

Abdelouahed Tounsi
Material and Hydrology Laboratory, Civil Engineering Department, Faculty of Technology, University of Sidi Bel Abbes, Sidi Bel Abbès, Algeria

Matteo Viscoti
Department of Innovation Engineering, University of Salento, Lecce, Italy

Chenxi Wei
State Key Laboratory of Nonlinear Mechanics (LNM), Institute of Mechanics, Chinese Academy of Sciences, Beijing, China

Yin Zhang
State Key Laboratory of Nonlinear Mechanics (LNM), Institute of Mechanics, Chinese Academy of Sciences, Beijing, China

Chapter 1

The static behavior of functionally graded beams

Sahmani Saeid and Safaei Babak

In the chapter, we investigate the nonlinear static characteristics of beam-type structures made of functionally graded composites at the microscale and the nanoscale. For this purpose, different unconventional continuum theories of elasticity, including the nonlocal strain gradient theory and the nonlocal couple stress theory, are applied to the classical beam formulations to analyze the size-dependent nonlinear static behaviors of beams at these small scales. Accordingly, unconventional load–deflection curves associated with the nonlinear stability of beams under axial compressive loads and the nonlinear flexural behavior of beams subjected to uniform distributed loads are plotted. For this purpose, the isogeometric kind of finite element approach is employed to solve the developed size-dependent beam models with the aid of appropriate nonuniform rational B-spline basis functions to achieve accurate geometrical descriptions.

1.1 Introduction

Beam-type structures have been employed in many fields of technology and various engineering applications at different scales. To mention some instances, Caddemi and Calio [1] exactly extracted the dynamic stiffness matrix of multiply cracked Euler–Bernoulli beams for frame damage applications. Shaat and Abdelkefi [2] simulated the material properties as well as the effect of couple stress on the pull-in response of nanocrystalline silicon beams employed for biomass sensing applications. Nadal *et al* [3] modeled the excitation characteristics of a beam-type piezoelectric actuator used for tactile applications. Sahmani *et al* [4–7] manufactured and analyzed the mechanical behaviors of beam-type bony implants used in various biological applications. Chadha and Todd [8] introduced a comprehensive kinematic model associated with single-manifold Cosserat beams to study the coupling effect of Poisson's ratio and warping effects for use in strain gauge applications. Sahmani *et al* [9–13] fabricated beam-type implants compatible with drug delivery and predicted their mechanical and biological performances in biomedical applications.

doi:10.1088/978-0-7503-5301-4ch1

Nuhu and Safaei [14] presented a comprehensive review of the oscillation responses of beam-type structures fabricated at small scales.

In the last decade, several research studies have employed non-classical continuum theories of elasticity to analyze the size-dependent mechanical features of beam-type structures and other types of structure at the microscale or the nanoscale. Wang [15] proposed a size-dependent beam model based on the modified couple stress theory of elasticity to analyze the oscillations of fluid-carrying microtubes. Akgoz and Civalek [16] constructed strain-gradient-based beam formulations to extract the critical buckling loads of axially compressed microbeams. Ansari *et al* [17, 18] employed the Rayleigh–Ritz technique to analyze the stability of axially compressed single-walled carbon nanotubes under different boundary conditions. Ansari and Sahmani [19, 20] combined the surface stress continuum mechanics with the classical beam and plate models in order to perform size-dependent bending and buckling analyses of nanobeams and nanoplates. Fu *et al* [21] carried out a nonlinear analysis of the dynamic stability characteristics of functionally graded piezoelectric beams in a thermal environment. Murmu and Adhikari [22] utilized nonlocal continuum elasticity to capture the natural frequencies of prestressed nanobeams. Ansari and Sahmani [23] developed nonlocal beam models based upon various beam theories to analyze the size-dependent vibration of single-walled carbon nanotubes. Wang *et al* [24] introduced a couple-stress-based Euler–Bernoulli beam model for the nonlinear oscillations of microbeams. Ansari and Sahmani [25] calibrated a developed nonlocal plate model for the biaxial buckling analysis of single-layered graphene sheets using a molecular dynamics simulation. Shen and Xiang [26] used a two-step perturbation method to perform nonlinear mechanical analyses of nanocomposite beams embedded in an elastic medium. Sahmani and Ansari [27] predicted the critical buckling loads of functionally graded composite microbeams with third-order shear flexibility using the strain gradient elasticity theory. Eltaher *et al* [28] extracted the natural frequencies of Timoshenko composite beams with nonlinear graded inhomogeneity and included the neutral axis position in their study. Darijani and Mohammadabadi [29] presented a new beam theory that included shear deformation in a way which satisfied couple-free conditions on the lower and upper surfaces. Sahmani *et al* [30–32] employed the surface stress and strain gradient elasticity theories to analyze different nonlinear mechanical characteristics of functionally graded third-order shear deformable beams at the nanoscale and the microscale. Kiani [33] studied the nonlocal stress and shear flexibility effects on the column buckling response of single-layered membranes, including both the in-plane and out-of-plane buckling modes. Karparvarfard *et al* [34] extracted the geometrically nonlinear governing differential equations of motion for Euler–Bernoulli beams based on the second strain gradient continuum mechanics. Sahmani *et al* [35–37] investigated the effect of the surface stress type of size dependency on the postbuckling of nanoplates and nanoshells and the vibrations of postbuckled nanobeams that had third-order shear deformation. Togun and Bagdatli [38] explored the nonlinear oscillations of Euler–Bernoulli beams at the nanoscale based upon the modified couple stress formulations. Li and Hu [39] incorporated the nonlocal strain gradient continuum mechanics into both the

Euler–Bernoulli and Timoshenko beam theories to analyze the nonlinear bending of functionally graded composite microbeams. Sahmani *et al* [40, 41] examined the surface stress effect on the postbuckling characteristics of nanoshells subjected to axial compressive loads together with a temperature change and an external electric field. Akgoz and Civalek [42] integrated the modified strain gradient elasticity theory to develop a size-dependent beam model which was used to analyze the stability of axially compressed carbon nanotubes. Wu *et al* [43] investigated the sensitivity to imperfections of the thermal postbuckling behavior of functionally graded nanocomposite beams using a generic imperfection model. Sahmani and Aghdam [44–46] established nonlocal shell models for the nonlinear stability analysis of nanoshells made of functionally graded composites with piezoelectric facings. Mohamed *et al* [47] predicted the nonlinear free as well as steady-state forced oscillations of buckled curved beams surrounded by a nonlinear elastic foundation. Chen and Li [48] analyzed the buckling and postbuckling responses of geometrically imperfect microbeams, including midplane stretching. Sahmani and Aghdam [49, 50] introduced nonlocal strain gradient beam models for the size-dependent postbuckling and nonlinear resonant behaviors of microscale beam-type structures made of biomedical porous materials. Fang *et al* [51] investigated the size-dependent three-dimensional free oscillations of rotating functionally graded Euler–Bernoulli microbeams. Sheng and Wang [52] analyzed the nonlinear free vibrations of functionally graded inhomogeneous Euler–Bernoulli beams under parametric external excitations with the aid of a multiscale technique. Sahmani *et al* [53–55] studied the effects of the nonlocal stress tensor and the strain gradient on the nonlinear mechanical characteristics of functionally graded porous plates at the microscale.

More recently, Aria and Friswell [56] anticipated the effect of the nonlocal stress tensor on the hygrothermal oscillations and buckling responses of multilayer functionally graded composite microbeams with temperature-dependent properties. Hashemian *et al* [57] analyzed the bending and stability behaviors of nanobeams modeled by size-dependent Euler–Bernoulli, Timoshenko, and Reddy beam formulations incorporating the nonlocal and surface stresses. Amiri *et al* [58] explored the effects of flexoelectric and surface elasticity on the natural frequencies of fluid-induced and fluid-conveying nanotubes modeled via the Euler–Bernoulli beam theory. Sahmani *et al* [59–62] employed different size-dependent continuum theories for the resonant characteristics of beams at the microscale and the nanoscale subjected to periodic hard excitations. Sahmani and Safaei [63–65] developed nonlocal strain gradient beam models for the nonlinear mechanical responses of bidirectional functionally graded composite microbeams. Fang *et al* [66] investigated the thermal stability of rotating functionally graded composite Euler–Bernoulli beams at the nanoscale exposed to temperature changes. Gao *et al* [67] studied the snap-buckling phenomenon in functionally graded laminated nanocomposite curved nanobeams in the presence of geometrical imperfection. Yuan *et al* [68–70] and Yang *et al* [71] predicted the nonlinear static and dynamic stability characteristics of functionally graded inhomogeneous truncated conical microshells, including various size effects. Esen [72] captured the effect of couple stress on the dynamics of Timoshenko microbeams subjected to moving loads using the finite element method.

Fan *et al* [73–76] and Rao *et al* [77] performed an isogeometric analysis of the nonlinear mechanical behaviors of microplates with a central cutout and through-the-thickness graded porosity in the presence of various types of size dependency. Liu *et al* [78] extracted the critical buckling loads and natural frequencies of axially graded inhomogeneous nanocomposite Euler–Bernoulli beams via the state-space technique. Tang *et al* [79], Sahmani *et al* [80], Yang *et al* [81], Hou *et al* [82], and Wang *et al* [83] established size-dependent plate formulations that included variable thickness for the nonlinear mechanical characteristics of quasi-three-dimensional composite microplates with arbitrary shapes. Yin *et al* [84] carried out an isogeometric analysis based on strain gradient elasticity of the size-dependent behavior of an Euler–Bernoulli beam model. Xu *et al* [85] calculated the natural frequency of rotating functionally graded nanocomposite beams exposed to temperature changes under general boundary conditions. Sun *et al* [86], Sahmani and Safaei [87], Lu *et al* [88], Yang *et al* [89], Liu *et al* [90], and Ma *et al* [91] formulated moving Kriging mesh-free shell and panel models based upon different unconventional continuum mechanics for the nonlinear stability behaviors of randomly reinforced nanocomposite microshells. Zhao *et al* [92] obtained the critical buckling loads of hydrogen-functionalized nanocomposite beams using a machine-learning-assisted homogenization scheme. Fu *et al* [93] applied the general strain gradient continuum elasticity to analyze the stability of partially covered multilayer beams at the microscale. Rao *et al* [94] and Alshenawy *et al* [95, 96] investigated the buckling mode transition phenomenon in the nonlinear postbuckling response of functionally graded piezoelectric microshells under electro-thermo-mechanical loading conditions.

1.2 Mathematical formulations

1.2.1 Constitutive equations

As depicted in figure 1.1, consider a functionally graded composite beam at the microscale or the nanoscale with length L and thickness h. The functionally graded composite material is assumed to be metal-rich and ceramic-rich at the top surface ($x_3 = -h/2$) and the bottom surface ($x_3 = h/2$), respectively. The effective materials properties for the FGM substrate can be estimated as shown below:

$$E(x_3) = (E_c - E_m)V_f(x_3) + E_m \tag{1.1a}$$

$$\nu(x_3) = (\nu_c - \nu_m)V_f(x_3) + \nu_m, \tag{1.1b}$$

in which the subscripts m and c denote the metal and ceramic phases, respectively. Furthermore, in accordance with the power law function, the volume fraction can be written as [97]

$$V_f(x_3) = \left(\frac{1}{2} + \frac{x_3}{h}\right)^k, \tag{1.2}$$

where k represents the material gradient index.

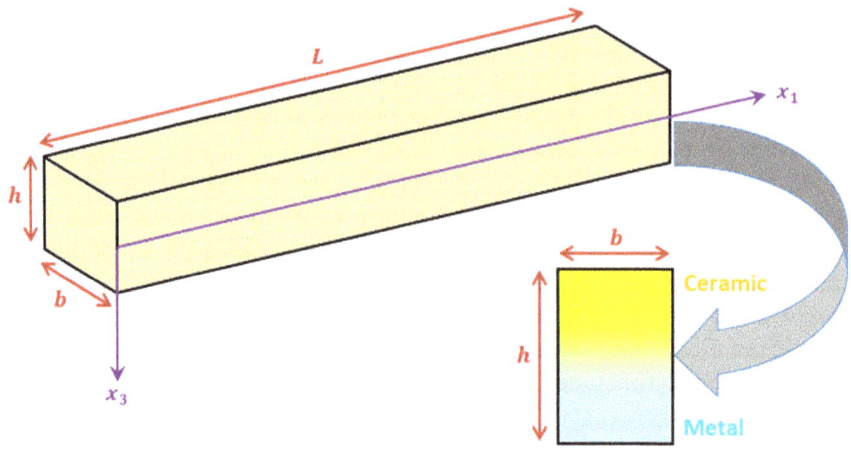

Figure 1.1. A schematic representation of a functionally graded composite beam.

According to the first-order shear deformation beam theory, the displacement field for a beam-type structure can be expressed as

$$u_1(x_1, x_3) = u(x_1) + x_3\psi(x_1) \tag{1.3a}$$

$$u_2(x_1, x_3) = 0 \tag{1.3b}$$

$$u_3(x_1, x_3) = w(x_1), \tag{1.3c}$$

where u and w stand for the displacement components along the x_1-axis and the x_3-axis, respectively. In addition, ψ refers to the face rotation with respect to the x_1–x_3 plane.

In this regard, the nonzero von Kármán nonlinear strain components for a beam at the microscale or the nanoscale can be expressed as follows:

$$\varepsilon_{11} = \frac{\partial u}{\partial x_1} + \frac{1}{2}\left(\frac{\partial w}{\partial x_1}\right)^2 + x_3\frac{\partial \psi}{\partial x_1} \tag{1.4a}$$

$$\gamma_{13} = \psi + \frac{\partial w}{\partial x_1}. \tag{1.4b}$$

1.2.2 The nonlocal strain gradient continuum beam model

In order to capture the size dependency features more efficiently, the physical coupling effect between the strain gradient stress tensor and nonlocal stress tensor can be taken into account via the nonlocal strain gradient continuum elasticity in the following form [98]:

$$\sigma_{ij} = \sigma_{ij}^{(0)} - \frac{\partial \sigma_{ij}^{(1)}}{\partial x_1}, \tag{1.5}$$

in which $\sigma_{ij}^{(0)}$ and $\sigma_{ij}^{(1)}$ denote the stress tensors associated with the conventional strain and strain gradient tensors as shown below

$$\sigma_{ij}^{(0)} = \int_0^L \left\{C_{ijmn}\beta_0(x, x', e_0a)\frac{\partial \varepsilon_{mn}(x')}{\partial x}\right\}dx' \tag{1.6a}$$

$$\sigma_{ij}^{(1)} = l_1^2 \int_0^L \left\{C_{ijmn}\beta_0(x, x', e_1a)\nabla\left(\frac{\partial \varepsilon_{mn}(x')}{\partial x}\right)\right\}dx', \tag{1.6b}$$

where C_{ijmn} denotes the elastic parameter, e_0a and e_1a represent the nonlocal parameters related to the conventional and higher-order stress tensors, and l_1 denotes an internal length scale parameter for the strain gradient size effect. In addition, $\beta_0(x, x', e_0a)$ and $\beta_1(x, x', e_1a)$ are the nonlocal kernel functions corresponding to the conditions introduced by Eringen [99].

Therefore, the constitutive equation for a beam-type structure modeled via the nonlocal strain gradient continuum mechanics can be written as

$$\begin{aligned}\left[1 - (e_1a)^2\frac{\partial^2}{\partial x_1^2}\right]\left[1 - (e_0a)^2\frac{\partial^2}{\partial x_1^2}\right]\sigma_{ij} &= C_{ijmn}\left[1 - (e_1a)^2\frac{\partial^2}{\partial x_1^2}\right] \\ &\quad \varepsilon_{mn} - C_{ijmn}l_1^2\left[1 - (e_0a)^2\frac{\partial^2}{\partial x_1^2}\right]\frac{\partial^2 \varepsilon_{mn}}{\partial x_1^2}.\end{aligned} \tag{1.7}$$

Given the assumption of $e = e_0 = e_1$ made by Lim *et al* [100], the derived constitutive equation of a nonlocal strain gradient beam can be rewritten as

$$\left[1 - (ea)^2\frac{\partial^2}{\partial x_1^2}\right]\sigma_{ij} = C_{ijmn}\left[1 - l_1^2\frac{\partial^2}{\partial x_1^2}\right]\varepsilon_{mn}. \tag{1.8}$$

In this regard, the stress–strain constitutive relations can be stated in the following form

$$\left(1 - (ea)^2\frac{\partial^2}{\partial x_1^2}\right)\begin{Bmatrix}\sigma_{11}\\ \tau_{13}\end{Bmatrix} = \left(1 - l_1^2\frac{\partial^2}{\partial x_1^2}\right)\begin{bmatrix}E(x_3) & 0\\ 0 & G(x_3)\end{bmatrix}\begin{Bmatrix}\varepsilon_{11}\\ \gamma_{13}\end{Bmatrix}, \tag{1.9}$$

where

$$G(x_3) = \frac{E(x_3)}{2(1 + \nu(x_3))}. \tag{1.10}$$

As a result, the strain energy stored in first-order shear flexible functionally graded composite beams at the microscale or the nanoscale can be presented in terms of the stress resultants as follows:

$$\begin{aligned}\Pi_s &= \frac{1}{2}\int_L\int_{-\frac{h}{2}}^{\frac{h}{2}} \sigma_{ij}\varepsilon_{ij}dx_3dx_1 \\ &= \frac{1}{2}\int_L\left\{N_{11}^*\left[\frac{\partial u}{\partial x_1} + \frac{1}{2}\left(\frac{\partial w}{\partial x_1}\right)^2\right] - M_{11}^*\frac{\partial \psi}{\partial x_1} + Q_1^*\left(\frac{\partial w}{\partial x_1} - \psi\right)\right\}dx_1,\end{aligned} \tag{1.11}$$

where

$$N_{11}^* - (ea)^2\frac{\partial^2 N_{11}^*}{\partial x_1^2} = A_{11}^*\left[\frac{\partial u}{\partial x_1} + \frac{1}{2}\left(\frac{\partial w}{\partial x_1}\right)^2 - l_1^2\frac{\partial^3 u}{\partial x_1^3} - l_1^2\frac{\partial w}{\partial x_1}\frac{\partial^3 w}{\partial x_1^3} - l_1^2\left(\frac{\partial^2 w}{\partial x_1^2}\right)^2\right]$$
$$- B_{11}^*\left(\frac{\partial \psi}{\partial x_1} - l_1^2\frac{\partial^3 \psi}{\partial x_1^3}\right)$$

$$M_{11}^* - (ea)^2\frac{\partial^2 M_{11}^*}{\partial x_1^2} = B_{11}^*\left[\frac{\partial u}{\partial x_1} + \frac{1}{2}\left(\frac{\partial w}{\partial x_1}\right)^2 - l_1^2\frac{\partial^3 u}{\partial x_1^3} - l_1^2\frac{\partial w}{\partial x_1}\frac{\partial^3 w}{\partial x_1^3} - l_1^2\left(\frac{\partial^2 w}{\partial x_1^2}\right)^2\right]$$
$$- D_{11}^*\left(\frac{\partial \psi}{\partial x_1} - l_1^2\frac{\partial^3 \psi}{\partial x_1^3}\right)$$

$$Q_1^* - (ea)^2\frac{\partial^2 Q_1^*}{\partial x_1^2} = k_s A_{55}^*\left(\frac{\partial w}{\partial x_1} - \psi - l_1^2\frac{\partial^3 w}{\partial x_1^3} + l_1^2\frac{\partial^2 \psi}{\partial x_1^2}\right), \tag{1.12}$$

k_s represents the shear correction factor, and

$$\begin{Bmatrix} A_{11}^* \\ B_{11}^* \\ D_{11}^* \end{Bmatrix} = \int_{-\frac{h}{2}}^{\frac{h}{2}} E(x_3)\begin{Bmatrix} 1 \\ x_3 \\ x_3^2 \end{Bmatrix} dx_3, \quad A_{55}^* = \int_{-\frac{h}{2}}^{\frac{h}{2}} G(x_3)dx_3. \tag{1.13}$$

Consequently, the nonlocal strain-gradient-based static governing differential equations can be expressed as follows:

$$\frac{\partial N_{11}^*}{\partial x_1} = 0$$
$$\frac{\partial}{\partial x_1}\left[N_{11}^*\frac{\partial w}{\partial x_1}\right] + \frac{\partial Q_1^*}{\partial x_1} = 0 \tag{1.14}$$
$$Q_1^* - \frac{\partial M_{11}^*}{\partial x_1} = 0.$$

1.2.3 The nonlocal couple stress continuum beam model

Using the nonlocal couple stress continuum elasticity [100], the physical coupling effect between the nonlocal stress tensor and the couple stress tensor can be incorporated. Accordingly, the rotation gradient tensor for a functionally graded composite beam at the microscale or the nanoscale can be derived as shown below:

$$\chi_{ij} = \frac{1}{4}\left(e_{imn}\frac{\partial^2 u_n}{\partial x_m \partial x_i} + e_{jmn}\frac{\partial^2 u_n}{\partial x_m \partial x_j}\right), \tag{1.15}$$

where δ_{ij} denotes the Kronecker delta and e_{ijk} refers to the permutation symbol.

Consequently, the components of the rotation gradient tensor for a functionally graded composite beam at the microscale or the nanoscale can be written as

$$\chi_{12} = -\frac{1}{4}\left(\frac{\partial^2 w}{\partial x_1^2} + \frac{\partial \psi}{\partial x_1}\right). \tag{1.16}$$

In this regard, the classical stress–strain constitutive relations can be stated in the following form:

$$\left(1 - (ea)^2\frac{\partial^2}{\partial x_1^2}\right)\begin{Bmatrix}\sigma_{11}\\ \tau_{13}\end{Bmatrix} = \begin{bmatrix}E(x_3) & 0\\ 0 & G(x_3)\end{bmatrix}\begin{Bmatrix}\varepsilon_{11}\\ \gamma_{13}\end{Bmatrix}. \tag{1.17}$$

In addition, the non-classical constitutive relation associated with the rotation gradient tensor can be expressed as

$$\mathcal{M}_{12} - (ea)^2\frac{\partial^2 \mathcal{M}_{12}}{\partial x_1^2} = \frac{l_2^2 E(x_3)}{(1 + \nu(x_3))}\chi_{12}, \tag{1.18}$$

where l_2 denotes the material length scale parameter related to the couple stress tensor.

Consequently, the strain energy stored in first-order shear flexible functionally graded composite beams at the microscale or the nanoscale can be written in terms of the classical and non-classical stress resultants as follows:

$$\begin{aligned}\Pi_s &= \frac{1}{2}\int_L\int_{-\frac{h}{2}}^{\frac{h}{2}}\left\{\sigma_{ij}\varepsilon_{ij} + \mathcal{M}_{ij}\chi_{ij}\right\}dx_3dx_1\\ &= \frac{1}{2}\int_L\left\{N_{11}^{**}\left[\frac{\partial u}{\partial x_1} + \frac{1}{2}\left(\frac{\partial w}{\partial x_1}\right)^2\right] - M_{11}^{**}\frac{\partial \psi}{\partial x_1} + Q_1^{**}\left(\frac{\partial w}{\partial x_1} - \psi\right)\right.\\ &\quad\left. - \frac{Y_{12}^{**}}{2}\left(\frac{\partial^2 w}{\partial x_1^2} + \frac{\partial \psi}{\partial x_1}\right)\right\}dx_1\end{aligned} \tag{1.19}$$

in which

$$\begin{aligned}N_{11}^{**} - (ea)^2\frac{\partial^2 N_{11}^{**}}{\partial x_1^2} &= A_{11}^{*}\left[\frac{\partial u}{\partial x_1} + \frac{1}{2}\left(\frac{\partial w}{\partial x_1}\right)^2\right] - B_{11}^{*}\frac{\partial \psi}{\partial x_1}\\ M_{11}^{**} - (ea)^2\frac{\partial^2 M_{11}^{**}}{\partial x_1^2} &= B_{11}^{*}\left[\frac{\partial u}{\partial x_1} + \frac{1}{2}\left(\frac{\partial w}{\partial x_1}\right)^2\right] - D_{11}^{*}\frac{\partial \psi}{\partial x_1}\\ Q_1^{**} - (ea)^2\frac{\partial^2 Q_1^{**}}{\partial x_1^2} &= k_s A_{55}^{*}\left(\frac{\partial w}{\partial x_1} - \psi\right)\end{aligned} \tag{1.20}$$

$$Y_{12}^{**} - (ea)^2 \frac{\partial^2 Y_{12}^{**}}{\partial x_1^2} = -\frac{A_{55}^{**}}{4}\left(\frac{\partial^2 w}{\partial x_1^2} + \frac{\partial \psi}{\partial x_1}\right)$$

and

$$A_{55}^{**} = \int_{-\frac{h}{2}}^{\frac{h}{2}} 2l_2^2 G(x_3) dx_3. \tag{1.21}$$

Consequently, the nonlocal couple-stress-based static governing differential equations can be expressed as follows:

$$\frac{\partial N_{11}^{**}}{\partial x_1} = 0$$

$$\frac{\partial}{\partial x_1}\left[N_{11}^{**}\frac{\partial w}{\partial x_1}\right] + \frac{\partial Q_1^{**}}{\partial x_1} + \frac{1}{2}\frac{\partial^2 Y_{12}^{**}}{\partial x_1^2} = 0 \tag{1.22}$$

$$Q_1^{**} - \frac{\partial M_{11}^{**}}{\partial x_1} - \frac{1}{2}\frac{\partial Y_{12}^{**}}{\partial x_1} = 0.$$

1.3 Isogeometric analysis

With the aid of the isogeometric analysis that has basis functions of the non-uniform rational B-spline (NURBS) type, it is possible to model any arbitrary geometry without the need to enforce kinematic constraints via the Lagrange multipliers. In this regard, several research studies have been carried out in recent years using isogeometric analysis of the mechanical behaviors of structures at different scales [101–103].

In accordance with this approach, the NURBS basis functions employed are constructed via an appropriate higher-order continuous approximation function. For this purpose, a knot vector $\varphi = \left[\xi_1, \xi_2, \ldots, \xi_{n+p+1}\right]$ that has a non-decreasing sequence denotes the parameter domain. In this regard, n denotes the number of the basis function, while p represents the order of the B-spline basis function employed.

Therefore, the B-spline basis functions can be introduced in the following form:

$$p = 0: \quad \mathcal{T}_i^0(\xi) = \begin{cases} 1 & \text{if } \xi_i \leqslant \xi \leqslant \xi_{i+1} \\ 0 & \text{otherwise} \end{cases}$$

$$p \geqslant 1: \quad \mathcal{T}_i^p(\xi) = \frac{\xi - \xi_i}{\xi_{i+p} - \xi_i}\mathcal{T}_i^{p-1}(\xi) + \frac{\xi_{i+p+1} - \xi}{\xi_{i+p+1} - \xi_{i+1}}\mathcal{T}_{i+1}^{p-1}(\xi). \tag{1.23}$$

The B-spline curve can be introduced as shown below:

$$\mathfrak{Y}(\xi) = \sum_{i=1}^{n} \mathcal{T}_i^p(\xi)\mathbf{S}_i, \tag{1.24}$$

where $\mathbf{S}_i$ denotes the ith control point.

Using the appropriate B-spline basis function, the ith NURBS basis function that has the order p can be expressed as follows:

$$\mathfrak{Z}_i^p(\xi) = \frac{\mathcal{T}_i^p(\xi)\mathbb{W}_i}{\sum_{j=1}^{n}\mathcal{T}_i^p(\xi)\mathbb{W}_j}, \tag{1.25}$$

where $\mathbb{W}_i$ denotes the weight coefficient associated with the control node $\mathbf{S}_i$.

As in the case of the B-spline behavior, the NURBS curve related to the control point $\mathbf{S}_i$ and the basis function $\mathfrak{Z}_i^p(\xi)$ can be introduced as follows:

$$\mathfrak{Y}(\xi) = \sum_{i=1}^{n}\mathfrak{Z}_i^p(\xi)\mathbf{S}_i, \tag{1.26}$$

in which ξ represents the parametric coordinates employed.

After the order related to the basis function is validated, it is possible to use the default $\mathfrak{Z}_i(\xi) = \mathfrak{Z}_i^p(\xi)$. For the first-order shear deformable beam formulations, the associated displacement approximation can be given using the one-dimensional NURBS basis function in the following form

$$u = \sum_{j=1}^{m}\mathfrak{Z}_j(\xi)u_j, \quad w = \sum_{j=1}^{m}\mathfrak{Z}_j(\xi)w_j, \quad \psi = \sum_{j=1}^{m}\mathfrak{Z}_j(\xi)\psi_j, \tag{1.27}$$

where m stands for the number of the control points employed in the associated element.

In accordance with equation (1.4) and considering equation (1.27), the strain components can be rewritten as shown below:

$$\mathfrak{P}_b = \mathfrak{P}_b^L + \mathfrak{P}_b^{NL} = \sum_{i=1}^{m}\mathcal{T}_{Lb}^i\mathbb{X} + \sum_{i=1}^{m}\frac{1}{2}\mathcal{T}_{NLb}^i\mathbb{X}, \quad \mathfrak{P}_s = \sum_{i=1}^{m}\mathcal{T}_s^i\mathbb{X}, \tag{1.28}$$

in which

$$\mathcal{T}_{Lb}^i = \left\{\mathcal{T}_{b1}^i \;\; \mathcal{T}_{b2}^i\right\}^T, \quad \mathcal{T}_{NLb}^i = \left\{\mathcal{T}_{b3}^i \;\; \mathbf{0}\right\}^T\mathcal{T}_G^i, \; \mathbb{X} = \begin{Bmatrix} u_i \\ w_i \\ \psi_i \end{Bmatrix}$$

$$\mathcal{T}_{b1}^i = \begin{bmatrix} \dfrac{\partial\mathfrak{Z}_i(x_1)}{\partial x_1} & 0 & 0 \end{bmatrix}, \quad \mathcal{T}_{b1}^i = \begin{bmatrix} 0 & 0 & \dfrac{\partial\mathfrak{Z}_i(x_1)}{\partial x_1} \end{bmatrix}, \quad \mathcal{T}_{b3}^i = \begin{bmatrix} \dfrac{\partial w}{\partial x} & 0 \\ 0 & \dfrac{\partial w}{\partial x} \end{bmatrix} \tag{1.29}$$

$$\mathcal{T}_s^i = \begin{bmatrix} 0 & \frac{\partial\mathfrak{Z}_i(x_1)}{\partial x_1} & \mathfrak{Z}_i(x_1) \end{bmatrix}.$$

1.4 Numerical results and discussion

In this section, we analyze the nonlocal strain-gradient-based and nonlocal couple-stress-based nonlinear bending responses of uniformly laterally loaded microbeams and the nonlinear postbuckling of axially compressed microbeams made of functionally graded composites. The material gradient of the functionally graded composite microsized beams is considered because there is a ceramic-rich surface at the top and a metal-rich surface at the bottom of the microbeams, the properties of which are taken to be $E_c = 210$ GPa, $\nu_c = 0.24$ for the ceramic phase and $E_m = 70$ GPa, $\nu_m = 0.35$ for the metal phase [104].

Figure 1.2 illustrates the effect of the nonlocal stress tensor on the postbuckling load–deflection stability paths of axially compressed functionally graded composite first-order shear deformable microbeams under simply supported and clamped boundary conditions. It can be observed that if the nonlocal type of size dependency is taken into account, the critical buckling loads of functionally graded beams at the microscale are reduced. In addition, this figure reveals that in deeper parts of the postbuckling territory, the gap between the load–deflection stability paths associated with various values of the nonlocal parameter decreases, which means that the effect of the nonlocal stress tensor becomes smaller for regions of deeper postbuckling.

Figure 1.3 illustrates the effect of the strain gradient on the postbuckling load–deflection stability paths of axially compressed functionally graded composite

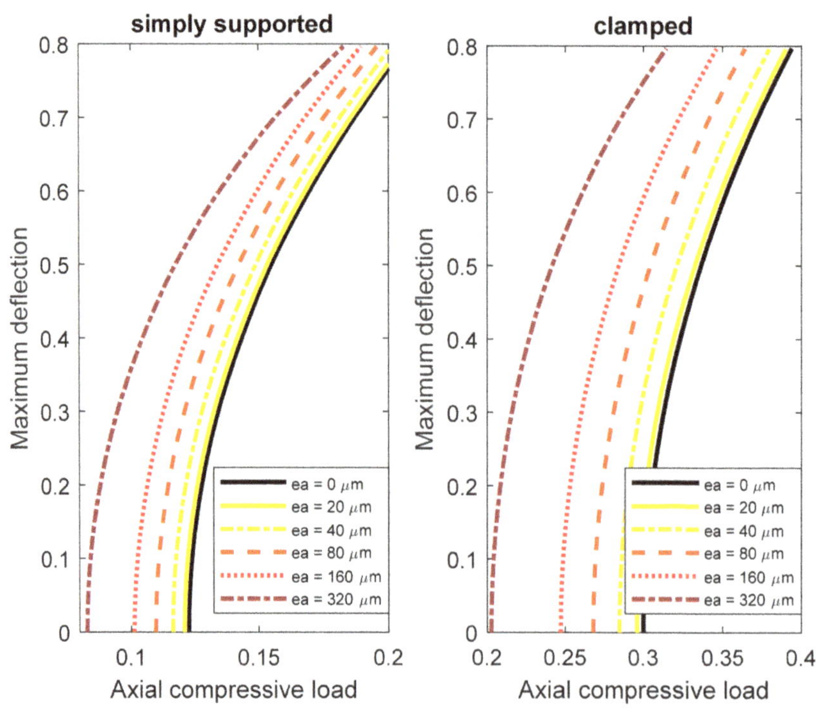

Figure 1.2. The effect of the nonlocal stress tensor on the dimensionless nonlinear postbuckling stability response of axially compressed functionally graded beams at the microscale ($k = 0.5$, $l_1 = l_2 = 0$ μm).

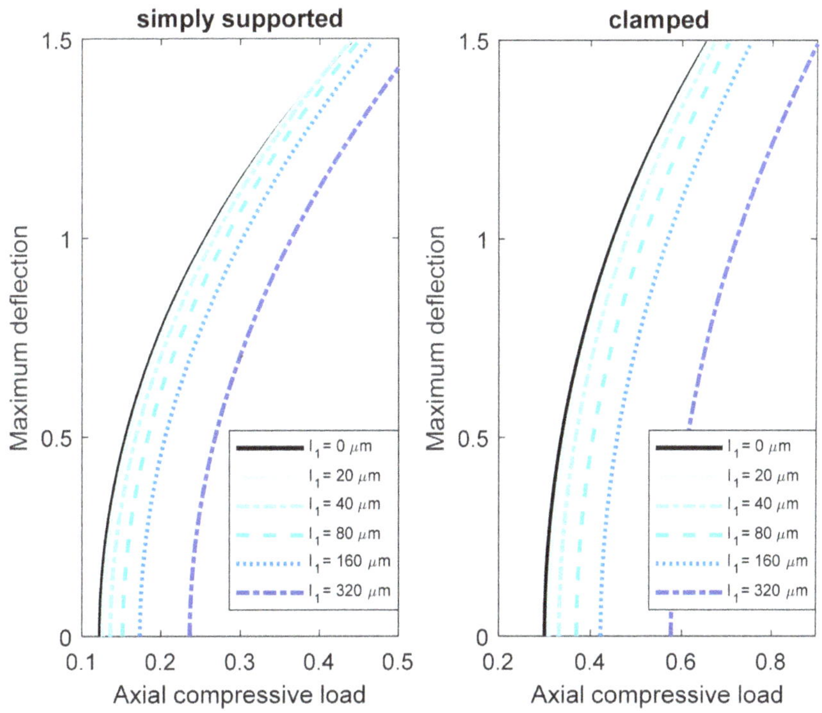

Figure 1.3. The effect of the strain gradient on the dimensionless nonlinear postbuckling stability response of axially compressed functionally graded beams at the microscale ($k = 0.5$, $ea = l_2 = 0$ μm).

first-order shear deformable microbeams under simply supported and clamped boundary conditions. It shows that if the strain gradient type of size dependency is taken into consideration, the critical buckling loads of functionally graded beams at microscale are enhanced. Furthermore, it can be seen that in deeper parts of the postbuckling territory, the gap between the load–deflection stability paths associated with various values of the strain gradient parameter reduces, which means that the effect of the strain gradient size effect becomes smaller in deeper parts of the postbuckling domain. These predictions are seen for both simply supported and clamped boundary conditions.

Figure 1.4 depicts the effect of the couple stress tensor on the postbuckling load–deflection stability paths of axially compressed functionally graded composite first-order shear deformable microbeams under simply supported and clamped boundary conditions. It shows that if the couple stress type of size dependency is taken into account, the critical buckling loads of functionally graded beams at the microscale are increased. However, this enhancement in the value of the critical buckling load is less than that caused by the strain gradient size effect. Moreover, it can be seen that in deeper parts of the postbuckling territory, the gap between the load–deflection stability paths associated with various values of the couple stress parameter decreases, which means that the effect of the couple stress tensor becomes smaller in deeper regions of the postbuckling regime.

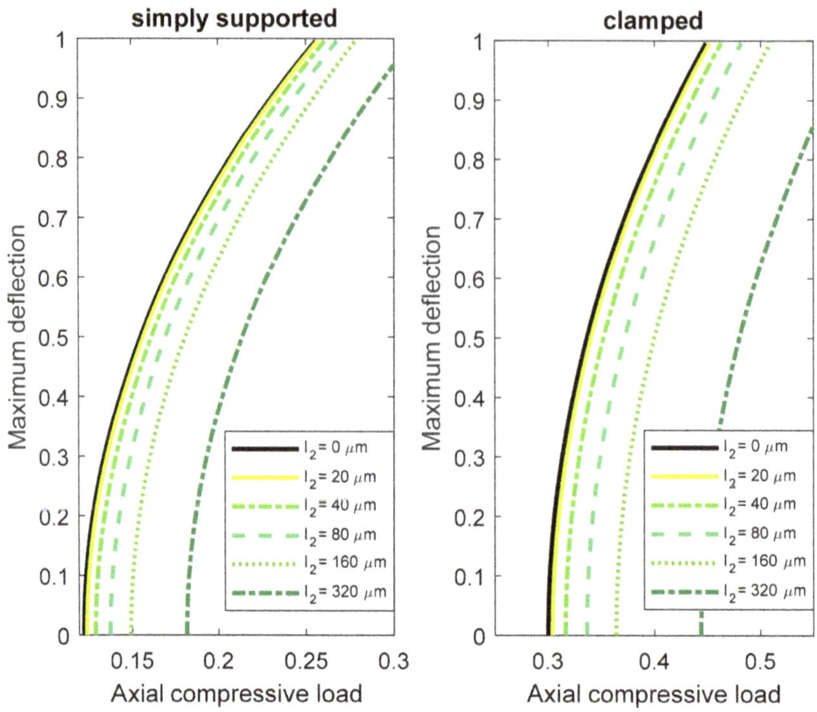

Figure 1.4. The effect of the couple stress tensor on the dimensionless nonlinear postbuckling stability response of axially compressed functionally graded beams at the microscale ($k = 0.5$, $ea = l_1 = 0$ μm).

Figure 1.5 shows the effect of the nonlocal stress tensor on the nonlinear bending load–deflection curves of uniformly laterally loaded functionally graded composite first-order shear deformable microbeams under simply supported and clamped boundary conditions. It shows that if the nonlocal type of size dependency is considered, the slope of the nonlinear bending curve becomes shallower in such a way that a smaller distributed lateral load is required to induce a specific maximum deflection in a microbeam. In other words, this means that when the nonlocal stress tensor is taken into account, the stiffness associated with the flexibility of a functionally graded composite beam at the microscale reduces. This prediction can be observed for both simply supported and clamped boundary conditions, as shown in the figure.

Figure 1.6 displays the effect of the strain gradient on the nonlinear bending load–deflection curves of uniformly laterally loaded functionally graded composite first-order shear deformable microbeams under simply supported and clamped boundary conditions. It can be observed that if the strain gradient type of size dependency is considered, the slope of the nonlinear bending curve becomes steeper in such a way that a larger distributed lateral load is necessary to induce a specific maximum deflection in a microbeam. In other words, this means that when the strain gradient tensor is considered, the stiffness associated with the flexibility of a functionally graded composite beam at the microscale is enhanced. For both simply supported

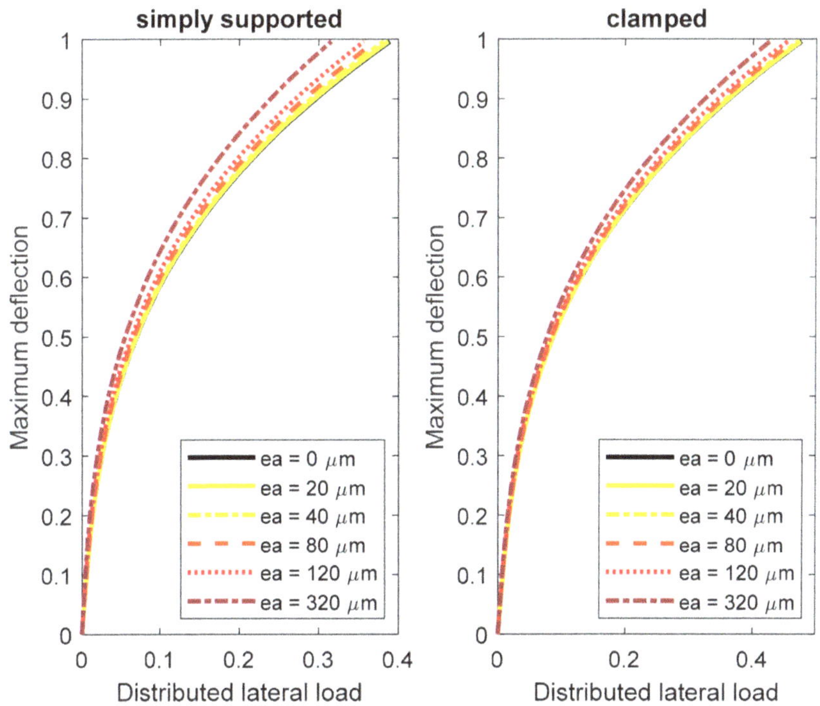

Figure 1.5. The effect of the nonlocal stress tensor on the dimensionless nonlinear bending curve of uniformly lateral loaded functionally graded beams at the microscale ($k = 0.5$, $l_1 = l_2 = 0$ μm).

and clamped boundary conditions, the predictions are repeated, as represented in the figure.

Figure 1.7 represents the effect of the couple stress tensor on the nonlinear bending load–deflection curves of uniformly lateral loaded functionally graded composite first-order shear deformable microbeams under simply supported and clamped boundary conditions. It demonstrates that if the couple stress type of size dependency is considered, the slope of the nonlinear bending curve steepens in such a way that a larger distributed lateral load is required to induce a specific maximum deflection in a microbeam. In other words, this means that when the couple stress tensor is taken into account, the stiffness associated with the flexibility of a functionally graded composite beam at the microscale increases. However, this enhancement in the stiffness associated with flexibility is lower than that achieved by the strain gradient size effect for both simply supported and clamped boundary conditions.

1.5 Concluding remarks

In this chapter, the static characteristics of first-order shear flexible beams at the microscale made of functionally graded composites were studied, including the nonlinear bending of uniformly laterally loaded microbeams and the nonlinear postbuckling of axially compressed microbeams. In this regard, the isogeometric

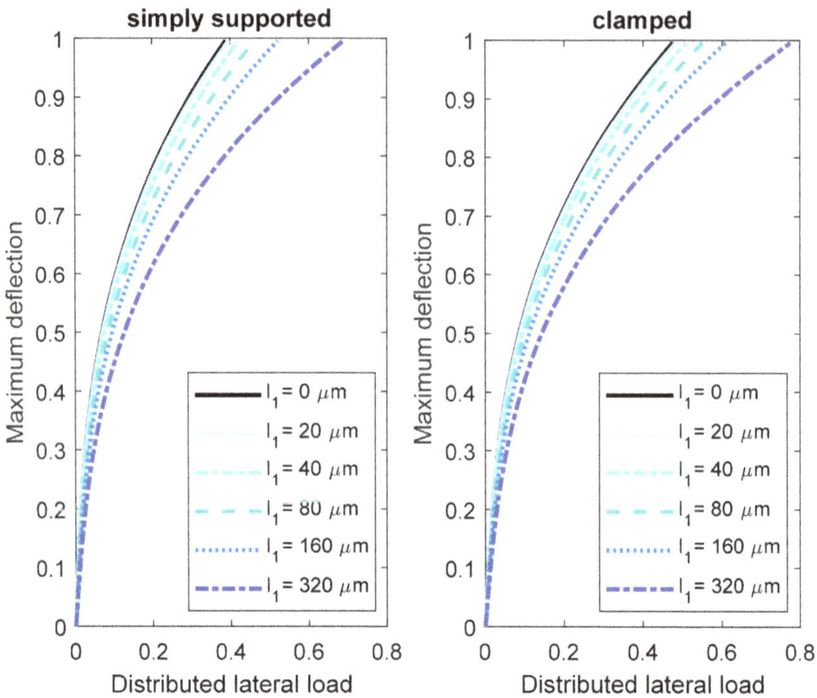

Figure 1.6. The effect of the strain gradient on the dimensionless nonlinear bending curve of uniformly lateral loaded functionally graded beams at the microscale ($k = 0.5$, $ea = l_2 = 0$ μm).

kind of finite element approach was employed to solve the developed size-dependent beam models with the aid of appropriate nonuniform rational B-spline basis functions in order to obtain accurate geometrical descriptions.

This chapter showed that when the nonlocal type of size dependency is taken into account, the critical buckling loads of functionally graded beams at the microscale are reduced; conversely, it was seen that when the couple stress type of size dependency is taken into account, the critical buckling loads of functionally graded beams at the microscale are increased. In addition, this enhancement in the value of the critical buckling load is less than that caused by the strain gradient size effect. It was also revealed that in deeper parts of the postbuckling territory, the gap between the load–deflection stability paths associated with various values of the size effect parameters decreases, meaning that the effect of all the considered size dependencies is reduced in deeper regions of the postbuckling regime.

This chapter reported that when the nonlocal type of size dependency is considered, the slops of the nonlinear bending curve becomes shallower in such a way that a lower distributed lateral load is required to induce a specific maximum deflection in a microbeam. In other words, this means that when the nonlocal stress tensor is taken into account, the stiffness associated with the flexibility of a functionally graded composite beam at the microscale reduces. On the other hand, if couple stress tensor is taken into account, the stiffness associated with the

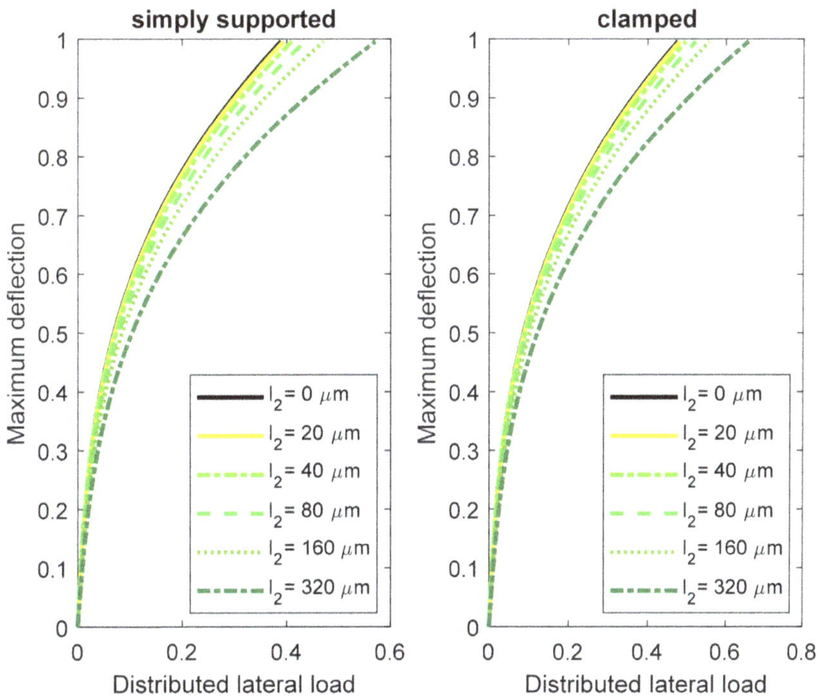

Figure 1.7. The effect of the couple stress tensor on the dimensionless nonlinear bending curve of uniformly lateral loaded functionally graded beams at the microscale ($k = 0.5$, $ea = l_1 = 0$ μm).

flexibility of a functionally graded composite beam at the microscale increases. However, this enhancement in the stiffness associated with flexibility is lower than that achieved by the strain gradient size effect for both simply supported and clamped boundary conditions.

References

[1] Caddemi S and Calio I 2013 The exact explicit dynamic stiffness matrix of multi-cracked Euler–Bernoulli beam and applications to damaged frame structures *J. Sound Vib.* **332** 3049–63

[2] Shaat M and Abdelkefi A 2015 Modeling the material structure and couple stress effects of nanocrystalline silicon beams for pull-in and bio-mass sensing applications *Int. J. Mech. Sci.* **102** 280–91

[3] Nadal C, Giraud-Audine C, Giraud F, Amberg M and Lemaire-Semail B 2017 Modelling of a beam excited by piezoelectric actuators in view of tactile applications *Math. Comput. Simul.* **131** 295–315

[4] Sahmani S, Khandan A, Saber-Samandari S and Aghdam M M 2018 Vibrations of beam-type implants made of 3D printed bredigite–magnetite bio-nanocomposite scaffolds under axial compression: application, communication and simulation *Ceram. Int.* **44** 11282–91

[5] Sahmani S, Shahali M, Khandan A, Saber-Samandari S and Aghdam M M 2018 Analytical and experimental analyses for mechanical and biological characteristics of novel

nanoclay bio-nanocomposite scaffolds fabricated via space holder technique *Appl. Clay Sci.* **165** 112–23

[6] Sahmani S, Khandan A, Saber-Samandari S and Aghdam M M 2018 Nonlinear bending and instability analysis of bioceramics composed with magnetite nanoparticles: fabrication, characterization, and simulation *Ceram. Int.* **44** 9540–9

[7] Sahmani S, Saber-Samandari S, Aghdam M M and Khandan A 2018 Nonlinear resonance response of porous beam-type implants corresponding to various morphology shapes for bone tissue engineering applications *J. Mater. Eng. Perform.* **27** 5370–83

[8] Chadha M and Todd M D 2019 A comprehensive kinematic model of single-manifold Cosserat beam structures with application to a finite strain measurement model for strain gauges *Int. J. Solids Struct.* **159** 58–76

[9] Sahmani S, Shahali M, Ghadiri Nejad M, Khandan A, Aghdam M M and Saber-Samandari S 2019 Effect of copper oxide nanoparticles on electrical conductivity and cell viability of calcium phosphate scaffolds with improved mechanical strength for bone tissue engineering *Eur. Phys. J. Plus* **134** 7

[10] Sahmani S, Saber-Samandari S, Khandan A and Aghdam M M 2019 Nonlinear resonance investigation of nanoclay based bio-nanocomposite scaffolds with enhanced properties for bone substitute applications *J. Alloys Compd.* **773** 636–53

[11] Sahmani S, Saber-Samandari S, Khandan A and Aghdam M M 2019 Influence of MgO nanoparticles on the mechanical properties of coated hydroxyapatite nanocomposite scaffolds produced via space holder technique: fabrication, characterization *J. Mech. Behav. Biomed. Mater.* **95** 76–88

[12] Sahmani S, Khandan A, Esmaeili S, Saber-Samandari S, Nejad M G and Aghdam M M 2020 Calcium phosphate-PLA scaffolds fabricated by fused deposition modeling technique for bone tissue applications: fabrication, characterization and simulation *Ceram. Int.* **46** 2447–56

[13] Sahmani S, Khandan A, Saber-Samandari S and Aghdam M M 2020 Effect of magnetite nanoparticles on the biological and mechanical properties of hydroxyapatite porous scaffolds coated with ibuprofen drug *Mater. Sci. Eng.* C **111** 110835

[14] Nuhu A A B S 2022 State-of-the-art of vibration analysis of small-sized structures by using nonclassical continuum theories of elasticity *Arch. Comput. Meth. Eng.* **29** 4959–5147

[15] Wang L 2010 Size-dependent vibration characteristics of fluid-conveying microtubes *J. Fluids Struct.* **26** 675–84

[16] Akgoz B O C 2011 Strain gradient elasticity and modified couple stress models for buckling analysis of axially loaded micro-scaled beams *Int. J. Eng. Sci.* **49** 1268–80

[17] Ansari R, Sahmani S and Rouhi H 2011 Rayleigh–Ritz axial buckling analysis of single-walled carbon nanotubes with different boundary conditions *Phys. Lett.* A **375** 1255–63

[18] Ansari R, Sahmani S and Rouhi H 2011 Axial buckling analysis of single-walled carbon nanotubes in thermal environments via the Rayleigh–Ritz technique *Comput. Mater. Sci.* **50** 3050–5

[19] Ansari R S S 2011 Bending behavior and buckling of nanobeams including surface stress effects corresponding to different beam theories *Int. J. Eng. Sci.* **49** 1244–55

[20] Ansari R and Sahmani S 2011 Surface stress effects on the free vibration behavior of nanoplates *Int. J. Eng. Sci.* **49** 1204–15

[21] Fu Y, Wang J and Mao Y 2012 Nonlinear analysis of buckling, free vibration and dynamic stability for the piezoelectric functionally graded beams in thermal environment *Appl. Math. Model.* **36** 4324–40

[22] Murmu T and Adhikari S 2012 Nonlocal elasticity based vibration of initially pre-stressed coupled nanobeam systems *Eur. J. Mech.* A **34** 52–61

[23] Ansari R and Sahmani S 2012 Small scale effect on vibrational response of single-walled carbon nanotubes with different boundary conditions based on nonlocal beam models *Commun. Nonlinear Sci. Numer. Simul.* **17** 1965–79

[24] Wang Y-G, Lin W-H and Liu N 2013 Nonlinear free vibration of a microscale beam based on modified couple stress theory *Physica* E **47** 80–5

[25] Ansari R and Sahmani S 2013 Prediction of biaxial buckling behavior of single-layered graphene sheets based on nonlocal plate models and molecular dynamics simulations *Appl. Math. Model.* **37** 7338–51

[26] Shen H-S Y X 2013 Nonlinear analysis of nanotube-reinforced composite beams resting on elastic foundations in thermal environments *Eng. Struct.* **56** 698–708

[27] Sahmani S R A 2013 Size-dependent buckling analysis of functionally graded third-order shear deformable microbeams including thermal environment effect *Appl. Math. Model.* **37** 9499–515

[28] Eltaher M A, Abdelrahman A A, Al-Nabawy A, Khater M and Mansour A 2014 Vibration of nonlinear graduation of nano-Timoshenko beam considering the neutral axis position *Appl. Math. Comput.* **235** 512–29

[29] Darijani H and Mohammadabadi H 2014 A new deformation beam theory for static and dynamic analysis of microbeams *Int. J. Mech. Sci.* **89** 31–9

[30] Sahmani S, Bahrami M and Ansari R 2014 Surface energy effects on the free vibration characteristics of postbuckled third-order shear deformable nanobeams *Compos. Struct.* **116** 552–61

[31] Sahmani S, Bahrami M and Ansari R 2014 Nonlinear free vibration analysis of functionally graded third-order shear deformable microbeams based on the modified strain gradient elasticity theory *Compos. Struct.* **110** 219–30

[32] Sahmani S, Bahrami M, Aghdam M M and Ansari R 2014 Surface effects on the nonlinear forced vibration response of third-order shear deformable nanobeams *Compos. Struct.* **118** 149–58

[33] Kiani K 2015 Nonlocal and shear effects on column buckling of single-layered membranes from stocky single-walled carbon nanotubes *Composites* B **79** 535–51

[34] Karparvarfard S M H, Asghari M and Vatankhah R 2015 A geometrically nonlinear beam model based on the second strain gradient theory *Int. J. Eng. Sci.* **91** 63–75

[35] Sahmani S, Bahrami M, Aghdam M M and Ansari R 2015 Postbuckling behavior of circular higher-order shear deformable nanoplates including surface energy effects *Appl. Math. Model.* **39** 3678–89

[36] Sahmani S, Aghdam M M and Bahrami M 2015 Nonlinear buckling and postbuckling behavior of cylindrical nanoshells subjected to combined axial and radial compressions incorporating surface stress effects *Composites* B **79** 676–91

[37] Sahmani S, Aghdam M M and Bahrami M 2015 On the free vibration characteristics of postbuckled third-order shear deformable FGM nanobeams including surface effects *Compos. Struct.* **121** 377–85

[38] Togun N and Bagdatli S M 2016 Size dependent nonlinear vibration of the tensioned nanobeam based on the modified couple stress theory *Composites* B **97** 255–61

[39] Li L and Hu Y 2016 Nonlinear bending and free vibration analyses of nonlocal strain gradient beams made of functionally graded material *Int. J. Eng. Sci.* **107** 77–97

[40] Sahmani S, Aghdam M M and Bahrami M 2016 Size-dependent axial buckling and postbuckling characteristics of cylindrical nanoshells in different temperatures *Int. J. Mech. Sci.* **107** 170–9

[41] Sahmani S, Aghdam M M and Akbarzadeh A H 2016 Size-dependent buckling and postbuckling behavior of piezoelectric cylindrical nanoshells subjected to compression and electrical load *Mater. Des.* **105** 341–51

[42] Akgoz B and Civalek O 2017 A size-dependent beam model for stability of axially loaded carbon nanotubes surrounded by Pasternak elastic foundation *Compos. Struct.* **176** 1028–38

[43] Wu H, Kitipornchai S and Yang J 2017 Imperfection sensitivity of thermal post-buckling behaviour of functionally graded carbon nanotube-reinforced composite beams *Appl. Math. Model.* **42** 735–51

[44] Sahmani S and Aghdam M M 2017 Temperature-dependent nonlocal instability of hybrid FGM exponential shear deformable nanoshells including imperfection sensitivity *Int. J. Mech. Sci.* **122** 129–41

[45] Sahmani S and Aghdam M M 2017 Size dependency in axial postbuckling behavior of hybrid FGM exponential shear deformable nanoshells based on the nonlocal elasticity theory *Compos. Struct.* **166** 104–13

[46] Sahmani S and Aghdam M M 2017 Nonlinear instability of hydrostatic pressurized hybrid FGM exponential shear deformable nanoshells based on nonlocal continuum elasticity *Composites* B **114** 404–17

[47] Mohamed N, Eltaher M A, Mohamed S A and Seddek L F 2018 Numerical analysis of nonlinear free and forced vibrations of buckled curved beams resting on nonlinear elastic foundations *Int. J. Non Linear Mech.* **101** 157–73

[48] Chen X and Li Y 2018 Size-dependent post-buckling behaviors of geometrically imperfect microbeams *Mech. Res. Commun.* **88** 25–33

[49] Sahmani S and Aghdam M M 2018 Nonlocal strain gradient beam model for postbuckling and associated vibrational response of lipid supramolecular protein micro/nano-tubules *Math. Biosci.* **295** 24–35

[50] Sahmani S and Aghdam M M 2018 Nonlinear primary resonance of micro/nano-beams made of nanoporous biomaterials incorporating nonlocality and strain gradient size dependency *Results Phys.* **8** 879–91

[51] Fang J, Gu J and Wang H 2018 Size-dependent three-dimensional free vibration of rotating functionally graded microbeams based on a modified couple stress theory *Int. J. Mech. Sci.* **136** 188–99

[52] Sheng G G and Wang X 2018 Nonlinear vibration of FG beams subjected to parametric and external excitations *Eur. J. Mech.* A **71** 224–34

[53] Sahmani S, Aghdam M M and Rabczuk T 2018 Nonlocal strain gradient plate model for nonlinear large-amplitude vibrations of functionally graded porous micro/nano-plates reinforced with GPLs *Compos. Struct.* **198** 51–61

[54] Sahmani S, Aghdam M M and Rabczuk T 2018 A unified nonlocal strain gradient plate model for nonlinear axial instability of functionally graded porous micro/nano-plates reinforced with graphene platelets *Mater. Res. Express* **5** 045048

[55] Sahmani S, Aghdam M M and Rabczuk T 2018 Nonlinear bending of functionally graded porous micro/nano-beams reinforced with graphene platelets based upon nonlocal strain gradient theory *Compos. Struct.* **186** 68–78

[56] Aria A I and Friswell M I 2019 Computational hygro-thermal vibration and buckling analysis of functionally graded sandwich microbeams *Composites* B **165** 785–97

[57] Hashemian M, Foroutan S and Toghraie D 2019 Comprehensive beam models for buckling and bending behavior of simple nanobeam based on nonlocal strain gradient theory and surface effects *Mech. Mater.* **139** 103209

[58] Amiri A, Vesal R and Telabitooti R 2019 Flexoelectric and surface effects on size-dependent flow-induced vibration and instability analysis of fluid-conveying nanotubes based on flexoelectricity beam model *Int. J. Mech. Sci.* **156** 474–85

[59] Sahmani S, Fattahi A M and Ahmed N A 2019 Size-dependent nonlinear forced oscillation of self-assembled nanotubules based on the nonlocal strain gradient beam model *J. Braz. Soc. Mech. Sci. Eng.* **41** 239

[60] Sahmani S, Fotouhi M and Aghdam M M 2019 Size-dependent nonlinear secondary resonance of micro-/nano-beams made of nano-porous biomaterials including truncated cube cells *Acta Mech.* **230** 1077–103

[61] Sarafraz A, Sahmani S and Aghdam M M 2019 Nonlinear secondary resonance of nanobeams under subharmonic and superharmonic excitations including surface free energy effects *Appl. Math. Model.* **66** 195–226

[62] Fattahi A M, Sahmani S and Ahmed N A 2020 Nonlocal strain gradient beam model for nonlinear secondary resonance analysis of functionally graded porous micro/nano-beams under periodic hard excitations *Mech. Based Des. Struct. Mach.* **48** 403–31

[63] Sahmani S and Safaei B 2019 Nonlinear free vibrations of bi-directional functionally graded micro/nano-beams including nonlocal stress and microstructural strain gradient size effects *Thin-Walled Struct.* **140** 342–56

[64] Sahmani S and Safaei B 2019 Nonlocal strain gradient nonlinear resonance of bi-directional functionally graded composite micro/nano-beams under periodic soft excitation *Thin-Walled Struct.* **143** 106226

[65] Sahmani S and Safaei B 2020 Influence of homogenization models on size-dependent nonlinear bending and postbuckling of bi-directional functionally graded micro/nano-beams *Appl. Math. Model.* **82** 336–58

[66] Fang J, Zheng S, Xiao J and Zhang X 2020 Vibration and thermal buckling analysis of rotating nonlocal functionally graded nanobeams in thermal environment *Aerosp. Sci. Technol.* **106** 106146

[67] Gao Y, Xiao W-S and Zhu H 2020 Snap-buckling of functionally graded multilayer graphene platelet-reinforced composite curved nanobeams with geometrical imperfections *Eur. J. Mech.* A **82** 103993

[68] Yuan Y, Zhao K, Zhao Y, Sahmani S and Safaei B 2020 Couple stress-based nonlinear buckling analysis of hydrostatic pressurized functionally graded composite conical micro-shells *Mech. Mater.* **148** 103507

[69] Yuan Y, Zhao K, Han Y, Sahmani S and Safaei B 2020 Nonlinear oscillations of composite conical microshells with in-plane heterogeneity based upon a couple stress-based shell model *Thin-Walled Struct.* **154** 106857

[70] Yuan Y, Zhao X, Zhao Y, Sahmani S and Safaei B 2021 Dynamic stability of nonlocal strain gradient FGM truncated conical microshells integrated with magnetostrictive face-sheets resting on a nonlinear viscoelastic foundation *Thin-Walled Struct.* **159** 107249

[71] Yang Y, Sahmani S and Safaei B 2021 Couple stress-based nonlinear primary resonant dynamics of FGM composite truncated conical microshells integrated with magneto-strictive layers *Appl. Math. Mech.* **42** 209–21

[72] Esen I 2020 Dynamics of size-dependent Timoshenko micro beams subjected to moving loads *Int. J. Mech. Sci.* **175** 105501

[73] Fan F, Xu Y, Sahmani S and Safaei B 2020 Modified couple stress-based geometrically nonlinear oscillations of porous functionally graded microplates using NURBS-based isogeometric approach *Comput. Meth. Appl. Mech. Eng.* **372** 113400

[74] Fan F, Cai X, Sahmani S and Safaei B 2021 Isogeometric thermal postbuckling analysis of porous FGM quasi-3D nanoplates having cutouts with different shapes based upon surface stress elasticity *Compos. Struct.* **262** 113604

[75] Fan F, Safaei B and Sahmani S 2021 Buckling and postbuckling response of nonlocal strain gradient porous functionally graded micro/nano-plates via NURBS-based isogeometric analysis *Thin-Walled Struct.* **159** 107231

[76] Fan F, Sahmani S and Safaei B 2021 Isogeometric nonlinear oscillations of nonlocal strain gradient PFGM micro/nano-plates via NURBS-based formulation *Compos. Struct.* **255** 112969

[77] Rao R, Sahmani S and Safaei B 2021 Isogeometric nonlinear bending analysis of porous FG composite microplates with a central cutout modeled by the couple stress continuum quasi-3D plate theory *Arch. Civ. Mech. Eng.* **21** 98

[78] Liu D, Chen D, Yang J and Kitipornchai S 2021 Buckling and free vibration of axially functionally graded graphene reinforced nanocomposite beams *Eng. Struct.* **249** 113327

[79] Tang P, Sun Y, Sahmani S and Madyira D M 2021 Isogeometric small-scale-dependent nonlinear oscillations of quasi-3D FG inhomogeneous arbitrary-shaped microplates with variable thickness *J. Braz. Soc. Mech. Sci. Eng.* **43** 343

[80] Sahmani S, Safaei B and Aldakheel F 2021 Surface elastic-based nonlinear bending analysis of functionally graded nanoplates with variable thickness *Eur. Phys. J. Plus* **136** 676

[81] Yang Z, Lu H, Sahmani S and Safaei B 2021 Isogeometric couple stress continuum-based linear and nonlinear flexural responses of functionally graded composite microplates with variable thickness *Arch. Civ. Mech. Eng.* **21** 114

[82] Hou R, Sahmani S and Safaei B 2021 Nonlinear oscillations of elliptical and sector prefabricated nanoplate-type structures made of functionally graded building material *Phys. Scr.* **96** 115704

[83] Wang P, Yuan P, Sahmani S and Safaei B 2021 Surface stress size dependency in nonlinear free oscillations of FGM quasi-3D nanoplates having arbitrary shapes with variable thickness using IGA *Thin-Walled Struct.* **166** 108101

[84] Yin S, Xiao Z, Deng Y, Zhang G, Liu J and Gu S 2021 Isogeometric analysis of size-dependent Bernoulli–Euler beam based on a reformulated strain gradient elasticity theory *Comput. Struct.* **253** 106577

[85] Xu J, Yang Z, Yang J and Li Y 2021 Free vibration analysis of rotating FG-CNT reinforced composite beams in thermal environments with general boundary conditions *Aerosp. Sci.Technol.* **118** 107030

[86] Sun J H, Zhou Z D, Sahmani S and Safaei B 2021 Microstructural size dependency in nonlinear lateral stability of random reinforced microshells via meshfree-based applied mathematical modeling *Int. J. Struct. Stab. Dyn.* **21** 2150164
[87] Sahmani S B S 2021 Microstructural-dependent nonlinear stability analysis of random checkerboard reinforced composite micropanels via moving Kriging meshfree approach *Eur. Phys. J. Plus* **136** 806
[88] Lu H, Zhou J, Sahmani S and Safaei B 2021 Nonlinear stability of axially compressed couple stress-based composite micropanels reinforced with random checkerboard nanofillers *Phys. Scr.* **96** 125703
[89] Yang Z, Safaei B, Sahmani S and Zhang Y 2022 A couple-stress-based moving Kriging meshfree shell model for axial postbuckling analysis of random checkerboard composite cylindrical microshells *Thin-Walled Struct.* **170** 108631
[90] Liu H, Safaei B and Sahmani S 2022 Combined axial and lateral stability behavior of random checkerboard reinforced cylindrical microshells via a couple stress-based moving Kriging meshfree model *Arch. Civ. Mech. Eng.* **22** 15
[91] Ma X, Sahmani S and Safaei B 2022 Probabilistic-based nonlinear stability analysis of randomly reinforced microshells under combined axial-lateral load using meshfree strain gradient formulations *Eng. Struct.* **262** 114344
[92] Zhao S, Zhang Y, Zhang Y, Zhang W, Yang J and Kitipornchai S 2022 Buckling of functionally graded hydrogen-functionalized graphene reinforced beams based on machine learning-assisted micromechanics models *Eur. J. Mech.* A **96** 104675
[93] Fu G, Zhang Z, Fu J and Zheng H 2022 On the strain gradient effects on buckling of the partially covered laminated microbeam *Appl. Math. Model.* **102** 472–91
[94] Rao R, Ye Z, Yang Z, Sahmani S and Safaei B 2022 Nonlinear buckling mode transition analysis of axial–thermal–electrical-loaded FG piezoelectric nanopanels incorporating nonlocal and couple stress tensors *Arch. Civ. Mech. Eng.* **22** 125
[95] Alshenawy R, Safaei B, Sahmani S, Elmoghazy Y, Al-Alwan A and Al Nuwairan M 2022 Buckling mode transition in nonlinear strain gradient-based stability behavior of axial–thermal–electrical loaded FG piezoelectric cylindrical panels at microscale *Eng. Anal. Boundary Elem.* **141** 36–64
[96] Alshenawy R, Sahmani S, Safaei B, Elmoghazy Y, Al-Alwan A and Al Nuwairan M 2023 Three-dimensional nonlinear stability analysis of axial–thermal–electrical loaded FG piezoelectric microshells via MKM strain gradient formulations *Appl. Math. Comput.* **439** 127623
[97] Phung-Van P, Thai C H, Nguyen-Xuan H and Abdel-Wahab M 2019 An isogeometric approach of static and free vibration analyses for porous FG nanoplates *Eur. J. Mech.* A **78** 103851
[98] Yang F, Chong A C M and Lam D C C *et al* 2002 Couple stress based strain gradient theory for elasticity *Int. J. Solids Struct.* **39** 2731–43
[99] Eringen A C 1972 Linear theory of nonlocal elasticity and dispersion of plane waves *Int. J. Eng. Sci.* **10** 425–35
[100] Yang F, Chong A C M, Lam D C C and Tong P 2002 Couple stress-based strain gradient theory for elasticity *Int. J. Solids Struct.* **39** 2731–43
[101] Chu J, Wang Y, Sahmani S and Safaei B 2022 Nonlinear large-amplitude oscillations of PFG composite rectangular microplates based upon the modified strain gradient elasticity theory *Int. J. Struct. Stab. Dyn.* **22** 2250068

[102] Wang J, Ma B, Gao J, Liu H, Safaei B and Sahmani S 2022 Nonlinear stability characteristics of porous graded composite microplates including various microstructural-dependent strain gradient tensors *Int. J. Appl. Mech.* **14** 2150129

[103] Zuo D, Safaei B, Sahmani S and Ma G 2022 Nonlinear free vibrations of porous composite microplates incorporating various microstructural-dependent strain gradient tensors *Appl. Math. Mech.* **43** 825–44

[104] Ma X, Sahmani S and Safaei B 2022 Quasi-3D large deflection nonlinear analysis of isogeometric FGM microplates with variable thickness via nonlocal stress–strain gradient elasticity *Eng. Comput.* **38** 3691–704

Chapter 2

The static behavior of functionally graded sandwich beams

Akgöz Bekir and Civalek Ömer

Sandwich structures are a type of composite structure and are basically obtained by combining two thin face layers with high strength and a core material of low density that is relatively thicker than the face layers. The weight savings offered by sandwich structures are very important for materials that require bending stiffness. Sandwich structures are found in many industrial applications due to their high flexural rigidity-to-weight and high strength-to-weight ratios. Sandwich structures, whose applications have expanded considerably thanks to developing technologies, are used in the aerospace, aviation, construction, biomedical, and automotive sectors in particular. Considering the usage areas, it is important to determine their mechanical behaviors and characteristics in order to create safe and optimum designs.

In this chapter, we analyze the bending of functionally graded (FG) sandwich beams subjected to uniformly distributed loads on an elastic foundation. The interaction between the FG sandwich beam and the elastic foundation is addressed through the use of a Winkler foundation model. Two types of sandwich beams are considered, namely those with a homogeneous core and FG skins (Type 1) and those with a FG core and homogeneous skins (Type 2). The material properties of the beam are assumed to change gradually through its thickness. The governing differential equations are derived by implementing the principle of minimum total potential energy based on Bernoulli–Euler beam theory. Navier's solution method is employed for the solution of simply supported sandwich beams.

2.1 Introduction

A relatively new category of composite materials called FG materials (FGMs) has been extensively studied over the last few years due to its desirable properties. FGMs are materials consisting of a combination of at least two different discrete components with continuous gradient properties, i.e. both mechanical and physical.

doi:10.1088/978-0-7503-5301-4ch2

FGMs have numerous benefits over other pure and laminated composite structures due to their continuous gradient properties. Normally, these properties change in certain direction(s), for example in the thickness and/or axial directions. Usually, FGMs are composed of a metal and a ceramic; the metal prevents the material from breaking when subjected to thermal stress, while the ceramic has resistance at extreme temperatures. Together, these materials exhibit hardness, high strength, and resistance to high temperature and corrosion. The idea of FGMs was first proposed by scientists in 1984 during a space project in Japan [1]. In conventional composite materials, marked property differences in the stresses of the constituent materials can occur at the interfaces between two different component materials due to material incompatibilities. FGMs overcome this problem and have made composites applicable in aircraft, vehicles, military and defense projects, space shuttles, biomedical fields, electronics, energy, and engineering structures. Due to the numerous benefits of FGMs, many studies of the dynamic stability and the dynamic and buckling properties of FGMs have been performed in recent years in an effort to better understand their mechanical performance [2–13].

Among the various configurations, one of the important forms of composite structures is called the sandwich form. Because they have high strength for their light weight, sandwich structures are widely used in aircraft, spacecraft, and marine and many other engineering applications. Normal sandwich structures usually consist of a soft core bonded to two sheaths of hard material. These structures often exhibit delamination problems at the interconnecting surfaces due to abrupt changes in the properties of the different constituent materials. To overcome this problem, sandwich FG structures are used. In sandwich FG structures, the core or coatings are formed from FGMs. In this case, the properties of the composite materials change smoothly from one surface to another, thus avoiding the stress concentration problem usually encountered in layered structures.

A sandwich structure consists of three basic elements. It consists of the upper and lower surfaces on the outside, the material called the core in the middle, and the adhesive layers that provide the connection. Each part fulfills its specific function in the unit. The aim here is to use materials with maximum efficiency. The surface materials are placed at a certain distance from each other to increase the moment of inertia, thus increasing the bending stiffness. A sandwich beam has considerably higher stiffness compared to a normal beam due to its high moment of inertia relative to a normal beam of the same width and weight. Sandwich beams are used in places where the weight-to-strength ratio of honeycomb structures is important (such as automotive, railway, and aircraft applications). They are also used to absorb energy in collisions. For example, honeycomb structures used in the front parts of high-speed trains and automobile chassis absorb the impact in the event of an accident, forming a safety cell.

Recently, a number of studies have been performed on the mechanical characteristics of sandwich beams, such as their bending [14–30], buckling [31–46], and vibration [47–60] responses. In this chapter, we investigate the static bending behavior of FG sandwich beams on elastic foundations under uniformly distributed loads. The interaction between the structure and the elastic foundation is simulated using the Winkler elastic foundation model. Two types of sandwich beam are considered, namely those with a homogeneous core and FG skins (Type 1) and those with an FG core and homogeneous skins (Type 2). The material properties of the beam are assumed to change gradually through its thickness. The equilibrium equations are derived by implementing principle of minimum total potential energy based on Bernoulli–Euler beam theory. Navier's solution method is used to solve the equilibrium equations for simply supported embedded sandwich beams.

2.2 Theory and formulation

Two types of sandwich beams (see figure 2.1) are taken into consideration in this chapter, namely sandwich beams with a homogeneous core and two FG skins (Type 1) and sandwich beams with an FG core and two homogeneous skins (Type 2).

The material properties of the beam, such as its Young's modulus (E) and Poisson's ratio (v), are assumed to vary smoothly in the thickness direction. The effective material properties can be calculated by the classical rule of mixtures, as follows:

$$E(z) = E_{\mathrm{m}} + (E_{\mathrm{c}} - E_{\mathrm{m}})V_{\mathrm{c}}(z) \tag{2.1}$$

$$v(z) = v_{\mathrm{m}} + (v_{\mathrm{c}} - v_{\mathrm{m}})V_{\mathrm{c}}(z), \tag{2.2}$$

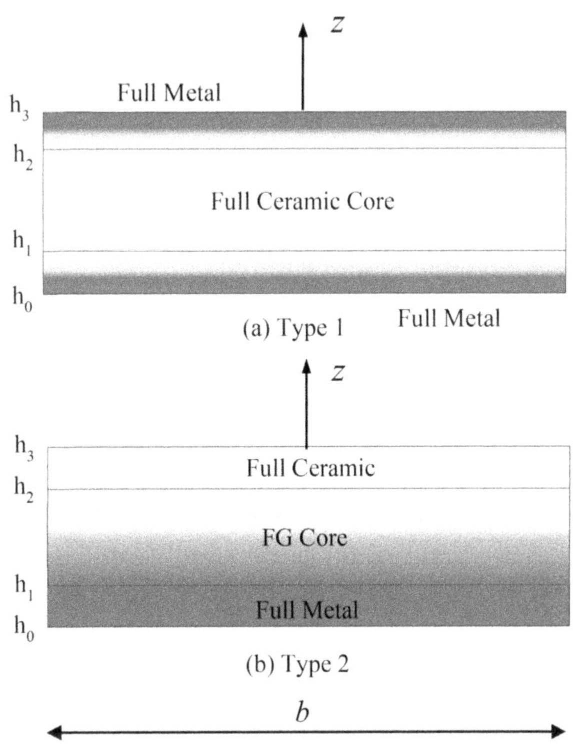

Figure 2.1. Cross sections of two types of FG sandwich beam: (a) Type 1 and (b) Type 2.

where V is the volume fraction of the material and the subscripts m and c represent the metal and ceramic phases, respectively. The volume fractions of the metal and ceramic constituents are formulated as

$$V_{\mathrm{m}} + V_{\mathrm{c}} = 1. \tag{2.3}$$

The volume fraction of the ceramic constituent can be defined as

Type 1:

$$V_{\mathrm{c}}(z) = \begin{cases} \left(\dfrac{z - h_o}{h_1 - h_0}\right)^k & z \in [h_0, h_1] \\ 1 & z \in [h_1, h_2] \\ \left(\dfrac{z - h_3}{h_2 - h_3}\right)^k & z \in [h_2, h_3] \end{cases} \tag{2.4}$$

Type 2:

$$V_{\mathrm{c}}(z) = \begin{cases} 0 & z \in [h_0, h_1] \\ \left(\dfrac{z - h_1}{h_2 - h_1}\right)^k & z \in [h_1, h_2], \\ 1 & z \in [h_2, h_3] \end{cases} \tag{2.5}$$

where k is the power-law index. Variations of the volume fraction of the ceramic phase V_{c} through the thickness are illustrated in figures 2.2–2.4 with various power-law indices for different schemes.

The strain energy in a linear elastic isotropic body can be written as

$$U = \frac{1}{2}\int_0^L \int_A \sigma_{ij}\varepsilon_{ij} dA dx, \tag{2.6}$$

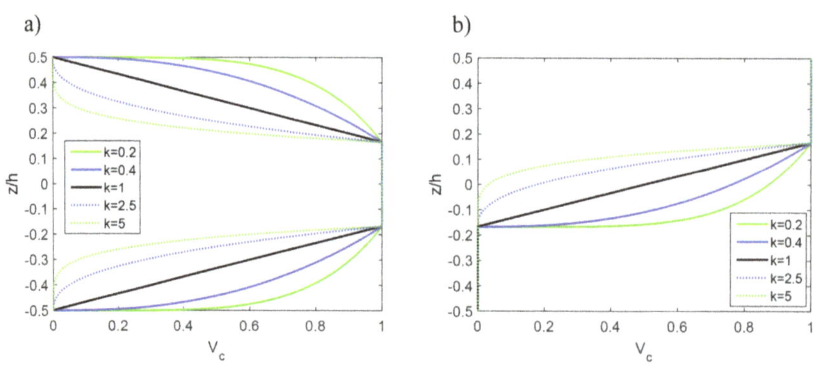

Figure 2.2. Variations in V_{c} through the thickness with various power-law indices for the scheme (1-1-1): (a) Type 1 and (b) Type 2.

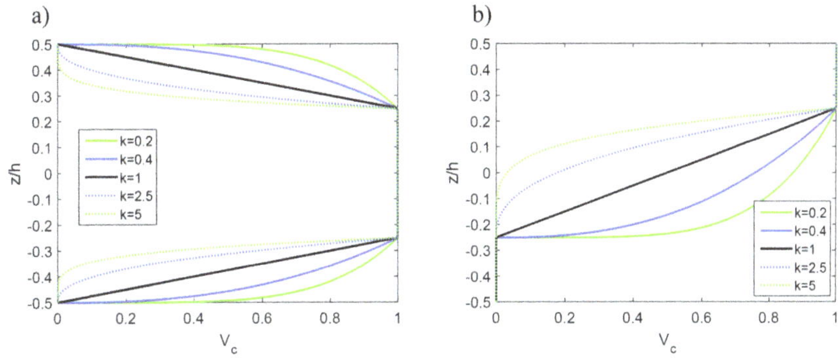

Figure 2.3. Variations of V_c through the thickness with various power-law indices for the scheme (1-2-1): (a) Type 1 and (b) Type 2.

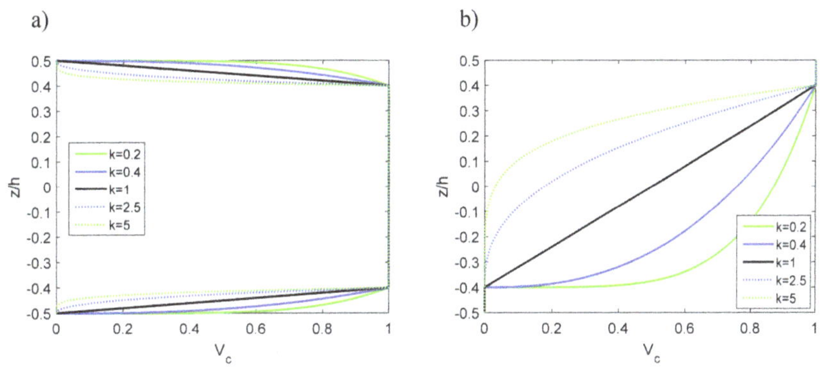

Figure 2.4. Variations of V_c through the thickness with various power-law indices for the scheme (1-8-1): (a) Type 1 and (b) Type 2.

where $\sigma_{ij}$and ε_{ij} are the components of the stress and strain tensors, respectively, defined by

$$\sigma_{ij} = \lambda \varepsilon_{kk} \delta_{ij} + 2\mu \varepsilon_{ij} \tag{2.7}$$

$$\varepsilon_{ij} = \frac{1}{2}(u_{i,\,j} + u_{j,\,i}), \tag{2.8}$$

where λ and μ are Lamé constants and δ_{ij} denotes the Kronecker delta.

According to Bernoulli–Euler beam theory, the components of displacement vector can be written as

$$u_1 = u - z\frac{dw(x)}{dx},\; u_2 = 0,\; u_3 = w(x), \tag{2.9}$$

where u_1, u_2, and u_3 are the displacements in the x-, y-, and z-directions and u and w denote the axial and transverse displacements of any point on the neutral axis, respectively. The nonzero components of the strain and stress tensors are obtained as follows (by neglecting the Poisson effect):

$$\varepsilon_{xx} = \frac{du}{dx} - z\frac{d^2w(x)}{dx^2} \tag{2.10}$$

$$\sigma_{xx} = E(z)\varepsilon_{xx}. \tag{2.11}$$

The governing equations of an FG sandwich beam subjected to bending can be derived with the aid of a variational principle, as follows:

$$\delta(U - W) = 0, \tag{2.12}$$

where U is the strain energy and W is the work done by the external loads. The first variation of the strain energy can be expressed as

$$\delta U = \int_0^L \int_A \sigma_{xx}\delta\varepsilon_{xx} dA dx = \int_0^L \left[N\frac{d\delta u}{dx} - M\frac{d^2\delta w}{dx^2}\right]dx. \tag{2.13}$$

The stress resultants are described as

$$N = A_x\frac{du}{dx} - B_x\frac{d^2w}{dx^2} \tag{2.14}$$

$$M = B_x\frac{du}{dx} - D_x\frac{d^2w}{dx^2}, \tag{2.15}$$

where

$$(A_x, B_x, D_x) = \int_A E(z)(1, z, z^2)dA \tag{2.16}$$

The first variation of the work done by the external loads is

$$\delta W = \int_0^L [f\delta u + (q - k_w w)\delta w]dx, \tag{2.17}$$

where f and q are the axial and transverse loads, respectively, and k_w is the Winkler foundation stiffness. If we use the above equations, integrate by parts, and set the coefficients δu and δw, the following equilibrium equations can be obtained:

$$\frac{dN}{dx} + f = 0 \tag{2.18}$$

$$\frac{d^2M}{dx^2} + q - k_w w = 0. \tag{2.19}$$

The corresponding boundary conditions ($x = 0, L$) can be expressed as

$$N = 0 \text{ or } u = 0 \tag{2.20}$$

$$\frac{dM}{dx} = 0 \text{ or } w = 0 \tag{2.21}$$

$$M = 0 \text{ or } \frac{dw}{dx} = 0. \tag{2.22}$$

Equations (2.18) and (2.19) can be rewritten in terms of displacements as follows:

$$A_x \frac{d^2u}{dx^2} - B_x \frac{d^3w}{dx^3} + f = 0 \tag{2.23}$$

$$-B_x \frac{d^3u}{dx^3} + D_x \frac{d^4w}{dx^4} - q + k_w w = 0. \tag{2.24}$$

2.3 Navier's solution technique

In this section, the governing equations under bending conditions are analytically solved for simply supported FG sandwich beams. Navier's solution technique is used to solve the related equations. The boundary conditions of a simply supported sandwich beam are expressed as follows:

$$N = 0,\ w = 0,\ M = 0 \tag{2.25}$$

The following expansions of the displacement field are assumed:

$$u(x) = \sum_{n=1}^{\infty} U_n \cos\beta x \tag{2.26}$$

$$w(x) = \sum_{n=1}^{\infty} W_n \sin\beta x, \tag{2.27}$$

where U_n and W_n are the unknown Fourier coefficients and $\beta = n\pi/L$. The applied transverse load q is written in Fourier series as

$$q(x) = \sum_{m=1}^{\infty} Q_n \sin\beta x \tag{2.28}$$

$$Q_n = \frac{2}{L}\int_0^L q(x)\sin\beta x dx, \tag{2.29}$$

in which Q_n is the Fourier coefficient and is given for a uniformly distributed load as follows:

$$q(x) = q_0;\ Q_n = \frac{4q_0}{n\pi} n = 1, 3, 5, \ldots . \tag{2.30}$$

Substituting equations (2.26)–(2.28) into equations (2.23) and (2.24) yields the following system of algebraic equations:

$$\begin{bmatrix} k_{11} & k_{12} \\ k_{21} & k_{22} \end{bmatrix} \begin{Bmatrix} U_n \\ W_n \end{Bmatrix} = \begin{Bmatrix} 0 \\ Q_n \end{Bmatrix}, \tag{2.31}$$

where

$$k_{11} = \beta^2 A_{xx},\ k_{12} = k_{21} = -\beta^3 B_{xx},\ k_{22} = \beta^4 D_{xx} + k_{\mathrm{w}}. \tag{2.32}$$

2.4 Results and discussion

In this section, some numerical examples of the bending behavior of FG embedded sandwich beams are given. The material properties for the metal and ceramic constituents are assumed to be [61]: $E_{\mathrm{m}} = 70$ GPa, $\upsilon_{\mathrm{m}} = 0.3$, $E_{\mathrm{c}} = 380$ GPa, $\upsilon_{\mathrm{c}} = 0.3$. The dimensionless deflection and Winkler parameter are defined as follows:

$$\overline{w} = 100w\frac{E_{\mathrm{m}}I}{q_0L^4} \tag{2.33}$$

$$K_{\mathrm{w}} = \frac{k_{\mathrm{w}}L^4}{E_{\mathrm{m}}I} \tag{2.34}$$

Figures 2.5–2.7 display the variation of the nondimensional deflection along the length with respect to various Winkler foundation parameters and schemes (1-1-1), (1-2-1), and (1-8-1) for $k = 1$, respectively. It is obvious from these figures that the biggest deflections occur at the mid-span of beam, that they are obtained for $K_{\mathrm{w}} = 0$, and that the deflections decrease as the Winkler parameter increases. In addition, it can be observed that the smallest deflections occur for the scheme (1-8-1), whereas the greatest ones occur for the scheme (1-1-1) for both Type 1 and Type 2.

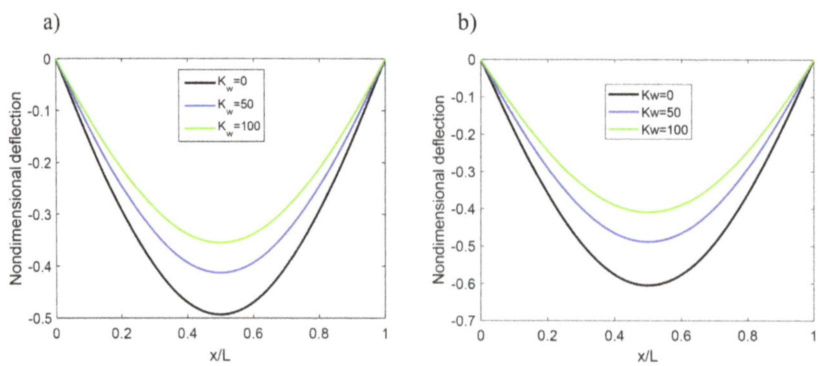

Figure 2.5. The effect of the Winkler parameter on the deflection of a simply supported FG sandwich beam for $k = 1$ and the scheme (1-1-1): (a) Type 1 and (b) Type 2.

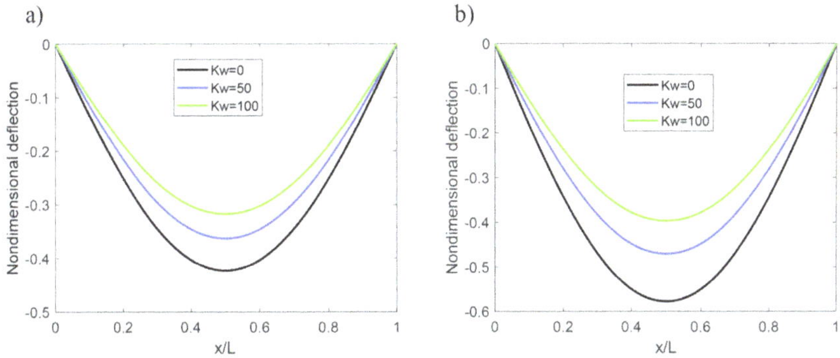

Figure 2.6. The effect of the Winkler parameter on the deflection of a simply supported FG sandwich beam for $k = 1$ and the scheme (1-2-1): (a) Type 1 and (b) Type 2.

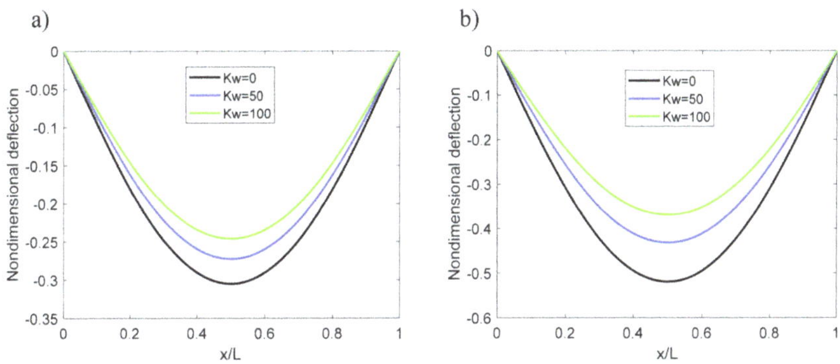

Figure 2.7. The effect of the Winkler parameter on the deflection of a simply supported FG sandwich beam for $k = 1$ and the scheme (1-8-1): (a) Type 1 and (b) Type 2.

Similarly, the variations of the nondimensional deflection along the length with respect to various Winkler foundation parameters and schemes (1-1-1), (1-2-1), and (1-8-1) for $k = 10$ are plotted in figures 2.8–2.10, respectively. It can be found from these figures that the deflections for $k = 10$ are bigger than those for $k = 1$. As stated above, an increase in the power-law index (k) leads to an increase in the metal constituent. Since the Young's modulus of the metal is smaller than the Young's modulus of the ceramic, the bending rigidity tends to decrease and the deflections tend to increase for higher power-law indices.

2.5 Conclusions

This chapter examined the static bending behavior of FG sandwich beams under a uniformly distributed load on an elastic foundation. The Winkler elastic foundation model was employed to consider the effect of the foundation. Two types of sandwich beam with different schemes were considered, namely one with a full ceramic core and FG skins and also one with an FG core and full ceramic and metal skins. Navier's solution method was employed for the solution of simply supported

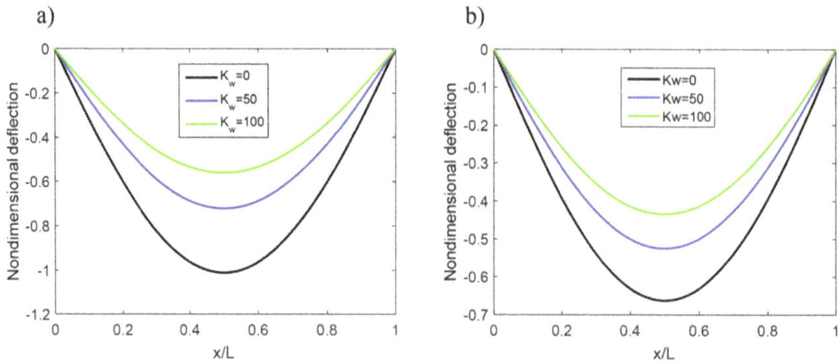

Figure 2.8. The effect of the Winkler parameter on the deflection of a simply supported FG sandwich beam for $k = 10$ and the scheme (1-1-1): (a) Type 1 and (b) Type 2.

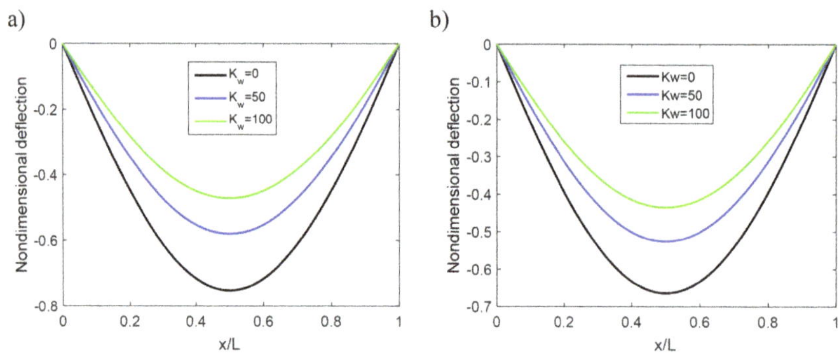

Figure 2.9. The effect of the Winkler parameter on the deflection of a simply supported FG sandwich beam for $k = 10$ and the scheme (1-2-1): (a) Type 1 and (b) Type 2.

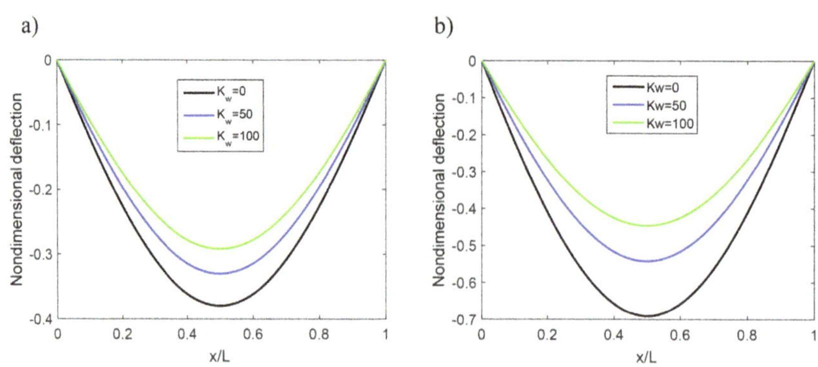

Figure 2.10. The effect of the Winkler parameter on the deflection of a simply supported FG sandwich beam for $k = 10$ and the scheme (1-8-1): (a) Type 1 and (b) Type 2.

sandwich beams. The effects of the Winkler foundation parameter, power-law index, and sandwich schemes on the deflections of FG sandwich beams were investigated in detail. It can be concluded that an increase in the Winkler parameter gives rise to a considerable decrease in deflection. It can also be confirmed that the smallest deflections were obtained for the (1-8-1) scheme of Type 1 sandwich structures. The reason for this is that this type of sandwich beam has the highest stiff core (ceramic) thickness. Moreover, the deflections tend to increase for increasing values of the power-law index (k).

References

[1] Koizumi M 1997 FGM activities in Japan *Composites* B **28** 1–4

[2] Kanu N J, Vates U K, Singh G K and Chavan S 2019 Fracture problems, vibration, buckling, and bending analyses of functionally graded materials: a state-of-the-art review including smart FGMS *Part. Sci. Technol.* **37** 579–604

[3] Naebe M and Shirvanimoghaddam K 2016 Functionally graded materials: a review of fabrication and properties *Appl. Mater. Today* **5** 223–45

[4] Nikbakht S, Kamarian S and Shakeri M 2019 A review on optimization of composite structures. Part II: functionally graded materials *Compos. Struct.* **214** 83–102

[5] Shukla A, Jain N and Chona R 2007 A review of dynamic fracture studies in functionally graded materials *Strain* **43** 76–95

[6] Xu F X, Zhang X and Zhang H 2018 A review on functionally graded structures and materials for energy absorption *Eng. Struct.* **171** 309–25

[7] Zhao S Y, Zhao Z, Yang Z C, Ke L L, Kitipornchai S and Yang J 2020 Functionally graded graphene reinforced composite structures: a review *Eng. Struct.* **210** 110339

[8] Zhang H, Gao C, Li H C, Pang F Z, Zou T D, Wang H F and Wang N 2020 Analysis of functionally graded carbon nanotube-reinforced composite structures: a review *Nanotechnol. Rev.* **9** 1408–26

[9] Zahedinejad P, Zhang C, Zhang H F and Ju S 2020 A comprehensive review on vibration analysis of functionally graded beams *Int. J. Struct. Stab. Dyn.* **20** 2030002

[10] Wu H L, Yang J and Kitipornchai S 2020 Mechanical analysis of functionally graded porous structures: a review *Int. J. Struct. Stab. Dyn.* **20** 2041015

[11] Sinha G P and Kumar B 2021 Review on vibration analysis of functionally graded material structural components with cracks *J. Vib. Eng. Technol.* **9** 23–49

[12] Sayyad A S and Ghugal Y M 2019 Modeling and analysis of functionally graded sandwich beams: a review *Mech. Adv. Mater. Struct.* **26** 1776–95

[13] Ghatage P S, Kar V R and Sudhagar P E 2020 On the numerical modelling and analysis of multi-directional functionally graded composite structures: a review *Compos. Struct.* **236** 111837

[14] Xu G D, Yang F, Zeng T, Cheng S and Wang Z H 2016 Bending behavior of graded corrugated truss core composite sandwich beams *Compos. Struct.* **138** 342–51

[15] Sobhy M 2021 Differential quadrature method for magneto-hygrothermal bending of functionally graded graphene/Al sandwich-curved beams with honeycomb core via a new higher-order theory *J. Sandw. Struct. Mater.* **23** 1662–700

[16] Sayyad A S and Ghugal Y M 2017 A unified shear deformation theory for the bending of isotropic, functionally graded, laminated and sandwich beams and plates *Int. J. Appl. Mech.* **9** 1750007

[17] Patil R, Joladarashi S and Kadoli R 2022 Bending and vibration studies of FG porous sandwich beam with viscoelastic boundary conditions: FE approach *Mech. Adv. Mater. Struct.*

[18] Nguyen N D, Vo T P and Nguyen T K 2020 An improved shear deformable theory for bending and buckling response of thin-walled FG sandwich I-beams resting on the elastic foundation *Compos. Struct.* **254** 112823

[19] Nguyen N D, Nguyen T K, Vo T P and Nguyen L B 2020 Bending, buckling and free vibration behaviors of thin-walled functionally graded sandwich and composite channel-section beams *Mech. Based Des. Struct. Mach.* **51** 932–60

[20] Magnucki K, Lewinski J, Far M and Michalak P 2020 Three-point bending of an expanded-tapered sandwich beam—analytical and numerical FEM study *Mech. Res. Commun.* **103** 103471

[21] Magnucki K 2019 Bending of symmetrically sandwich beams and I-beams—analytical study *Int. J. Mech. Sci.* **150** 411–9

[22] Li C, Shen H S and Wang H 2019 Nonlinear bending of sandwich beams with functionally graded negative Poisson's ratio honeycomb core *Compos. Struct.* **212** 317–25

[23] Koutoati K, Mohri F and Daya E 2021 Finite element approach of axial bending coupling on static and vibration behaviors of functionally graded material sandwich beams *Mech. Adv. Mater. Struct.* **28** 1537–53

[24] Karamanli A 2017 Bending behaviour of two directional functionally graded sandwich beams by using a quasi-3D shear deformation theory *Compos. Struct.* **174** 70–86

[25] Hadji L and Safa A 2020 Bending analysis of softcore and hardcore functionally graded sandwich beams *Earthq. Struct.* **18** 481–92

[26] Garg A, Chalak H D, Zenkour A M, Belarbi M O and Sahoo R 2022 Bending and free vibration analysis of symmetric and unsymmetric functionally graded CNT reinforced sandwich beams containing softcore *Thin-Walled Struct.* **170** 108626

[27] Garg A, Chalak H D, Belarbi M O and Zenkour A M 2022 A parametric analysis of free vibration and bending behavior of sandwich beam containing an open-cell metal foam core *Arch. Civ. Mech. Eng.* **22** 56

[28] Fereidoon A, Andalib M and Hemmatian H 2015 Bending analysis of curved sandwich beams with functionally graded core *Mech. Adv. Mater. Struct.* **22** 564–77

[29] Draiche K, Bousahla A A, Tounsi A and Hussain M 2021 An integral shear and normal deformation theory for bending analysis of functionally graded sandwich curved beams *Arch. Appl. Mech.* **91** 4669–91

[30] Belarbi M O, Khechai A, Bessaim A, Houari M S A, Garg A, Hirane H and Chalak H D 2021 Finite element bending analysis of symmetric and non-symmetric functionally graded sandwich beams using a novel parabolic shear deformation theory *Proc. Inst. Mech. Eng.* L **235** 2482–504

[31] Vo T P, Thai H T, Nguyen T K, Maheri A and Lee J 2014 Finite element model for vibration and buckling of functionally graded sandwich beams based on a refined shear deformation theory *Eng. Struct.* **64** 12–22

[32] Vo T P, Thai H T, Nguyen T K, Inam F and Lee J H 2015 A quasi-3D theory for vibration and buckling of functionally graded sandwich beams *Compos. Struct.* **119** 1–12

[33] Poortabib A and Maghsoudi M 2014 The analytical solution for buckling of curved sandwich beams with a transversely flexible core subjected to uniform load *Struct. Eng. Mech.* **52** 323–49

[34] Nguyen T K, Vo T P, Nguyen B D and Lee J 2016 An analytical solution for buckling and vibration analysis of functionally graded sandwich beams using a quasi-3D shear deformation theory *Compos. Struct.* **156** 238–52

[35] Nguyen T K and Nguyen B D 2015 A new higher-order shear deformation theory for static, buckling and free vibration analysis of functionally graded sandwich beams *J. Sandw. Struct. Mater.* **17** 613–31

[36] Nejadi M M, Mohammadimehr M and Mehrabi M 2021 Free vibration and buckling of functionally graded carbon nanotubes/graphene platelets Timoshenko sandwich beam resting on variable elastic foundation *Adv. Nano. Res.* **10** 525–34

[37] Nejadiand M M and Mohammadimehr M 2020 Buckling analysis of nano composite sandwich Euler–Bernoulli beam considering porosity distribution on elastic foundation using DQM *Adv. Nano. Res.* **8** 59–68

[38] Lanc D, Vo T P, Turkalj G and Lee J 2015 Buckling analysis of thin-walled functionally graded sandwich box beams *Thin-Walled Struct.* **86** 148–56

[39] Kahya V 2016 Buckling analysis of laminated composite and sandwich beams by the finite element method *Composites* B **91** 126–34

[40] Hamed M A, Mohamed S A and Eltaher M A 2020 Buckling analysis of sandwich beam rested on elastic foundation and subjected to varying axial in-plane loads *Steel Compos. Struct.* **34** 75–89

[41] Grygorowicz M, Magnucki K and Malinowski M 2015 Elastic buckling of a sandwich beam with variable mechanical properties of the core *Thin-Walled Struct.* **87** 127–32

[42] Eltaher M A and Mohamed S A 2020 Buckling and stability analysis of sandwich beams subjected to varying axial loads *Steel Compos. Struct.* **34** 241–60

[43] Douville M A and Le Grognec P 2013 Exact analytical solutions for the local and global buckling of sandwich beam-columns under various loadings *Int. J. Solids Struct.* **50** 2597–609

[44] Derikv M, Farhatnia F and Hodges D H 2021 Functionally graded thick sandwich beams with porous core: buckling analysis via differential transform method *Mech. Based Des. Struct. Mach.* **51** 3650–77

[45] Chakrabarti A, Chalak H D, Iqbal M A and Sheikh A H 2012 Buckling analysis of laminated sandwich beam with soft core *Lat. Am. J. Solids Struct.* **9** 367–81

[46] Cao P Y and Niu K M 2020 New unified model of composite sandwich panels/beams buckling introducing interlayer shear effects *Compos. Struct.* **252** 112722

[47] Lai Z D, Li Z W, Lin B C and Tang H P 2022 Free vibration analysis of rotating sandwich beams with FG-CNTRC face sheets in thermal environments with general boundary conditions *Z. Naturforsch.* A **77** 1153–73

[48] Fallah Y and Mohammadimehr M 2022 On the free vibration behavior of Timoshenko sandwich beam model with honeycomb core and nano-composite face sheet layers integrated by sensor and actuator layers *Eur. Phys. J. Plus* **137** 741

[49] Sayyad A S and Avhad P V 2022 A new higher order shear and normal deformation theory for the free vibration analysis of sandwich curved beams *Compos. Struct.* **280** 114948

[50] Chen C D and Su P W 2021 An analytical solution for vibration in a functionally graded sandwich beam by using the refined zigzag theory *Acta Mech.* **232** 4645–68

[51] Nguyen D K, Vu A N T, Pham V N and Truong T T 2021 Vibration of a three-phase bidirectional functionally graded sandwich beam carrying a moving mass using an enriched beam element *Eng. Comput.* **38** 4629–50

[52] Yuan L H and Liang S 2021 Free vibration analysis of embedded and co-cured perforated damping sandwich composite beam *J. Compos. Mater.* **55** 2605–15

[53] Senthilkumaran P, Venkatachalam R and Raja K V 2021 Vibration Analysis of a multi core sandwich composite beam with cutouts—a critical investigation *Mater. Res. Express* **8** 075301

[54] Kohsaka K, Ushijima K and Cantwell W J 2021 Study on vibration characteristics of sandwich beam with BCC lattice core *Mater. Sci. Eng.* B **264** 114986

[55] Liu B and Sun Y G 2022 Free vibration analysis of carbon-fiber-reinforced Miura-ori core sandwich beam: theoretical prediction and numerical simulation *Mech. Adv. Mater. Struct.* **29** 1266–75

[56] Keshtegar B, Motezaker M, Kolahchi R and Trung N T 2020 Wave propagation and vibration responses in porous smart nanocomposite sandwich beam resting on Kerr foundation considering structural damping *Thin-Walled Struct.* **154** 106820

[57] Li M, Du S J, Li F M and Jing X J 2020 Vibration characteristics of novel multilayer sandwich beams: modelling, analysis and experimental validations *Mech. Syst. Sig. Process.* **142** 106799

[58] Si H, Shen D M, Xia J H and Tahouneh V 2020 Vibration behavior of functionally graded sandwich beam with porous core and nanocomposite layers *Steel Compos. Struct.* **36** 1–16

[59] Selvaraj R and Ramamoorthy M 2022 Experimental and finite element vibration analysis of CNT reinforced MR elastomer sandwich beam *Mech. Based Des. Struct. Mach.* **50** 2414–26

[60] Alambeigi K, Mohammadimehr M, Bamdad M and Rabczuk T 2020 Free and forced vibration analysis of a sandwich beam considering porous core and SMA hybrid composite face layers on Vlasov's foundation *Acta Mech.* **231** 3199–218

[61] Vo T P, Thai H T, Nguyen T K, Inam F and Lee J 2015 Static behaviour of functionally graded sandwich beams using a quasi-3D theory *Composites* B **68** 59–74

Chapter 3

The static behavior of functionally graded plates

Shahriar Dastjerdi and Bekir Akgöz

This chapter investigates the static deformations of various structures made of functionally graded materials (FGMs). Different plate and shell structures that have geometric shapes are considered under symmetric and asymmetric static loadings with various boundary conditions. In addition, the impact of environmental conditions on this problem is considered using, e.g. thermoelastic analysis. The strain field is caused by large deformations. Therefore, the governing equations obtained using the energy method are a system of nonlinear partial differential equations. The governing equations are derived using first-order shear deformation theory (FSDT) and solved using an efficient semianalytical method called the semianalytical polynomial method (SAPM). The results thus obtained are evaluated against other results from various available references. In addition, the comprehensive nature of the analysis allows the impacts of various factors on the results, such as the loading conditions, the type of FGM, the boundary conditions, and the environmental factors, to be carefully studied. Overall, this chapter provides a comprehensive and complete static mechanical analysis of FGM structures in terms of the type of theory, loading, boundary conditions, and geometry of the structure, so that the reader will gain a general understanding of the subject and will not need to study different sources.

3.1 Introduction

The meaning of the word 'composite' is something that is made of various parts. Composite materials are obtained by combining two or more independent materials in a macrostructure. These materials can be classified according to the matrix material used or the shape and placement of the reinforcement elements. They are classified as metal, polymer, or ceramic composites according to the matrix material used. Depending on the shape and placement of the reinforcement elements, they may be classified as fibrous, particulate, layered, or mixed composites.

doi:10.1088/978-0-7503-5301-4ch3

The primary purpose of producing composite materials is to obtain materials with the desired mechanical properties. Carbon fiber/epoxy composites contain an epoxy polymer material, which has a very high tensile strength compared to steel materials and a low specific weight compared to carbon fiber. The epoxy holds the fibers together, and the result is a classic example of production based on the desired mechanical properties. From this combination, a material with high strength and low mass, composite materials with high strength in multiple directions can obtained using layered composite materials, in which composites oriented in different directions are brought together. Although layered composite materials exhibit only mechanical and thermal behaviors, adhesion problems between adjacent layers, damage, and fractures caused by interlayer bonding forces lead to a deterioration in the properties of such materials. At this point, it is helpful to use a functionally graded combination. For this purpose, FGMs are used. FGMs are composite materials in which the transition between two or more materials takes place according to a mathematical function. These materials eliminate the layers and thus the bonding forces and stress concentrations between them.

A group of Japanese scientists first proposed the idea of an FGM as a thermal barrier material in 1984 [1]. In this study, a functionally graded heat shield with ceramic materials on the outer surface and metal materials on the inner surface for use in high temperatures was designed and produced for space shuttles exposed to a temperature of 2100 K on the outer surface (a related project was active between 1987 and 1992). This temperature is too high for any industrial material to resist. FGMs reduce thermal and internal stresses and minimize damage. Consequently, the desired mechanical and thermal properties can be provided. In addition, the delamination damage caused by manufacturing defects and contact forces was greatly reduced through the use of FGMs.

When the adhesive strength of materials is increased, stress concentration is eliminated, and dynamic effects such as deformation, wear, corrosion, and dynamic effects can be controlled. For instance, there have been studies in which fracture resistance has been increased by adding a known ductile metal material to a brittle ceramic material. Some of the main fields of application of FGMs are defense technology, medicine, energy, aerospace, and mechanical and electronic engineering.

Many studies have investigated composites made of FGMs. They have also attracted attention in terms of both the production process and the determination of their mechanical behaviors, such as bending [2–23], buckling [24–42], and vibration [43–55]. As a result of rapid technological advances, FGMs have recently been used in micro- and nanoelectromechanical systems (MEMS and NEMS) such as shape memory alloy thin films, electrically actuated miniaturized devices, and atomic force microscopes. Many papers have been published on the static and dynamic responses of micro- and nanosized structures made of FGMs [56–75].

In this chapter, static deformations of various structures made of FGMs are investigated in a thermal environment using first-order shear deformation theory. Different plate and shell structures that have several geometric shapes are considered under symmetric and asymmetric static loadings with various boundary conditions.

The strain field is caused by large deformations. Therefore, the obtained governing equations obtained using the energy method are a system of nonlinear partial differential equations. The governing equations are solved using an efficient semi-analytical polynomial method. In addition, the comprehensive nature of the analysis allows the impacts of various factors on the results, such as the loading conditions, the type of FGM, the boundary conditions, and the environmental factors, to be carefully studied. Overall, this chapter provides a comprehensive and complete static mechanical analysis of FGM structures in terms of the type of theory, loading, boundary conditions, and geometry of the structure, so that the reader will gain a general understanding of the subject and will not need to study different sources.

3.2 Deriving the governing equations

As mentioned earlier, this chapter investigates the static behavior of structures made of FGMs. First, the structure's geometry is introduced. The structure's geometry determines the coordinate system and, as a result, the form of the governing equations. The optimal choice of geometry can make it possible to analyze one or more other geometrical structures simultaneously by choosing a specific geometry. The goal of this chapter is to investigate the static behavior of different FGM structures. Therefore, the geometry of the analyzed structure is divided into planar and curved geometries. In the following, the process of deriving the governing equations for each section is described in detail.

3.3 Planar FGM structures

The applicable planar structures often have rectangular, circular, annular/circular, and sectorial shapes. For a rectangular planar sheet, the Cartesian coordinate system (x, y, z) is suitable for analysis, and for circular, annular, and sectoral flat sheets, the cylindrical coordinate system (r, θ, z) is suitable for analysis. The schematic in figure 3.1 illustrates this point well.

However, as mentioned earlier, as a result of choosing a specific type of geometric structure, other geometric structures can be modeled by changing some parameters. For example, according to figure 3.1, other circular, annular/circular, and even rectangular shapes can be modeled by choosing a sectorial structure. The parameters used to introduce a sectorial geometric structure are R_i, R_o, θ_1, and θ_2, i.e. the inner and outer radiuses and the starting and ending angles, respectively. In the case of annular and circular structures, it is easy to model these geometric shapes by choosing the parameters mentioned below.

$$\begin{cases} \text{Annular: } R_i,\ R_o; \theta_1 = 0,\ \theta_2 = 360^\circ \\ \text{Circular: } R_i = 0,\ R_o; \theta_1 = 0,\ \theta_2 = 360^\circ \end{cases} \tag{3.1}$$

Nevertheless, the critical point is how to model a rectangular structure according to the shape of a sector. If the following parameters are chosen, a rectangular structure can be obtained, as shown in figure 3.2.

It can be seen that a rectangular structure can be obtained to an excellent approximation through the use of the sectorial shape (figure 3.2) [76]. Infinity (∞)

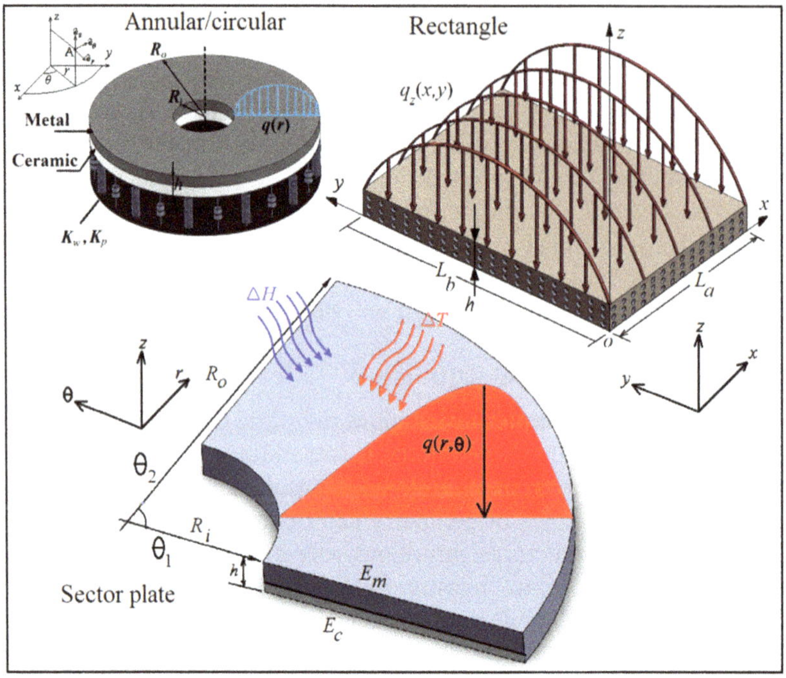

Figure 3.1. A schematic view of annular/circular, rectangular, and sectorial FGM plates.

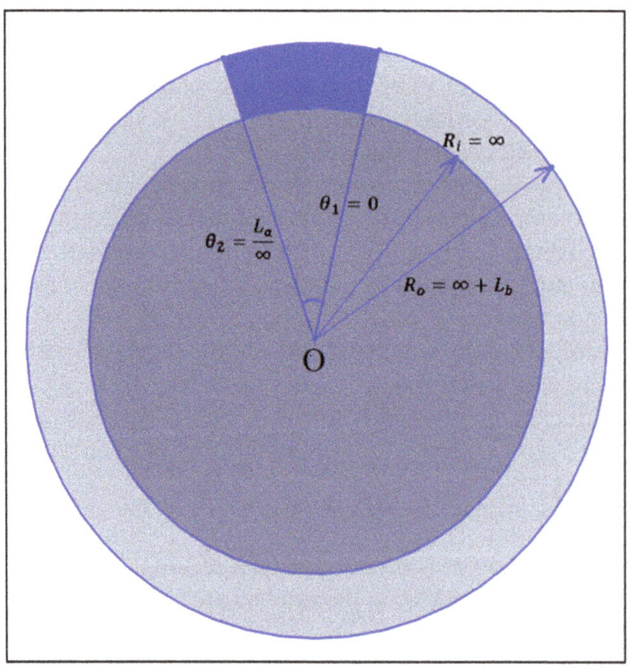

Figure 3.2. The parameters used to model a rectangle in sectorial geometry.

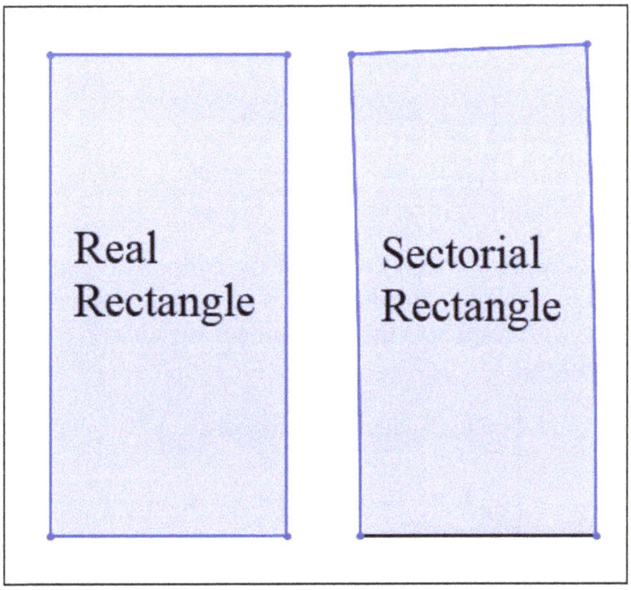

Figure 3.3. A comparison between actual and sectorial rectangular shapes.

has a physical meaning; it means that the inner radius should be chosen to be very large compared to the other parameters of the rectangular geometry, i.e. the length and the width, L_a and L_b, respectively. The larger the value of R_i chosen, the more accurate the simulated rectangle and the more consistent it is with a rectangular shape in the Cartesian coordinate system. Of course, it should be remembered that choosing a sizable value of R_i may cause calculation number errors while solving equations. For example, for a rectangle with the dimensions of length and width $L_a = 2$ m and $L_b = 1$ m, respectively, if $R_i = 50$ m is chosen, we have the following:

$$R_i = 50\,\text{m},\quad R_o = 51\text{ m};\quad \theta_1 = 0,\, \theta_2 = \left(\frac{2}{50} = 0.04\right)\text{rad} = \left(\frac{7.2}{\pi}\right)^{\circ}. \tag{3.2}$$

The difference between the two simulated shapes and the target rectangle can be seen schematically (to scale) in figure 3.3.

3.3.1 FGM sector sheet

It was mentioned earlier that due to the sectorial shape, the use of a cylindrical coordinate system is recommended. It is possible to convert the parameters of two cylindrical coordinates to the Cartesian system using the following equations (figure 3.2):

$$\begin{cases} x = r\cos(\theta) \\ y = r\sin(\theta) \\ z = z \end{cases} \begin{cases} r = \sqrt{x^2 + y^2} \\ \theta = \tan^{-1}\left(\dfrac{y}{x}\right) \end{cases} R_i \leqslant r \leqslant R_o; 0^\circ \leqslant \theta \leqslant 360^\circ. \tag{3.3}$$

Based on the above equations, the gradient of changes can be written using the following expressions:

$$\begin{cases} \vec{\nabla} = \dfrac{\partial}{\partial r}\hat{e}_r + \dfrac{1}{r}\dfrac{\partial}{\partial \theta}\hat{e}_\theta + \dfrac{\partial}{\partial z}\hat{e}_z \\ \nabla^2 = \dfrac{\partial^2}{\partial r^2} + \dfrac{1}{r}\dfrac{\partial}{\partial r} + \dfrac{1}{r^2}\dfrac{\partial^2}{\partial \theta^2}. \end{cases} \tag{3.4}$$

The relationships between the unit vectors of the cylindrical coordinate system and the Cartesian system as well as the derivatives of the unit vectors (in the cylindrical coordinate system) with respect to the independent variables r, θ, and z are given by the following equations:

$$\begin{cases} \left(\frac{\partial}{\partial r}\right) = \cos(\theta)\left(\frac{\partial}{\partial x}\right) + \sin(\theta)\left(\frac{\partial}{\partial y}\right) \\ \left(\frac{\partial}{\partial \theta}\right) = -\frac{1}{r}\sin(\theta)\left(\frac{\partial}{\partial x}\right) + \frac{1}{r}\cos(\theta)\left(\frac{\partial}{\partial y}\right) \\ \left(\frac{\partial}{\partial z}\right) = \left(\frac{\partial}{\partial z}\right) \end{cases}$$

$$\begin{bmatrix} \dfrac{\partial \hat{e}_r}{\partial r} = 0 & \dfrac{\partial \hat{e}_r}{\partial \theta} = \hat{e}_\theta & \dfrac{\partial \hat{e}_r}{\partial z} = 0 \\ \dfrac{\partial \hat{e}_\theta}{\partial r} = 0 & \dfrac{\partial \hat{e}_\theta}{\partial \theta} = -\hat{e}_r & \dfrac{\partial \hat{e}_\theta}{\partial z} = 0 \\ \dfrac{\partial \hat{e}_z}{\partial r} = 0 & \dfrac{\partial \hat{e}_z}{\partial \theta} = 0 & \dfrac{\partial \hat{e}_\theta}{\partial z} = 0 \end{bmatrix}. \tag{3.5}$$

3.3.2 Strain tensor

Using the equations given so far in the cylindrical coordinate system, the strain tensor can now be obtained using the following general equation.

$$\vec{\varepsilon} = \frac{1}{2}\left(\vec{\nabla}\vec{U} + \vec{\nabla}\vec{U}^T + \vec{\nabla}\vec{U}\cdot\vec{\nabla}\vec{U}^T\right) \tag{3.6}$$

$$\vec{\nabla}\vec{U} = \left[\hat{e}_r\left(\frac{\partial}{\partial r}\right) \;\; \hat{e}_\theta\left(\frac{1}{r}\frac{\partial}{\partial \theta}\right) \;\; \hat{e}_z\left(\frac{\partial}{\partial z}\right)\right]\begin{bmatrix} U_r\hat{e}_r \\ U_\theta\hat{e}_\theta \\ U_z\hat{e}_z \end{bmatrix} = \begin{bmatrix} \dfrac{\partial U_r}{\partial r} & \dfrac{\partial U_\theta}{\partial r} & \dfrac{\partial U_z}{\partial r} \\ \dfrac{1}{r}\left(\dfrac{\partial U_r}{\partial \theta} - U_\theta\right) & \dfrac{1}{r}\left(\dfrac{\partial U_\theta}{\partial \theta} + U_r\right) & \dfrac{1}{r}\dfrac{\partial U_z}{\partial \theta} \\ \dfrac{\partial U_r}{\partial z} & \dfrac{\partial U_\theta}{\partial z} & \dfrac{\partial U_z}{\partial z} \end{bmatrix}$$

$$\vec{\varepsilon}_{ij} = \begin{bmatrix} \varepsilon_{rr} & \varepsilon_{r\theta} & \varepsilon_{rz} \\ \varepsilon_{\theta r} & \varepsilon_{\theta\theta} & \varepsilon_{\theta z} \\ \varepsilon_{zr} & \varepsilon_{z\theta} & \varepsilon_{zz} \end{bmatrix}$$

$$
\begin{aligned}
\varepsilon_{rr} &= \left(\frac{\partial U_r}{\partial r}\right) + \frac{1}{2}\left(\left(\frac{\partial U_r}{\partial r}\right)^2 + \left(\frac{\partial U_\theta}{\partial r}\right)^2 + \left(\frac{\partial U_z}{\partial r}\right)^2\right);\ 2\varepsilon_{r\theta} = 2\varepsilon_{\theta r} \\
&= \left(\frac{\partial U_\theta}{\partial r}\right) + \frac{1}{r}\left(\frac{\partial U_r}{\partial \theta} - U_\theta\right) + \frac{1}{r}\left(\left(\frac{\partial U_r}{\partial r}\right)\left(\frac{\partial U_r}{\partial \theta} - U_\theta\right) + \left(\frac{\partial U_\theta}{\partial r}\right)\left(\frac{\partial U_\theta}{\partial \theta} + U_r\right)\right. \\
&\quad \left. + \left(\frac{\partial U_z}{\partial r}\right)\left(\frac{\partial U_z}{\partial \theta}\right)\right);\ 2\varepsilon_{rz} = 2\varepsilon_{zr} \\
&= \left(\frac{\partial U_r}{\partial z} + \frac{\partial U_z}{\partial r}\right) + \left(\frac{\partial U_r}{\partial r}\right)\left(\frac{\partial U_r}{\partial z}\right) + \left(\frac{\partial U_\theta}{\partial r}\right)\left(\frac{\partial U_\theta}{\partial z}\right) + \left(\frac{\partial U_z}{\partial r}\right)\left(\frac{\partial U_z}{\partial z}\right);\ \varepsilon_{\theta\theta} \\
&= \frac{1}{r}\left(\frac{\partial U_\theta}{\partial \theta} + U_r\right) + \frac{1}{2r^2}\left(\left(\frac{\partial U_r}{\partial \theta} - U_\theta\right)\left(\frac{\partial U_r}{\partial \theta} - U_\theta\right)\right. \\
&\quad \left. + \left(\frac{\partial U_\theta}{\partial \theta} + U_r\right)\left(\frac{\partial U_\theta}{\partial \theta} + U_r\right) + \left(\frac{\partial U_z}{\partial \theta}\right)^2\right);\ 2\varepsilon_{\theta z} = 2\varepsilon_{z\theta} \\
&= \left(\frac{1}{r}\frac{\partial U_z}{\partial \theta} + \frac{\partial U_\theta}{\partial z}\right) + \frac{1}{r}\left(\left(\frac{\partial U_r}{\partial \theta} - U_\theta\right)\left(\frac{\partial U_r}{\partial z}\right) + \left(\frac{\partial U_\theta}{\partial \theta} + U_r\right)\left(\frac{\partial U_\theta}{\partial z}\right)\right. \\
&\quad \left. + \left(\frac{\partial U_z}{\partial \theta}\right)\left(\frac{\partial U_z}{\partial z}\right)\right);\ \varepsilon_{zz} = \left(\frac{\partial U_z}{\partial z}\right) + \frac{1}{2}\left(\left(\frac{\partial U_r}{\partial z}\right)^2 + \left(\frac{\partial U_\theta}{\partial z}\right)^2 + \left(\frac{\partial U_z}{\partial z}\right)^2\right).
\end{aligned}
\tag{3.7}
$$

The above equations are completely nonlinear, and the strain components were introduced using the cylindrical coordinate system. This chapter tries to include environmental factors such as temperature and humidity in the calculations. Therefore, the strain generated by these factors can be rewritten as shown in the following equations. We note that α_T and β are the thermal conductivity coefficient (due to heat) and the humidity coefficient, respectively. ΔT and ΔH denote changes in the temperature and the humidity of the environment.

$$
\left(\vec{\varepsilon}_{ij}\right)_{\text{total}} = \begin{bmatrix} \varepsilon_{rr} & \varepsilon_{r\theta} & \varepsilon_{rz} \\ \varepsilon_{\theta r} & \varepsilon_{\theta\theta} & \varepsilon_{\theta z} \\ \varepsilon_{zr} & \varepsilon_{z\theta} & \varepsilon_{zz} \end{bmatrix} - (\alpha_T \Delta T + \beta \Delta H) \begin{bmatrix} 1 & 0 & 0 \\ 0 & 1 & 0 \\ 0 & 0 & 1 \end{bmatrix}
\tag{3.8}
$$

If we suppose that the strain tensor changes in the r and θ directions can be assumed to be insignificant (considering that the applied load is in the z-direction), the nonlinear terms in these two directions and in the thickness direction $((\varepsilon_{zz})_{\text{strain}} = \frac{\partial U_z}{\partial z})$ can be ignored. This hypothesis is the same as von Kàrmàn's theory. In other words, according to von Kàrmàn's theory of large deformations, expressions such as $\left(\frac{\partial U_\theta}{\partial r}\right)\left(\frac{\partial U_\theta}{\partial z}\right)$ in the strain tensor can be omitted, with the result that only the nonlinear terms of ∂U_z with respect to ∂r and $\partial \theta$ remain in the calculations. It should be noted that if loads are applied in directions other than the transverse direction (z), ignoring these terms may cause many errors in the results related to large non-transverse (radial and angular) loads. Therefore, in this chapter, all nonlinear expressions are included in the strain tensor so that the reader can understand this problem. However, due to the length of the fully nonlinear equations, von Kàrmàn's assumptions are used here. If the reader is interested in deriving fully nonlinear equations, he/she can follow the same process of deriving equations based on von Kàrmàn's assumptions as that used in this chapter and derive completely nonlinear static equations for sectorial sheets made of FGM. Therefore, according to von Kàrmàn's assumptions, the strain components obtained in equation (3.7) can be simplified and rewritten as the following equations:

$$\begin{aligned}\varepsilon_{rr} &= \frac{\partial U_r}{\partial r} + \frac{1}{2}\left(\frac{\partial U_z}{\partial r}\right)^2 - (\alpha_T \Delta T + \beta \Delta H); 2\varepsilon_{r\theta} = 2\varepsilon_{\theta r} \\ &= \frac{1}{r}\left(\frac{\partial U_r}{\partial \theta} - U_\theta\right) + \left(\frac{\partial U_\theta}{\partial r}\right) + \frac{1}{r}\left(\frac{\partial U_z}{\partial r}\right)\left(\frac{\partial U_z}{\partial \theta}\right); 2\varepsilon_{rz} = 2\varepsilon_{zr} \\ &= \left(\frac{\partial U_r}{\partial z} + \frac{\partial U_z}{\partial r}\right); \varepsilon_{\theta\theta} = \frac{1}{r}\left(\frac{\partial U_\theta}{\partial \theta} + U_r\right) + \frac{1}{2r^2}\left(\frac{\partial U_z}{\partial \theta}\right)^2 - (\alpha_T \Delta T + \beta \Delta H); \\ 2\varepsilon_{\theta z} &= 2\varepsilon_{z\theta} = \left(\frac{1}{r}\frac{\partial U_z}{\partial \theta} + \frac{\partial U_\theta}{\partial z}\right); (\varepsilon_{zz})_{\text{thermal}} = -(\alpha_T \Delta T + \beta \Delta H).\end{aligned} \tag{3.9}$$

3.3.3 FGMs

In the introduction to this chapter, the importance and unique properties of FGM materials were discussed. FGMs have attracted the attention of researchers due to their remarkable mechanical properties. The mechanical properties of FGMs (such as Young's modulus) change functionally from one surface to another surface of the structure [77]. These changes often take the form of changes from ceramic to metal properties. For example, in a sectorial sheet considered in this chapter (figure 3.1), the bottom has exclusively ceramic properties. However, with changes in the thickness direction under transverse load, its properties become completely metallic. These changes in the z-direction are introduced by the following functions for Young's modulus and Poisson's ratio.

$$\begin{cases} E(z) = (E_c - E_m)\left(\dfrac{z}{h} + \dfrac{1}{2}\right)^g + E_m\left(-\dfrac{h}{2} \leqslant z \leqslant \dfrac{h}{2}\right) \\ \nu(z) = (\nu_c - \nu_m)\left(\dfrac{z}{h} + \dfrac{1}{2}\right)^g + \nu_m\left(-\dfrac{h}{2} \leqslant z \leqslant \dfrac{h}{2}\right) \end{cases} \tag{3.10}$$

In the above equations, the power of g represents the intensity of the change from ceramic to metal. The higher the value, the more extreme the changes in material properties from ceramic to metal. Figure 3.4 is exemplifies this issue better for a sheet with a thickness of $h = 1$ cm and values of $E_m = 190$ GPa and E_c (Alumina–Al_2O_3) = 393 GPa (assuming different values of g).

Other properties, such as the thermal conductivity coefficient, the humidity coefficient, and Poisson's ratio can also be written using equation (3.10).

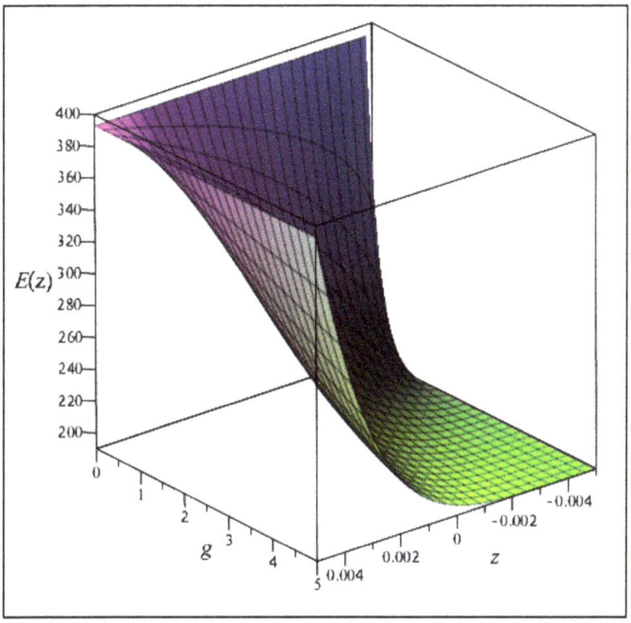

Figure 3.4. The variation of Young's modulus for an FGM structure.

Therefore, the mechanical properties of an analyzed FGM structure are not constant but rather depend on a function. Consequently, they affect the subsequent calculations used to describe the mechanical behavior of the FGM structure under applied loads.

3.3.4 The static governing equations for the FGM sectorial sheet

The static governing equations for the sectorial sheet can now be extracted. In the equations related to the components of the strain field, it can be seen that the functions U_r, U_θ, and U_z are functions of three independent variables r, θ, and z. It is expected that the governing equations are third-order partial differential equations. Solving third-order partial differential equations is very difficult and time-consuming. Therefore, it is possible to write the functions of the strain field in terms of two independent variables, r, and θ, using a simplification that does not subject the results to many errors. This simplification does not cause much error in the calculations, because R_i and R_o are much larger than the sheet thickness (h). In other words, the strains in the thickness direction can be ignored. One of the most popular theories, which provides good results for moderately thick sheets while being simple, is the FSDT. The FSDT introduces the displacement field in the form of the following equations in the cylindrical coordinate system.

$$\begin{cases} U_r(r,\theta,z) = u_0(r,\theta) + z.\,\varphi(r,\theta) \\ U_\theta(r,\theta,z) = v_0(r,\theta) + z.\,\psi(r,\theta) \\ U_z(r,\theta,z) = w_0(r,\theta) \end{cases} \tag{3.11}$$

In the above equations, u_0, v_0, and w_0 are the translational displacement functions, and φ and ψ are the functions for rotation around the θ and r axes, respectively. As can be seen, the independent variable z has been removed from the displacement functions. In other words, a two-dimensional analysis has been obtained instead of a three-dimensional analysis. As a result, the transverse displacement (i.e. the sheet's deflection) is only a function of r and θ and is constant through the thickness. The above statement is invalid if the thickness of the sheet is significant compared to its other dimensions, in which case three-dimensional or quasi-three-dimensional analyses should be used.

3.3.5 Two-dimensional static equations

As mentioned earlier, it is assumed that the thickness of the sheet is much smaller than its other dimensions. Therefore, the displacement field equations of the FSDT can be used, in which the shear force changes linearly along the thickness. If we substitute the FSDT displacement field (equation (3.11)) into the strain field (equation (3.9)), the components of the nonlinear strain field can be rewritten to give the following equations:

$$\begin{aligned} \varepsilon_{rr} &= \left(\frac{\partial u_0}{\partial r} + z\frac{\partial \varphi}{\partial r}\right) + \frac{1}{2}\left(\frac{\partial w_0}{\partial r}\right)^2 - (\alpha_T \Delta T + \beta \Delta H);\, 2\varepsilon_{r\theta} = 2\varepsilon_{\theta r} \\ &= \frac{1}{r}\left(\left(\frac{\partial u_0}{\partial r} + z\frac{\partial \varphi}{\partial r}\right) - v_0 - z.\,\psi\right) + \left(\frac{\partial v_0}{\partial r} + z\frac{\partial \psi}{\partial r}\right) + \frac{1}{r}\left(\frac{\partial w_0}{\partial r}\right)\left(\frac{\partial w_0}{\partial \theta}\right);\, 2\varepsilon_{rz} = 2\varepsilon_{zr} \\ &= \left(\varphi + \frac{\partial w_0}{\partial r}\right);\, \varepsilon_{\theta\theta} = \frac{1}{r}\left(\left(\frac{\partial v_0}{\partial \theta} + z\frac{\partial \psi}{\partial \theta}\right) + u_0 + z.\,\varphi\right) + \frac{1}{2r^2}\left(\frac{\partial w_0}{\partial \theta}\right)^2 \\ &\quad - (\alpha_T \Delta T + \beta \Delta H);\, 2\varepsilon_{\theta z} = 2\varepsilon_{z\theta} = \left(\frac{1}{r}\frac{\partial w_0}{\partial \theta} + \psi\right);\, (\varepsilon_{zz})_{\text{thermal}} \\ &= -(\alpha_T \Delta T + \beta \Delta H). \end{aligned} \tag{3.12}$$

All sections of this chapter use the energy method and the minimum potential energy variation principle to obtain the governing equations. The general equation used in this method is as follows:

$$\delta \Pi = \delta U_s + \delta U_p = 0. \tag{3.13}$$

In the above equations, the sign δ denotes variational calculus. δU_s and δU_p are the strain energy and the energy generated by external loads, respectively; these are given by the following equations ($i, j = r, \theta, z$):

$$\delta U_s = \iiint_V \left(\sigma_{ij}\delta\varepsilon_{ij}\right)dv$$
$$= \iiint_V \left(\sigma_{rr}\delta\varepsilon_{rr} + \sigma_{\theta\theta}\delta\varepsilon_{\theta\theta} + (\sigma_{zz}\delta\varepsilon_{zz})_{\text{thermal}} + \sigma_{r\theta}\delta\varepsilon_{r\theta} + 2\sigma_{rz}\delta\varepsilon_{rz} + 2\sigma_{\theta z}\delta\varepsilon_{\theta z}\right) \tag{3.14}$$
$$rdrd\theta dz$$

$$\delta U_p = -\iint_A \left(q_z + k_w w_0 + k_p \nabla^2 w_0\right)(\delta w_0)dA(dA = rdrd\theta). \tag{3.15}$$

In the above equation, σ_{ij} is the stress component in the FGM sector sheet, which is obtained using Hooke's law ($\sigma_{ij} = C\colon \varepsilon_{ij}$), resulting in the following tensor equation:

$$C = \frac{E(z)}{1-(\nu(z))^2}\begin{bmatrix} 1 & \nu(z) & 0 & 0 & 0 & 0 \\ \nu(z) & 1 & 0 & 0 & 0 & 0 \\ 0 & 0 & \frac{1-\nu(z)}{2} & 0 & 0 & 0 \\ 0 & 0 & 0 & \frac{1-\nu(z)}{2} & 0 & 0 \\ 0 & 0 & 0 & 0 & \frac{1-\nu(z)}{2} & 0 \\ 0 & 0 & 0 & 0 & 0 & \frac{1-\nu(z)}{2} \end{bmatrix}. \tag{3.16}$$

By substituting the strain components into equation (3.14) and calculating the variation, equation (3.14) can be rewritten as shown below:

$$\begin{aligned} \delta U_s = \iiint_V \Bigg(&\sigma_{rr}\left(\left(\frac{\partial \delta u_0}{\partial r} + z\frac{\partial \delta \varphi}{\partial r}\right) + \left(\frac{\partial w_0}{\partial r}\right)\left(\frac{\partial \delta w_0}{\partial r}\right)\right) \\ &+ \sigma_{\theta\theta}\left(\frac{1}{r}\left(\left(\frac{\partial \delta v_0}{\partial \theta} + z\frac{\partial \delta \psi}{\partial \theta}\right) + \delta u_0 + z.\ \delta\varphi\right) + \frac{1}{r^2}\left(\frac{\partial w_0}{\partial \theta}\right)\left(\frac{\partial \delta w_0}{\partial \theta}\right)\right) \\ &+ \sigma_{r\theta}\left(\frac{1}{r}\left(\left(\frac{\partial \delta u_0}{\partial r} + z\frac{\partial \delta \varphi}{\partial r}\right) - \delta v_0 - z.\ \delta\psi\right) + \left(\frac{\partial \delta v_0}{\partial r} + z\frac{\partial \delta \psi}{\partial r}\right)\right. \\ &+ \left.\frac{1}{r}\left(\left(\frac{\partial w_0}{\partial r}\right)\left(\frac{\partial \delta w_0}{\partial \theta}\right) + \left(\frac{\partial \delta w_0}{\partial r}\right)\left(\frac{\partial w_0}{\partial \theta}\right)\right)\right) + \sigma_{rz}\left(\delta\varphi + \frac{\partial \delta w_0}{\partial r}\right) \\ &+ \sigma_{\theta z}\left(\frac{1}{r}\frac{\partial \delta w_0}{\partial \theta} + \delta\psi\right)\Bigg) rdrd\theta dz. \end{aligned} \tag{3.17}$$

Finally, we convert the integral over the volume into the integral over the surface by integrating it in the z-direction. Applying integration introduces the stress and moment resultants as shown in the following equations, and equation (3.17) is also rewritten. The parameter κ_s is the shear correction factor.

$$\begin{cases} (N_{rr}, N_{\theta\theta}, N_{r\theta}) = \int_{-\frac{h}{2}}^{\frac{h}{2}} (\sigma_{rr}, \sigma_{\theta\theta}, \sigma_{r\theta})dz \\ (M_{rr}, M_{\theta\theta}, M_{r\theta}) = \int_{-\frac{h}{2}}^{\frac{h}{2}} (\sigma_{rr}, \sigma_{\theta\theta}, \sigma_{r\theta})zdz \\ \left(Q_{rz}, Q_{\theta z}\right) = \int_{-\frac{h}{2}}^{\frac{h}{2}} \kappa_s(\sigma_{rz}, \sigma_{\theta z})dz \end{cases} \tag{3.18}$$

$$\begin{aligned} \delta U_s \\ = \iint_A \Bigg(& \left(N_{rr}\frac{\partial \delta u_0}{\partial r} + M_{rr}\frac{\partial \delta\varphi}{\partial r}\right) + N_{rr}\left(\frac{\partial w_0}{\partial r}\right)\left(\frac{\partial \delta w_0}{\partial r}\right) \\ & + \frac{1}{r}\left(\left(N_{\theta\theta}\frac{\partial \delta v_0}{\partial \theta} + M_{\theta\theta}\frac{\partial \delta\psi}{\partial \theta}\right) + N_{\theta\theta}\delta u_0 + M_{\theta\theta}\delta\varphi\right) + \frac{N_{\theta\theta}}{r^2}\left(\frac{\partial w_0}{\partial \theta}\right)\left(\frac{\partial \delta w_0}{\partial \theta}\right) \\ & + \frac{1}{r}\left(\left(N_{r\theta}\frac{\partial \delta u_0}{\partial r} + M_{r\theta}\frac{\partial \delta\varphi}{\partial r}\right) - N_{r\theta}\delta v_0 - M_{r\theta}\delta\psi\right) \\ & + \left(N_{r\theta}\frac{\partial \delta v_0}{\partial r} + M_{r\theta}\frac{\partial \delta\psi}{\partial r}\right) + \frac{N_{r\theta}}{r}\left(\left(\frac{\partial w_0}{\partial r}\right)\left(\frac{\partial \delta w_0}{\partial \theta}\right) + \left(\frac{\partial \delta w_0}{\partial r}\right)\left(\frac{\partial w_0}{\partial \theta}\right)\right) \\ & + Q_{rz}\left(\delta\varphi + \frac{\partial \delta w_0}{\partial r}\right) + Q_{\theta z}\left(\frac{1}{r}\frac{\partial \delta w_0}{\partial \theta} + \delta\psi\right)\Bigg) r dr d\theta \end{aligned} \tag{3.19}$$

Some of the integrals in equation (3.19) are integrals by parts ($\int udv = uv - \int vdu$). We perform an integration by parts below, as an example.

$$\begin{aligned} \iint_A rN_{rr}\frac{\partial \delta u_0}{\partial r} dr d\theta &= rN_{rr}\delta u_0]_{R_i}^{R_o} - \iint_A \frac{\partial}{\partial r}(rN_{rr})\delta u_0 d\theta \\ &= rN_{rr}\delta u_0]_{R_i}^{R_o} - \iint_A \frac{\partial}{\partial r}\left(N_{rr} + r\frac{\partial \delta N_{rr}}{\partial r}\right)\delta u_0 d\theta \end{aligned} \tag{3.20}$$

The first part gives the mathematical definition of the boundary conditions ($rN_{rr}\delta u_0]_{R_i}^{R_o}$), and the governing equations are derived by considering the second part ($-\iint_A \frac{\partial}{\partial r}\left(N_{rr} + r\frac{\partial \delta N_{rr}}{\partial r}\right)\delta u_0 d\theta$). Finally, by collecting the similar values of δu_0, δv_0, δw_0, $\delta\varphi$, and $\delta\psi$ (and also considering δU_p), the two-dimensional static governing equations of the FGM sector sheet are extracted in the form of the following second-order partial differential equations:

$$\delta u_0: \left(\frac{\partial N_{rr}}{\partial r}\right) + \frac{1}{r}\left(\frac{\partial N_{r\theta}}{\partial \theta} + N_{rr} - N_{\theta\theta}\right) = 0 \tag{3.21}$$

$$\delta v_0: \left(\frac{\partial N_{r\theta}}{\partial r}\right) + \frac{1}{r}\left(\frac{\partial N_{\theta\theta}}{\partial \theta} + 2N_{r\theta}\right) = 0 \tag{3.22}$$

$$\begin{aligned}\delta w_0: &\left(\frac{\partial Q_{rz}}{\partial r}\right) + \frac{1}{r}\left(\frac{\partial Q_{\theta z}}{\partial \theta} + Q_{rz}\right)\\ &+ \left(q_z + k_w w_0 + k_p \nabla^2 w_0 + N_{rr}\left(\frac{\partial^2 w_0}{\partial r^2}\right) + N_{\theta\theta}\left(\frac{1}{r}\frac{\partial w_0}{\partial r} + \frac{1}{r^2}\frac{\partial^2 w_0}{\partial \theta^2}\right)\right.\\ &\left. + 2N_{r\theta}\left(\frac{1}{r}\frac{\partial^2 w_0}{\partial r \partial \theta} - \frac{1}{r^2}\frac{\partial w_0}{\partial \theta}\right)\right) = 0\end{aligned} \tag{3.23}$$

$$\delta \varphi: \left(\frac{\partial M_{rr}}{\partial r}\right) + \frac{1}{r}\left(\frac{\partial M_{r\theta}}{\partial \theta} + M_{rr} - M_{\theta\theta}\right) - Q_{rz} = 0 \tag{3.24}$$

$$\delta \psi: \left(\frac{\partial M_{r\theta}}{\partial r}\right) + \frac{1}{r}\left(\frac{\partial M_{\theta\theta}}{\partial \theta} + 2M_{r\theta}\right) - Q_{\theta z} = 0. \tag{3.25}$$

In addition, the mathematical definition of the boundary conditions can be expressed using the following equations, which are based on the first part of the integrals (e.g. equation (3.20)). The letters C, S, and F denote the clamped, simply supported, and free boundary conditions, respectively. It can be observed that there are four edges available for a sector sheet on $r = R_i$, R_o and $\theta = \theta_1$, θ_2 (figure 3.1). For example, the boundary conditions denoted by $CCCC$ refer to a sector sheet whose edges are all clamped. The first and second letters define the $r = R_i$, R_o boundaries, and the third and the fourth letters refer to the $\theta = \theta_1$, θ_2 boundaries.

$$C: u_0 = v_0 = w_0 = \varphi = \psi = 0 \quad r = R_i, R_o; \theta = \theta_1, \theta_2 \tag{3.26}$$

$$S: \begin{cases} u_0 = v_0 = w_0 = M_{rr} = \psi = 0 & r = R_i, R_o \\ u_0 = v_0 = w_0 = M_{\theta\theta} = \varphi = 0 & \theta = \theta_1, \theta_2 \end{cases} \tag{3.27}$$

$$F: \begin{cases} N_{rr} = N_{r\theta} = M_{rr} = M_{r\theta} = Q_{rz} = 0 & r = R_i, R_o \\ N_{\theta\theta} = N_{r\theta} = M_{\theta\theta} = M_{r\theta} = Q_{\theta z} = 0 & \theta = \theta_1, \theta_2 \end{cases}. \tag{3.28}$$

3.3.6 Static equations for quasi-three-dimensional sectorial FGM sheets

It was mentioned in an earlier section that if the thickness of the sheet is small compared to its other dimensions (R_i and R_o), the FSDT can be used to derive the governing partial differential equations in two dimensions, r and θ. However, if the

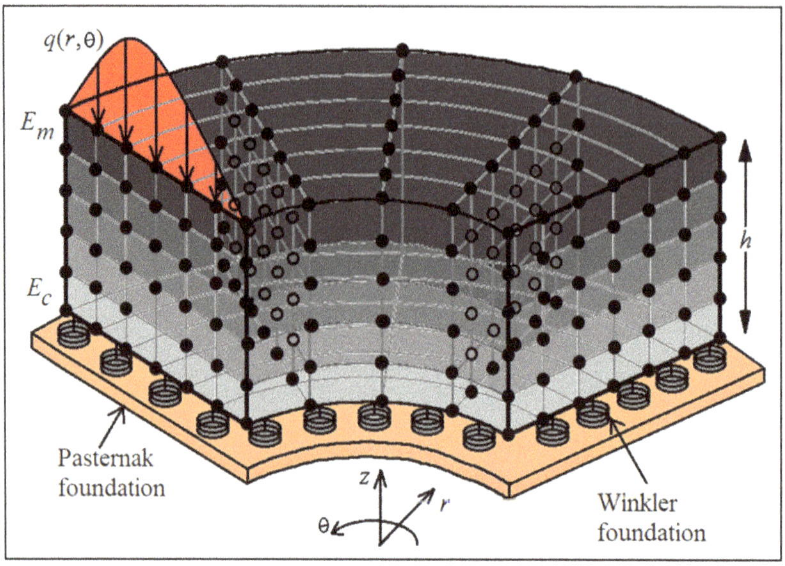

Figure 3.5. A thick sectorial FGM plate.

thickness of the sheet is significant, the strain changes in the z-direction should also be considered. A schematic view of the discussed geometry for a sectorial FGM plate is depicted in figure 3.5.

Let us consider the strain field in r, θ, and z. The governing partial differential equations are of the third order, which is very complicated to solve and requires many calculations that ordinary computers cannot solve. However, quasi-three-dimensional methods can be used instead. Quasi-three-dimensional methods do not ignore the strain in the z-direction and give complete three-dimensional results with appropriate accuracy. The authors of this chapter previously developed a quasi-3D method for the mechanical analysis of sheets [60, 78]. This method introduces the displacement field using a polynomial, as shown in the following equations:

$$\begin{cases} U_r(r,\theta,z) = \sum_{i=1}^{n} u_i(r,\theta) z^{(i-1)} \\ U_\theta(r,\theta,z) = \sum_{i=1}^{n} v_i(r,\theta) z^{(i-1)} \\ U_z(r,\theta,z) = \sum_{i=1}^{n} w_i(r,\theta) z^{(i-1)}. \end{cases} \tag{3.29}$$

If $n = 2$ is selected for U_r, U_θ, and $n = 1$ for U_z, the FSDT displacement field is obtained. The higher the value of n chosen, the closer the results of the quasi-3D analysis are to the three-dimensional results.

Based on the introduced quasi-three-dimensional displacement field, the same process is repeated to obtain the governing equations based on the FSDT. Suppose that $n = 4$ is selected in the quasi-three-dimensional displacement field. In this case, it is expected that 12 unknown functions and, as a result, 12 governing equations will be obtained. Of course, the number of independent variables of r and θ is similar to the number of independent variables in two-dimensional analysis. By solving the 12 equations thus obtained, it is possible to obtain similar results to those obtained through complete three-dimensional analysis (to an appropriate approximation).

The quasi-three-dimensional strain components and thus the variation of the potential energies generated by strains and external loads (δU_s, δU_p) can be written in the form of the following equations:

$$\varepsilon_{rr} = \frac{\partial\left(\sum_{i=1}^{4} u_i(r,\theta) z^{(i-1)}\right)}{\partial r} + \frac{1}{2}\left(\frac{\partial\left(\sum_{i=1}^{4} w_i(r,\theta) z^{(i-1)}\right)}{\partial r}\right)^2 - (\alpha_T \Delta T + \beta \Delta H);\, 2\varepsilon_{r\theta} = 2\varepsilon_{\theta r}$$

$$= \frac{1}{r}\left(\frac{\partial\left(\sum_{i=1}^{4} u_i(r,\theta) z^{(i-1)}\right)}{\partial \theta} - \left(\sum_{i=1}^{4} v_i(r,\theta) z^{(i-1)}\right)\right) + \left(\frac{\partial\left(\sum_{i=1}^{4} v_i(r,\theta) z^{(i-1)}\right)}{\partial r}\right)$$

$$+ \frac{1}{r}\left(\frac{\partial\left(\sum_{i=1}^{4} w_i(r,\theta) z^{(i-1)}\right)}{\partial r}\right)\left(\frac{\partial\left(\sum_{i=1}^{4} w_i(r,\theta) z^{(i-1)}\right)}{\partial \theta}\right);\, 2\varepsilon_{rz} = 2\varepsilon_{zr} \tag{3.30}$$

$$= \left(\sum_{i=1}^{3} (i) u_{i+1}(r,\theta) z^{(i-1)}\right) + \frac{\partial\left(\sum_{i=1}^{4} w_i(r,\theta) z^{(i-1)}\right)}{\partial r};\, \varepsilon_{\theta\theta}$$

$$= \frac{1}{r}\left(\frac{\partial\left(\sum_{i=1}^{4} v_i(r,\theta) z^{(i-1)}\right)}{\partial \theta} + \left(\sum_{i=1}^{4} u_i(r,\theta) z^{(i-1)}\right)\right) + \frac{1}{2r^2}\left(\frac{\partial\left(\sum_{i=1}^{4} w_i(r,\theta) z^{(i-1)}\right)}{\partial \theta}\right)^2 - (\alpha_T \Delta T + \beta \Delta$$

$$H);\, 2\varepsilon_{\theta z} = 2\varepsilon_{z\theta} = \frac{1}{r}\frac{\partial\left(\sum_{i=1}^{4} w_i(r,\theta) z^{(i-1)}\right)}{\partial \theta} + \left(\sum_{i=1}^{3} (i) v_{i+1}(r,\theta) z^{(i-1)}\right);\, (\varepsilon_{zz})_{\text{thermal}}$$

$$= -(\alpha_T \Delta T + \beta \Delta H)$$

$$\delta U_s = \iiint_V \left(\sigma_{rr} \left(\frac{\partial \delta\left(\sum_{i=1}^{4} u_i(r,\theta) z^{(i-1)}\right)}{\partial r} + \frac{1}{2}\left(\frac{\partial \delta\left(\sum_{i=1}^{4} w_i(r,\theta) z^{(i-1)}\right)}{\partial r} \right)^2 \right) \right.$$

$$+ \sigma_{\theta\theta} \left(\frac{1}{r} \left(\frac{\partial \delta\left(\sum_{i=1}^{4} v_i(r,\theta) z^{(i-1)}\right)}{\partial \theta} + \delta\left(\sum_{i=1}^{4} u_i(r,\theta) z^{(i-1)}\right) \right) + \frac{1}{2r^2}\left(\frac{\partial \delta\left(\sum_{i=1}^{4} w_i(r,\theta) z^{(i-1)}\right)}{\partial \theta} \right)^2 \right)$$

$$+ \sigma_{r\theta} \left(\frac{1}{r} \left(\frac{\partial \delta\left(\sum_{i=1}^{4} u_i(r,\theta) z^{(i-1)}\right)}{\partial \theta} - \delta\left(\sum_{i=1}^{4} v_i(r,\theta) z^{(i-1)}\right) \right) + \left(\frac{\partial \delta\left(\sum_{i=1}^{4} v_i(r,\theta) z^{(i-1)}\right)}{\partial r} \right) \right. \tag{3.31}$$

$$\left. + \frac{1}{r} \left(\frac{\partial \delta\left(\sum_{i=1}^{4} w_i(r,\theta) z^{(i-1)}\right)}{\partial r} \right) \left(\frac{\partial \delta\left(\sum_{i=1}^{4} w_i(r,\theta) z^{(i-1)}\right)}{\partial \theta} \right) \right)$$

$$+ \sigma_{rz} \left(\delta\left(\sum_{i=1}^{3} (i) u_{i+1}(r,\theta) z^{(i-1)}\right) + \frac{\partial \left(\sum_{i=1}^{4} w_i(r,\theta) z^{(i-1)}\right)}{\partial r} \right)$$

$$\left. + \sigma_{\theta z} \left(\frac{1}{r} \frac{\partial \delta\left(\sum_{i=1}^{4} w_i(r,\theta) z^{(i-1)}\right)}{\partial \theta} + \delta\left(\sum_{i=1}^{3} (i) v_{i+1}(r,\theta) z^{(i-1)}\right) \right) \right) r dr d\theta dz$$

$$\delta U_p = -\iint_A \left(q_z + k_w \left(\sum_{i=1}^{4} w_i(r,\theta) z^{(i-1)} \right) + k_p \nabla^2 \left(\sum_{i=1}^{4} w_i(r,\theta) z^{(i-1)} \right) \right) \left(\delta\left(\sum_{i=1}^{4} w_i(r,\theta) z^{(i-1)} \right) \right) dA (dA = r dr d\theta). \tag{3.32}$$

Finally, the 12 quasi-three-dimensional partial differential governing equations for the sectorial FGM sheet can be introduced as follows:

$$\delta u_1: \left(\frac{\partial N_{rr}}{\partial r} \right) + \frac{1}{r} \left(\frac{\partial N_{r\theta}}{\partial \theta} + N_{rr} - N_{\theta\theta} \right) = 0 \tag{3.33}$$

$$\delta u_2: \left(\frac{\partial M_{rr}}{\partial r}\right) + \frac{1}{r}\left(\frac{\partial M_{r\theta}}{\partial \theta} + M_{rr} - M_{\theta\theta}\right) - Q_{rz} = 0 \tag{3.34}$$

$$\delta u_3: \left(\frac{\partial P_{rr}}{\partial r}\right) + \frac{1}{r}\left(\frac{\partial P_{r\theta}}{\partial \theta} + P_{rr} - P_{\theta\theta}\right) - 2M_{rz} = 0 \tag{3.35}$$

$$\delta u_4: \left(\frac{\partial H_{rr}}{\partial r}\right) + \frac{1}{r}\left(\frac{\partial H_{r\theta}}{\partial \theta} + H_{rr} - H_{\theta\theta}\right) - 3P_{rz} = 0 \tag{3.36}$$

$$\delta v_1: \left(\frac{\partial N_{r\theta}}{\partial r}\right) + \frac{1}{r}\left(\frac{\partial N_{\theta\theta}}{\partial \theta} + 2N_{r\theta}\right) = 0 \tag{3.37}$$

$$\delta v_2: \left(\frac{\partial M_{r\theta}}{\partial r}\right) + \frac{1}{r}\left(\frac{\partial M_{\theta\theta}}{\partial \theta} + 2M_{r\theta}\right) - Q_{\theta z} = 0 \tag{3.38}$$

$$\delta v_3: \left(\frac{\partial P_{r\theta}}{\partial r}\right) + \frac{1}{r}\left(\frac{\partial P_{\theta\theta}}{\partial \theta} + 2P_{r\theta}\right) - 2M_{\theta z} = 0 \tag{3.39}$$

$$\delta v_4: \left(\frac{\partial H_{r\theta}}{\partial r}\right) + \frac{1}{r}\left(\frac{\partial H_{\theta\theta}}{\partial \theta} + 2H_{r\theta}\right) - 3P_{\theta z} = 0 \tag{3.40}$$

$$\begin{aligned}\delta w_1: &\left(\frac{\partial Q_{rz}}{\partial r}\right) + \frac{1}{r}\left(\frac{\partial Q_{\theta z}}{\partial \theta} + Q_{rz}\right) \\ &+ \left(q_z + k_w\left(\sum_{i=1}^{4} w_i\left(\frac{h}{2}\right)^{(i-1)}\right) + k_p\nabla^2\left(\sum_{i=1}^{4} w_i\left(\frac{h}{2}\right)^{(i-1)}\right) + N_{rr}\left(\frac{\partial^2 w_1}{\partial r^2}\right)\right. \\ &\left. + N_{\theta\theta}\left(\frac{1}{r}\frac{\partial w_1}{\partial r} + \frac{1}{r^2}\frac{\partial^2 w_1}{\partial \theta^2}\right) + 2N_{r\theta}\left(\frac{1}{r}\frac{\partial^2 w_1}{\partial r \partial \theta} - \frac{1}{r^2}\frac{\partial w_1}{\partial \theta}\right)\right) = 0\end{aligned} \tag{3.41}$$

$$\begin{aligned}\delta w_2: &\left(\frac{\partial M_{rz}}{\partial r}\right) + \frac{1}{r}\left(\frac{\partial M_{\theta z}}{\partial \theta} + M_{rz}\right) - N_{zz} \\ &+ \left(q_z\left(\frac{h}{2}\right) + k_w\left(\sum_{i=1}^{4} w_i\left(\frac{h}{2}\right)^{(i)}\right) + k_p\nabla^2\left(\sum_{i=1}^{4} w_i\left(\frac{h}{2}\right)^{(i)}\right) + M_{rr}\left(\frac{\partial^2 w_2}{\partial r^2}\right)\right. \\ &\left. + \left(M_{\theta\theta} + rQ_{rz}\right)\left(\frac{1}{r}\frac{\partial w_2}{\partial r} + \frac{1}{r^2}\frac{\partial^2 w_2}{\partial \theta^2}\right) + 2M_{r\theta}\left(\frac{1}{r}\frac{\partial^2 w_2}{\partial r \partial \theta} - \frac{1}{r^2}\frac{\partial w_2}{\partial \theta}\right)\right) = 0\end{aligned} \tag{3.42}$$

$$\delta w_3: \left(\frac{\partial P_{rz}}{\partial r}\right) + \frac{1}{r}\left(\frac{\partial P_{\theta z}}{\partial \theta} + P_{rz}\right) - 2M_{zz}$$
$$+ \left(q_z\left(\frac{h}{2}\right)^2 + k_w\left(\sum_{i=1}^{4} w_i\left(\frac{h}{2}\right)^{(i+1)}\right) + k_p\nabla^2\left(\sum_{i=1}^{4} w_i\left(\frac{h}{2}\right)^{(i+1)}\right) + P_{rr}\left(\frac{\partial^2 w_3}{\partial r^2}\right)\right. \tag{3.43}$$
$$\left. + (P_{\theta\theta} + 2rM_{rz})\left(\frac{1}{r}\frac{\partial w_3}{\partial r} + \frac{1}{r^2}\frac{\partial^2 w_3}{\partial \theta^2}\right) + 2P_{r\theta}\left(\frac{1}{r}\frac{\partial^2 w_3}{\partial r \partial \theta} - \frac{1}{r^2}\frac{\partial w_3}{\partial \theta}\right)\right) = 0$$

$$\delta w_4: \left(\frac{\partial H_{rz}}{\partial r}\right) + \frac{1}{r}\left(\frac{\partial H_{\theta z}}{\partial \theta} + H_{rz}\right) - 3P_{zz}$$
$$+ \left(q_z\left(\frac{h}{2}\right)^3 + k_w\left(\sum_{i=1}^{4} w_i\left(\frac{h}{2}\right)^{(i+2)}\right) + k_p\nabla^2\left(\sum_{i=1}^{4} w_i\left(\frac{h}{2}\right)^{(i+2)}\right) + H_{rr}\left(\frac{\partial^2 w_4}{\partial r^2}\right)\right. \tag{3.44}$$
$$\left. + (H_{\theta\theta} + 2rP_{rz})\left(\frac{1}{r}\frac{\partial w_4}{\partial r} + \frac{1}{r^2}\frac{\partial^2 w_4}{\partial \theta^2}\right) + 2H_{r\theta}\left(\frac{1}{r}\frac{\partial^2 w_4}{\partial r \partial \theta} - \frac{1}{r^2}\frac{\partial w_4}{\partial \theta}\right)\right) = 0$$

$$\begin{cases} \left(N_{rr}, N_{\theta\theta}, N_{zz}, N_{r\theta}, Q_{rz}, Q_{\theta z}\right) = \int_{-\frac{h}{2}}^{\frac{h}{2}} (\sigma_{rr}, \sigma_{\theta\theta}, \sigma_{zz}, \sigma_{r\theta}, \sigma_{rz}, \sigma_{\theta z})dz \\ (M_{rr}, M_{\theta\theta}, M_{zz}, M_{r\theta}, M_{rz}, M_{\theta z}) = \int_{-\frac{h}{2}}^{\frac{h}{2}} (\sigma_{rr}, \sigma_{\theta\theta}, \sigma_{zz}, \sigma_{r\theta}, \sigma_{rz}, \sigma_{\theta z})zdz \end{cases} \tag{3.45}$$

$$\begin{cases} (P_{rr}, P_{\theta\theta}, P_{zz}, P_{r\theta}, P_{rz}, P_{\theta z}) = \int_{-\frac{h}{2}}^{\frac{h}{2}} (\sigma_{rr}, \sigma_{\theta\theta}, \sigma_{zz}, \sigma_{r\theta}, \sigma_{rz}, \sigma_{\theta z})z^2dz \\ (H_{rr}, H_{\theta\theta}, H_{r\theta}, H_{rz}, H_{\theta z}) = \int_{-\frac{h}{2}}^{\frac{h}{2}} (\sigma_{rr}, \sigma_{\theta\theta}, \sigma_{r\theta}, \sigma_{rz}, \sigma_{\theta z})z^3dz. \end{cases} \tag{3.46}$$

3.4 Shell-type FGM structures

If the cross-section of the FGM structure is curved, a shell structure is formed. There are different shell structures in terms of geometric shape, the most common of which are cylindrical, spherical, and conical shell structures. In the case of conical structures, it can be concluded that if the equations related to conical structures can be obtained, the results for cylindrical structures can also be obtained. Figure 3.6 shows schematics of the shapes of spherical, cylindrical, and conical structures (along with the corresponding coordinate system). It should be noted that a cylindrical coordinate system can also be used for the analysis of conical structures.

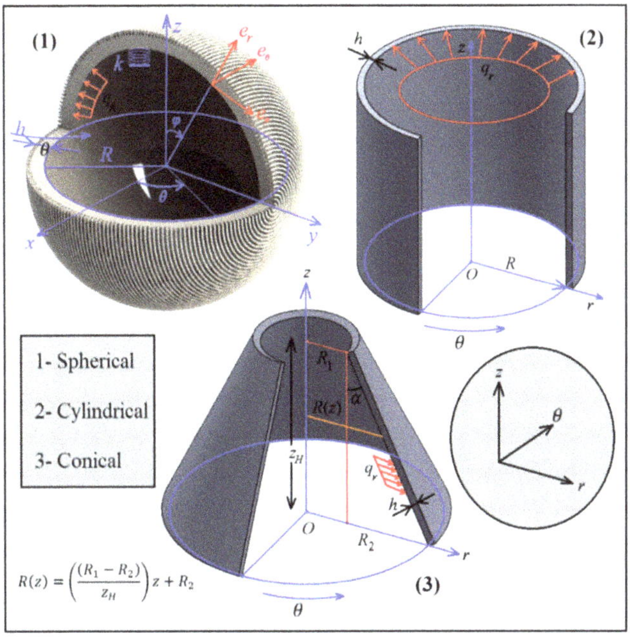

Figure 3.6. Spherical, cylindrical, and conical FGM structures.

3.4.1 The spherical FGM structure

First, the spherical FGM structure is considered and the governing equations are extracted. It is possible to convert spherical (r, φ, θ)and Cartesian (x, y, and z) coordinates according to the following equations:

$$\begin{cases} x = r \sin \varphi \cos \theta \\ y = r \sin \varphi \sin \theta \\ z = r \cos \varphi. \end{cases} \tag{3.47}$$

The method used for the planar sector FGM structure is repeated now. The gradient vector in the spherical coordinate system and the derivatives of the unit vectors with respect to the independent variables r, φ, and θ are given by the following equations:

$$\begin{cases} \vec{\nabla} = \frac{\partial}{\partial r}\hat{e}_r + \frac{1}{r}\frac{\partial}{\partial \varphi}\hat{e}_\varphi + \frac{1}{r\ \sin\varphi}\frac{\partial}{\partial \theta}\hat{e}_\theta \\ \nabla^2 = \left(\frac{\partial^2}{\partial r^2} + \frac{2}{r}\frac{\partial}{\partial r}\right) + \frac{1}{r^2 \sin\varphi}\frac{\partial}{\partial \varphi}\left(\sin\varphi\frac{\partial}{\partial \varphi}\right) + \frac{1}{r^2 \sin^2 \varphi}\frac{\partial^2}{\partial \theta^2} \end{cases} \tag{3.48}$$

$$\begin{bmatrix} \frac{\partial \hat{e}_r}{\partial r} = 0 & \frac{\partial \hat{e}_r}{\partial \varphi} = \hat{e}_\varphi & \frac{\partial \hat{e}_r}{\partial \theta} = \sin \varphi \hat{e}_\theta \\ \frac{\partial \hat{e}_\varphi}{\partial r} = 0 & \frac{\partial \hat{e}_\varphi}{\partial \varphi} = -\hat{e}_r & \frac{\partial \hat{e}_\varphi}{\partial \theta} = \cos \varphi \hat{e}_\theta \\ \frac{\partial \hat{e}_\theta}{\partial r} = 0 & \frac{\partial \hat{e}_\theta}{\partial \varphi} = 0 & \frac{\partial \hat{e}_\theta}{\partial \theta} = -\left(\sin \varphi \hat{e}_r + \cos \varphi \hat{e}_\varphi\right) \end{bmatrix}. \tag{3.49}$$

The nonlinear strain components (according to von Kàrmàn's theory) in the spherical coordinate system take the form of the following equations [79]. It is assumed that the sphere's radius is constant and equal to R.

$$\vec{\varepsilon} = \frac{1}{2}[\nabla \vec{U} + \nabla \vec{U}^T + \nabla \vec{U} \cdot \nabla \vec{U}^T] \tag{3.50}$$

$$\vec{\varepsilon}_{ij} = \begin{bmatrix} \varepsilon_{rr} & \varepsilon_{r\varphi} & \varepsilon_{r\theta} \\ \varepsilon_{\varphi r} & \varepsilon_{\varphi\varphi} & \varepsilon_{\varphi\theta} \\ \varepsilon_{\theta r} & \varepsilon_{\theta\varphi} & \varepsilon_{\theta\theta} \end{bmatrix} - (\alpha_T \Delta T + \beta \Delta H) \begin{bmatrix} 1 & 0 & 0 \\ 0 & 1 & 0 \\ 0 & 0 & 1 \end{bmatrix} \tag{3.51}$$

$$\begin{aligned} \nabla \vec{U} &= \begin{bmatrix} \frac{\partial}{\partial r}\hat{e}_r & \frac{1}{R}\frac{\partial}{\partial \varphi}\hat{e}_\varphi & \frac{1}{R\sin\varphi}\frac{\partial}{\partial \theta}\hat{e}_\theta \end{bmatrix} \begin{bmatrix} U_r\hat{e}_r \\ U_\varphi\hat{e}_\varphi \\ U_\theta\hat{e}_\theta \end{bmatrix} \\ &= \begin{bmatrix} \frac{\partial U_r}{\partial r} & \frac{\partial U_\varphi}{\partial r} & \frac{\partial U_\theta}{\partial r} \\ \frac{1}{R}\left(\frac{\partial U_r}{\partial \varphi} - U_\varphi\right) & \frac{1}{R}\left(\frac{\partial U_\varphi}{\partial \varphi} + U_r\right) & \frac{1}{R}\frac{\partial U_\theta}{\partial \varphi} \\ \frac{1}{R}\left(\frac{1}{\sin\varphi}\frac{\partial U_r}{\partial \theta} - U_\theta\right) & \frac{1}{R}\left(\frac{1}{\sin\varphi}\frac{\partial U_\varphi}{\partial \theta} - \cot\varphi U_\theta\right) & \frac{1}{R}\left(U_r + \cot\varphi U_\varphi + \frac{1}{\sin\varphi}\frac{\partial U_\theta}{\partial \theta}\right) \end{bmatrix} \end{aligned} \tag{3.52}$$

$$\nabla \vec{U}^T = \begin{bmatrix} \frac{\partial U_r}{\partial r} & \frac{1}{R}\left(\frac{\partial U_r}{\partial \varphi} - U_\varphi\right) & \frac{1}{R}\left(\frac{1}{\sin\varphi}\frac{\partial U_r}{\partial \theta} - U_\theta\right) \\ \frac{\partial U_\varphi}{\partial r} & \frac{1}{R}\left(\frac{\partial U_\varphi}{\partial \varphi} + U_r\right) & \frac{1}{R}\left(\frac{1}{\sin\varphi}\frac{\partial U_\varphi}{\partial \theta} - \cot\varphi U_\theta\right) \\ \frac{\partial U_\theta}{\partial r} & \frac{1}{R}\frac{\partial U_\theta}{\partial \varphi} & \frac{1}{R}\left(U_r + \cot\varphi U_\varphi + \frac{1}{\sin\varphi}\frac{\partial U_\theta}{\partial \theta}\right) \end{bmatrix} \tag{3.53}$$

$$\varepsilon_{rr} = \left(\frac{\partial U_r}{\partial r}\right) + \frac{1}{2}\left(\frac{\partial U_r}{\partial r}\right)^2 - (\alpha_T \Delta T + \beta \Delta H) \tag{3.54}$$

$$2\varepsilon_{r\varphi} = 2\varepsilon_{\varphi r} = \frac{\partial U_\varphi}{\partial r} + \frac{1}{R}\left(\frac{\partial U_r}{\partial \varphi} - U_\varphi\right) + \frac{1}{R}\left(\frac{\partial U_r}{\partial r}\right)\left(\frac{\partial U_r}{\partial \varphi}\right) \tag{3.55}$$

$$2\varepsilon_{r\theta} = 2\varepsilon_{\theta r} = \frac{\partial U_\theta}{\partial r} + \frac{1}{R}\left(\frac{1}{\sin\varphi}\frac{\partial U_r}{\partial \theta} - U_\theta\right) + \frac{1}{R\sin\varphi}\left(\frac{\partial U_r}{\partial r}\right)\left(\frac{\partial U_r}{\partial \theta}\right) \tag{3.56}$$

$$\varepsilon_{\varphi\varphi} = \frac{1}{R}\left(\frac{\partial U_\varphi}{\partial \varphi} + U_r\right) + \frac{1}{2R^2}\left(\frac{\partial U_r}{\partial \varphi}\right)^2 - (\alpha_T \Delta T + \beta \Delta H) \tag{3.57}$$

$$2\varepsilon_{\varphi\theta} = 2\varepsilon_{\theta\varphi} = \frac{1}{R}\left(\frac{\partial U_\theta}{\partial \varphi} + \frac{1}{\sin\varphi}\frac{\partial U_\varphi}{\partial \theta} - \cot\varphi U_\theta\right) + \frac{1}{R^2\sin\varphi}\left(\frac{\partial U_r}{\partial \varphi}\right)\left(\frac{\partial U_r}{\partial \theta}\right) \tag{3.58}$$

$$\varepsilon_{\theta\theta} = \frac{1}{R}\left(U_r + \cot\varphi U_\varphi + \frac{1}{\sin\varphi}\frac{\partial U_\theta}{\partial \theta}\right) + \frac{1}{2R^2\sin^2\varphi}\left(\frac{\partial U_r}{\partial \theta}\right)^2 - (\alpha_T \Delta T + \beta \Delta H) \tag{3.59}$$

We use the energy method explained in the previous section (section 3.1) and neglect the strain, i.e. $\varepsilon_{rr} = 0$. Other nonlinear terms (von Kàrmàn's nonlinear strains) and five linear static governing equations for the spherical FGM structure are extracted using the FSDT. A similar attempt can be made for the quasi-three-dimensional analysis, as shown in section 3.6.

$$\text{(FSDT displacement field)}\begin{cases} U_\varphi(\varphi, \theta, r) = u_0(\varphi, \theta) + r\psi_1(\varphi, \theta) \\ U_\theta(\varphi, \theta, r) = v_0(\varphi, \theta) + r\psi_2(\varphi, \theta) \\ U_r(\varphi, \theta, r) = w_0(\varphi, \theta) \end{cases} \tag{3.60}$$

$$\delta u_0\colon \left(\frac{\partial N_{\varphi\varphi}}{\partial \varphi}\right) + \cot\varphi\left(N_{\varphi\varphi} - N_{\theta\theta}\right) + N_{\varphi r} + \frac{1}{\sin\varphi}\frac{\partial N_{\varphi\theta}}{\partial \theta} = 0 \tag{3.61}$$

$$\delta v_0\colon \left(\frac{\partial N_{\varphi\theta}}{\partial \varphi}\right) + 2\cot\varphi N_{\varphi\theta} + N_{\theta r} + \frac{1}{\sin\varphi}\frac{\partial N_{\theta\theta}}{\partial \theta} = 0 \tag{3.62}$$

$$\begin{aligned}\delta w_0\colon &\frac{\partial}{\partial \varphi}\left(R\sin\varphi N_{\varphi r}\right) - R\sin\varphi\left(N_{\varphi\varphi} + N_{\theta\theta}\right) + R\frac{\partial N_{\theta r}}{\partial \theta} \\ &+ \left(q_r + k_w w_0 + k_p \nabla^2 w_0\right)R^2\sin\varphi = 0\end{aligned} \tag{3.63}$$

$$\begin{aligned}\delta\psi_1\colon &R\sin\varphi M_{\varphi r} + R^2\sin\varphi N_{\varphi r} - \frac{\partial}{\partial \varphi}\left(R\sin\varphi M_{\varphi\varphi}\right) + R\sin\varphi\cot\varphi M_{\theta\theta} - R\frac{\partial M_{\varphi\theta}}{\partial \theta} \\ &= 0\end{aligned} \tag{3.64}$$

$$\delta\psi_2\colon \frac{\partial M_{\varphi\theta}}{\partial \varphi} + 2\cot\varphi M_{\varphi\theta} + \frac{1}{\sin\varphi}\frac{\partial M_{\theta\theta}}{\partial \theta} - (M_{\theta r} + RN_{\theta r}) = 0 \tag{3.65}$$

$$\begin{cases} \left(N_{\varphi\varphi}, N_{\theta\theta}, N_{\varphi\theta}, N_{\varphi r}, N_{\theta r}\right) = \int_{-\frac{h}{2}}^{\frac{h}{2}} (\sigma_{\varphi\varphi}, \sigma_{\theta\theta}, \sigma_{\varphi\theta}, \kappa_s\sigma_{\varphi r}, \kappa_s\sigma_{\theta r})dr \\ \left(M_{\varphi\varphi}, M_{\theta\theta}, M_{\varphi\theta}, M_{\varphi r}, M_{\theta r}\right) = \int_{-\frac{h}{2}}^{\frac{h}{2}} (\sigma_{\varphi\varphi}, \sigma_{\theta\theta}, \sigma_{\varphi\theta}, \sigma_{\varphi r}, \sigma_{\theta r})rdr \end{cases} \tag{3.66}$$

3.4.2 Conical and cylindrical FGM structures

The static governing equations of the conical FGM structure can also be derived using a similar approach to that used to extract the static governing equations of the spherical FGM structure. Considering the geometry of a truncated cone, it is possible (figure 3.6) to model a cylindrical structure according to the introduced parameters. Therefore, the equations related to the truncated conical geometry are introduced, and the reader can quickly obtain the equations for a cylindrical structure by considering the governing equations of the conical FGM structure. Since the process of obtaining the governing equations for the conical structure is similar to that shown in the previous section (for a spherical FGM structure), only the final governing equations are given below to avoid repetition.

$$\text{(FSDT displacement field)} \begin{cases} U_\theta(\theta, z, r) = u_0(\theta, z) + r\psi_1(\theta, z) \\ U_z(\theta, z, r) = v_0(\theta, z) + r\psi_2(\theta, z) \\ U_r(\theta, z, r) = w_0(\theta, z) \end{cases} \tag{3.67}$$

$$\delta u_0\text{: } N_{r\theta} + \left(\frac{\partial N_{\theta\theta}}{\partial\theta}\right) + \frac{\partial}{\partial z}(R(z)N_{\theta z}) = 0 \tag{3.68}$$

$$\delta v_0\text{: } \left(\frac{\partial N_{\theta z}}{\partial\theta}\right) + \frac{\partial}{\partial z}(R(z)N_{zz}) = 0 \tag{3.69}$$

$$\begin{aligned} \delta w_0\text{: } & \left(\frac{\partial N_{r\theta}}{\partial\theta}\right) + \frac{\partial}{\partial z}(R(z)N_{rz}) - N_{\theta\theta} + \frac{1}{R(z)}\left(\frac{\partial}{\partial\theta}\left(N_{\theta\theta}\frac{\partial w_0}{\partial\theta}\right) - N_{\theta\theta}w_0\right) \\ & + \frac{\partial}{\partial\theta}\left(N_{\theta z}\frac{\partial w_0}{\partial z}\right) + \frac{\partial}{\partial z}\left(N_{\theta z}\frac{\partial w_0}{\partial\theta}\right) + \frac{\partial}{\partial z}\left(R(z)N_{zz}\frac{\partial w_0}{\partial z}\right) \\ & + R(z)\left(q_r + k_w w_0 + k_p\nabla^2 w_0\right) = 0 \end{aligned} \tag{3.70}$$

$$\delta\psi_1\text{: } \left(\frac{\partial M_{\theta\theta}}{\partial\theta}\right) + M_{r\theta} + \frac{\partial}{\partial z}(R(z)M_{\theta z}) - R(z)N_{r\theta} = 0 \tag{3.71}$$

$$\delta\psi_2\text{: } \left(\frac{\partial M_{\theta z}}{\partial\theta}\right) + \frac{\partial}{\partial z}(R(z)M_{zz}) - R(z)N_{rz} = 0 \tag{3.72}$$

$$\begin{cases} (N_{\theta\theta}, N_{zz}, N_{\theta z}, N_{r\theta}, N_{rz}) = \int_{-\frac{h}{2}}^{\frac{h}{2}} (\sigma_{\theta\theta}, \sigma_{zz}, \sigma_{\theta z}, \kappa_s\sigma_{r\theta}, \kappa_s\sigma_{rz})dr \\ (M_{\theta\theta}, M_{zz}, M_{\theta z}, M_{r\theta}) = \int_{-\frac{h}{2}}^{\frac{h}{2}} (\sigma_{\theta\theta}, \sigma_{zz}, \sigma_{\theta z}, \sigma_{r\theta})rdr \end{cases} \tag{3.73}$$

3.5 The solution method

The solution method that is applied in this chapter is the SAPM. This simple but efficient solution method has been presented and used by the authors of this chapter to solve various mechanical problems. To avoid a duplication of effort, we do not introduce the SAPM here; the interested reader can find the details in [80]. Efficiency and simplicity are the most significant advantages of the SAPM.

3.6 Numerical results and discussion

3.6.1 Validation

As mentioned earlier, the method used to solve the governing equations is the SAPM, which is based on a comprehensive polynomial based on the number of nodes in line with the independent variables of the problem. In figures 3.7(a) and (b), the convergence of the results can be seen for the sectorial and cylindrical structures with the characteristics given below. The boundary conditions considered are $CCCC$.

$$\begin{cases} \text{Sector: } R_i = 1\text{m},\ R_o = 2\text{m},\ h = 1\text{cm},\ \theta_2 = 90^\circ,\ E = 190\text{GPa}, \\ \qquad \nu = 0.29,\ q_z = 10\text{kPa} \\ \text{Cylinder: } R = 1\text{m},\ h = 1\text{cm},\ \varphi_1 = 20^\circ,\ \varphi_2 = 90^\circ,\ E = 190\text{GPa}, \\ \qquad \nu = 0.29,\ q_z = 1\text{MPa} \end{cases} \tag{3.74}$$

As can be seen, in the sectorial structure, the difference between the results for the $M = N = 5$ and $M = N = 7$ nodes in each direction of r and θ is not particularly large. If $N = M = 7$ nodes is selected, the results obtained have appropriate accuracy. Of course, it should be noted that choosing a large number of nodes in each direction (N and M) increases the calculation time and, in practice, does not affect the accuracy of the results. Therefore, the optimal and appropriate selection for the number of nodes (that provides sufficient assurance, regardless of the type of sheet or shell structure) is suggested here, namely $M = N = 9$. From now on, we use $N = M = 9$ nodes in each independent direction of the problem to simulate and obtain the results. The high convergence rate of SAPM is one of its strengths. Another point is that the boundary conditions do not affect the convergence rate.

A comparison and evaluation between the results of this study and those produced by the ABAQUS software for a sectorial structure and a spherical structure (with the same characteristics as those used for figure 3.7) are presented in figures 3.8 and 3.9.

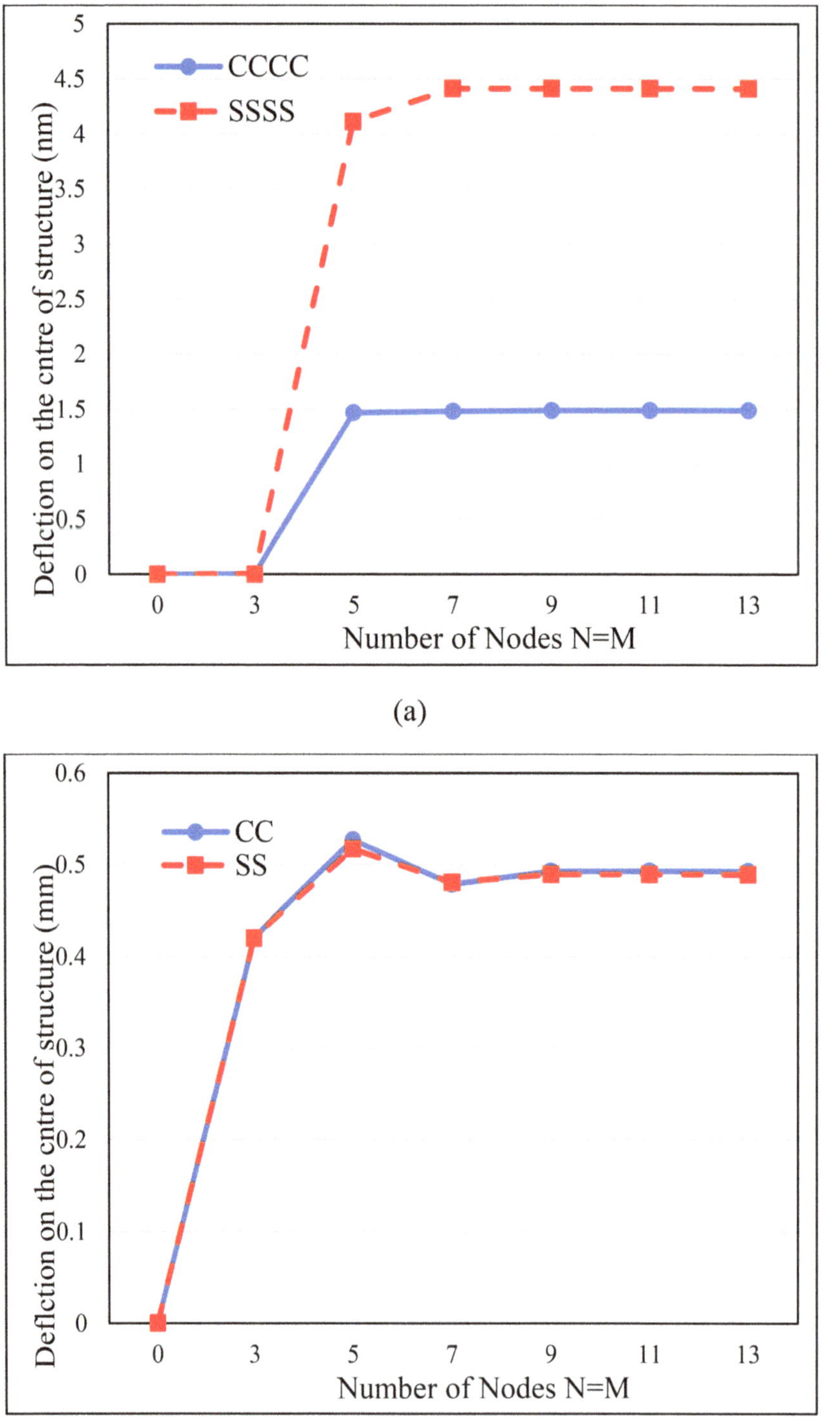

(b)

Figure 3.7. The convergence rates of the results for (a) a sectorial sheet and (b) a cylindrical panel.

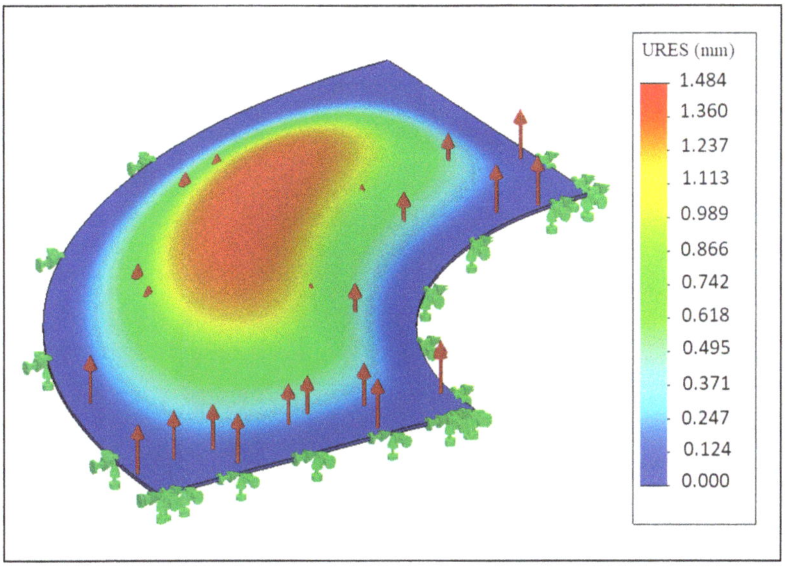

(a)

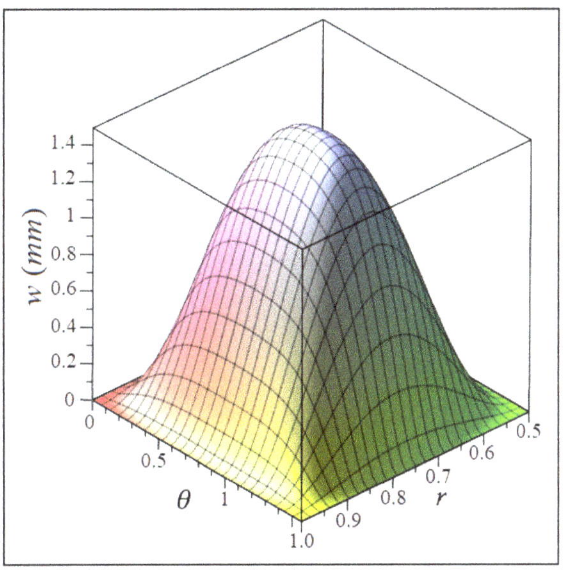

(b)

Figure 3.8. The deflection results for the sectorial sheet obtained using (a) ABAQUS and (b) the method outlined in this chapter.

As can be seen, the results produced by the theory and modeling described in this chapter show good agreement with the results produced by the ABAQUS software. The difference between the results is slight (for the structural deflection); therefore, the results can be used for subsequent simulations with sufficient confidence.

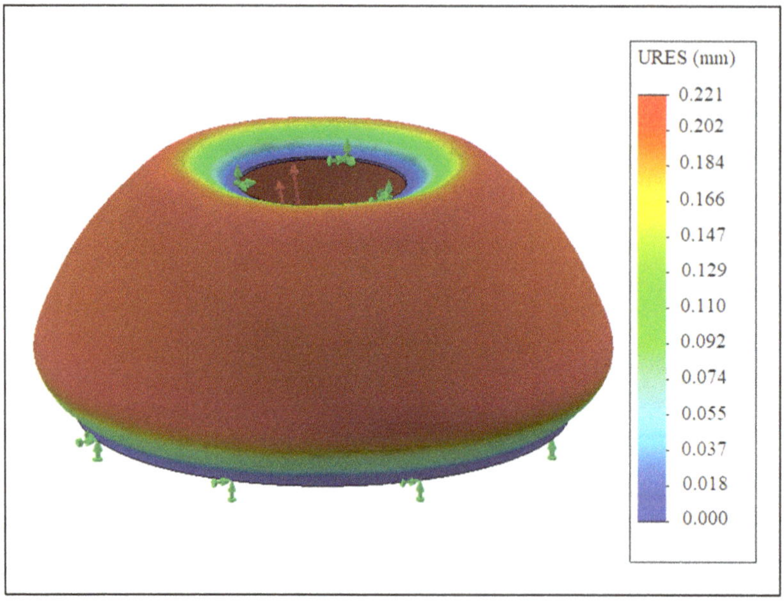

(a)

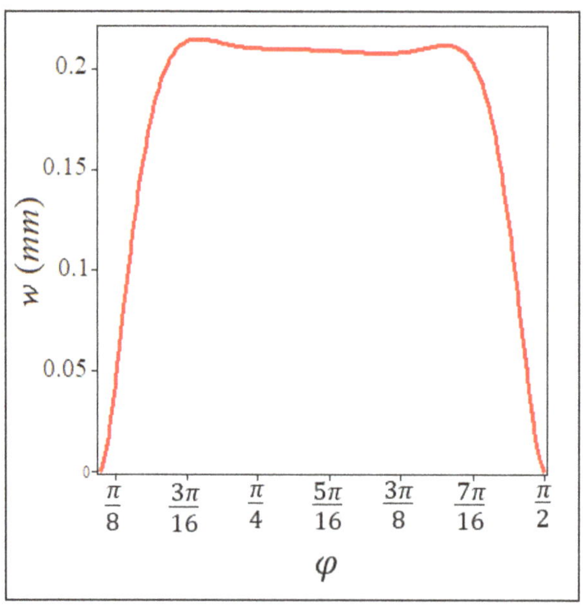

(b)

Figure 3.9. The deflection results for the spherical panel obtained using (a) ABAQUS and (b) the method outlined in this chapter.

3.6.2 Discussion

In this chapter, the material of the structure is considered to be an FGM. The properties of ceramic (98% alumina) and metal (type 304 steel) materials are assumed to be as follows:

$$E_c = 345\text{GPa}, \quad E_m = 190\text{GPa}, \quad \nu_c = 0.23, \quad \nu_m = 0.29. \tag{3.75}$$

Changes in Young's modulus in the direction of thickness, z, can be seen in figure 3.10.

As the value of g increases, this property of the material changes from ceramic to metallic more rapidly. In other words, for small values of g, this property of the material is closer to the properties of ceramics. As g increases, this property changes its nature more rapidly to that of a metal (for $g = 0$, it can be seen that this property of the material is entirely ceramic).

Figures 3.11(a) and (b) show the changes in the maximum deflection for two sectorial and spherical structures. Different values of the changing intensity of the FGM (g) have been considered. At the beginning of the changes, the maximum deflection is upward for increasing values of g. In other words, the deflection change from $g = 0$ (fully ceramic) to $g = 1$ is much greater than from $g = 1$ to $g = 10$ (i.e. ten times the value of g). Clearly, the lower the value of g, the more ceramic the behavior of the FGM. The deformation of the structure is indeed less for a full ceramic than for a full metal. However, if a fully ceramic structure is considered, its other properties, including high fragility, can be a significant defect. Therefore, the intensity of the change from ceramic to metal, which is represented by the value of g,

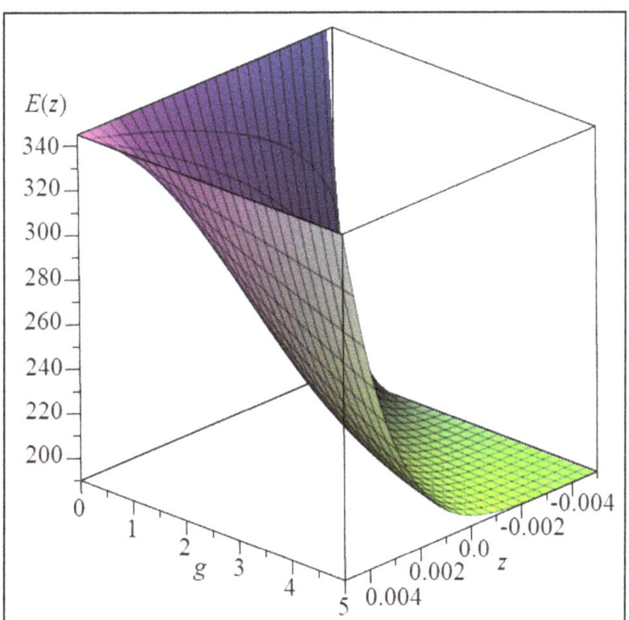

Figure 3.10. The variations of Young's modulus versus thickness and g parameter changes.

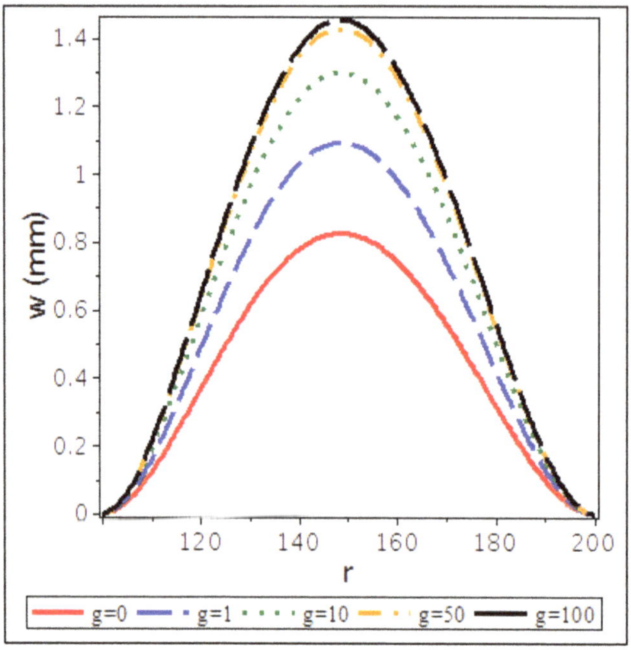

(a)

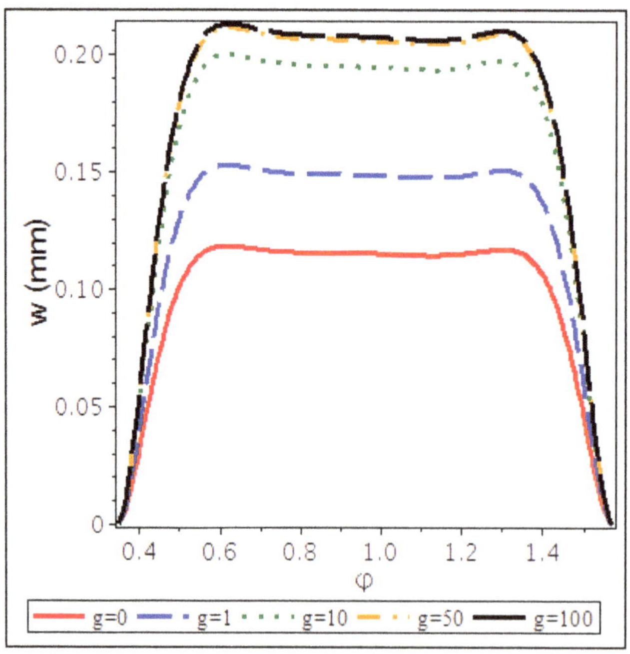

(b)

Figure 3.11. Deflection results versus different values of the *g* parameter for FGMs: (a) a sectorial sheet, (b) a spherical panel.

plays a significant role in determining the mechanical behavior of an FGM structure under applied loads. The trends of the changes shown in figures 3.11(a) and (b) are similar. We conclude that the results for FGM structures are not related to the types of their geometric shapes.

One of the strengths of modeling FGM structures in the form of sectorial geometry is the resulting ability to simulate a rectangular structure. This conclusion can be seen in figure 3.12. It can be seen that the results for a rectangular structure simulated by sectorial geometry are similar to the actual results for the rectangular structure (modeled in Cartesian coordinates). As mentioned earlier, the greater the value of the internal radius (∞) selected, the more the accuracy of the results of the rectangle simulated by sectorial geometry in the cylindrical coordinate system increases.

The use of nonlinear analysis (e.g. von Kàrmàn) causes large deformations. The results of the nonlinear analysis can be seen in figure 3.13 for a sectorial FGM sheet. As the transverse load on the sheet increases, the maximum deflection increases. If the linear analysis is used, then the changes in the maximum deflection are also linear. However, it can be seen that according to the nonlinear analysis, an increase in the transverse load applied to the structure also causes nonlinear changes in the maximum deflection (i.e. large deformations). Therefore, for a large transverse load applied to the structure, the linear analysis results are no longer valid, and the results of the nonlinear analysis should be used. In addition, the effects related to the property of the FGM are examined in figure 3.13. It can be seen that the higher the value of g, the steeper the slope of the nonlinear results. In other words, the closer

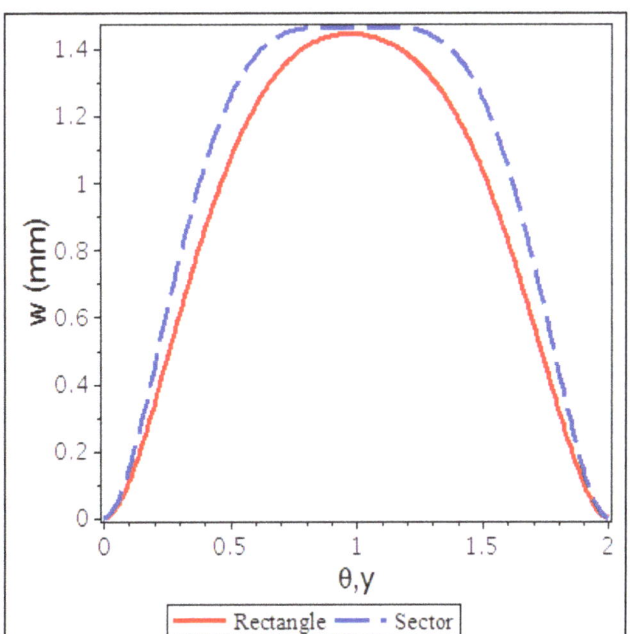

Figure 3.12. The modeling of a rectangular FGM sheet by a sectorial sheet.

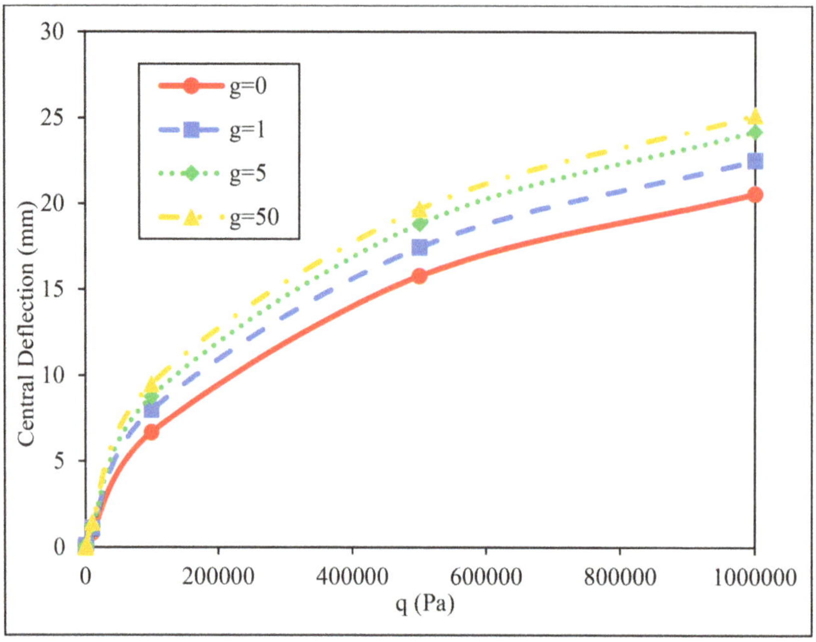

Figure 3.13. A large deformation analysis of a sectorial FGM sheet for different values of the g parameter.

the property of the FGM material is to the property of a metal, the more the effects related to nonlinear analysis increase. Finally, we conclude that for FGM structures, an increase in the metal properties causes the results of nonlinear analysis (large deformations) to depart further from the linear results. As a result, it is necessary to use nonlinear analysis to ensure the accuracy of the results, and the results of the linear analysis are no longer acceptable.

The effect of quasi-three-dimensional analysis can be seen in figure 3.14. It can be seen that the thickness of the analyzed structure, which is an annular/circular sheet, is significant compared to its other dimensions (in figure 3.14, the thickness of the sheet is $h = 1$ m). Therefore, according to figure 3.14, the amount of deflection of the sheet is not constant in the direction of its thickness, and its value is much higher on the surface under transverse load. In order to better understand the result that the sheet deflection value varies from $z = -\frac{h}{2}$ to $z = \frac{h}{2}$ according to the constant applied load, figure 3.15 shows the dimensionless sheet deflection value $w^{*}=\frac{w\,|_{z=\frac{h}{2}}}{w\,|_{z=-\frac{h}{2}}}$. The dimensionless thickness ($h^{*}=\frac{R_O}{h}$) is the ratio of the outer radius of the sheet (R_O) to its thickness (h). Visible changes in the dimensionless deflection difference occur for the dimensionless thickness values in the range of $0.15 < h^{*} < 0.2$.

$$\begin{aligned} &E_c = 345\text{GPa},\ E_m = 190\text{GPa},\ \nu_c = 0.23,\ \nu_m = 0.29,\ g = 5,\ R_i = 1\text{m}, \\ &\quad R_o = 2\text{m},\ h = 1\text{m},\ q_z = 1\text{GPa} \end{aligned} \tag{3.76}$$

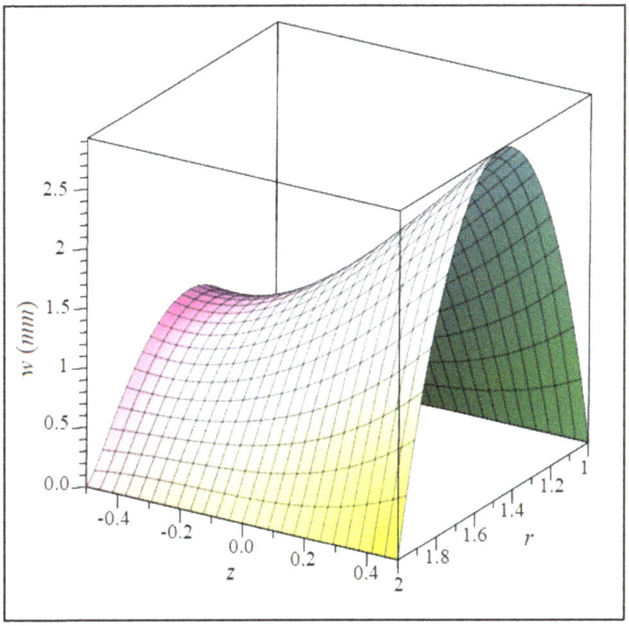

Figure 3.14. The quasi-three-dimensional deflection results for a thick annular/circular sheet in the radial (r) and thickness (z) directions.

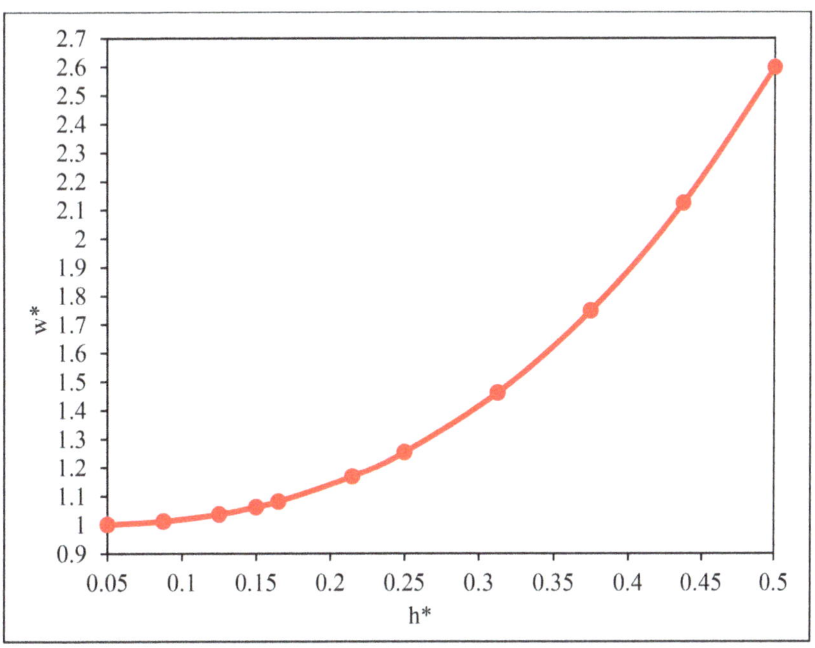

Figure 3.15. The dimensionless deflection w^* changes versus dimensionless thickness h^* for a thick annular/circular sheet.

Figures 3.16(a) and (b) show the deflection changes in the thickness direction for different values of the g coefficient of an FGM for two annular/circular and spherical structures. It can be seen that increasing the g value in the spherical

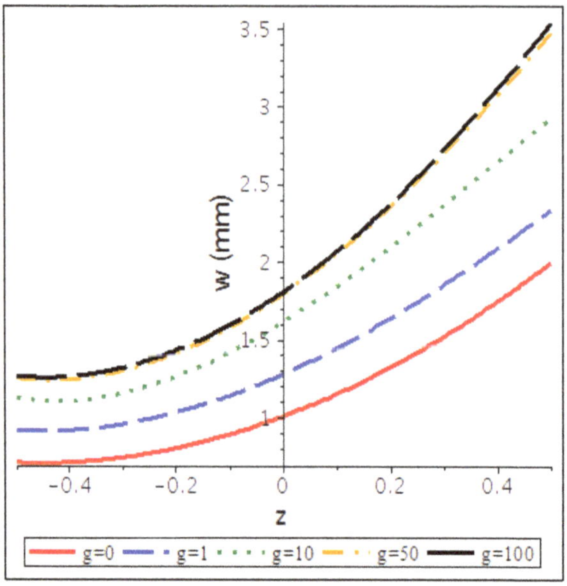

(a)

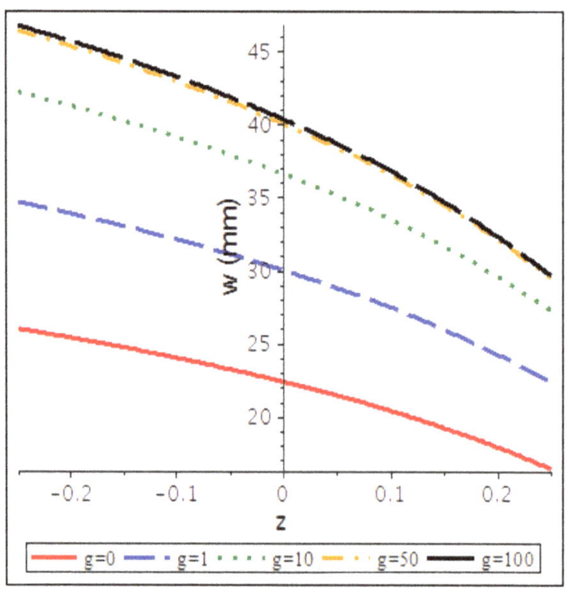

(b)

Figure 3.16. The quasi-3D deflection results for different values of the g parameter for an FGM: (a) an annular/circular sheet; (b) a spherical panel.

structure has no noticeable effect on the slope of the results, but in the case of the annular/circular structure, an increase in the g factor (in other words, an increase in the metallic property of the FGM structure) increases the deformation of the sheet on the surface that is under transverse load ($z = \frac{h}{2}$). The deformation is larger than its value at ($z = -\frac{h}{2}$). Therefore, we conclude that if the quasi-three-dimensional analysis is applied to FGM structures, its impact is greater for planar structures. To obtain results with reasonable accuracy, the quasi-three-dimensional analysis must be considered.

As mentioned earlier, environmental factors, including the effect of ambient temperature on the results, have been considered in the governing equations. The effect of these factors can be seen in figure 3.17. It can be seen that with an increase in the ambient temperature, the difference in the quasi-3D results for an annular/circular sheet increases in line with the thickness of the sheet. Therefore, we conclude that the effect of ambient temperature on the results is significant. If this factor is ignored, the mechanical performance of the sheet under external transverse loads will not be reliable. According to figure 3.17, it can be seen that an ambient temperature difference of $\Delta T = 600$ °C causes an deflection increase of about four times in comparison with the deflection at $\Delta T = 0$. Therefore, if the ambient temperature difference is significant in the analysis, this issue and the results shown in figure 3.17 must be considered.

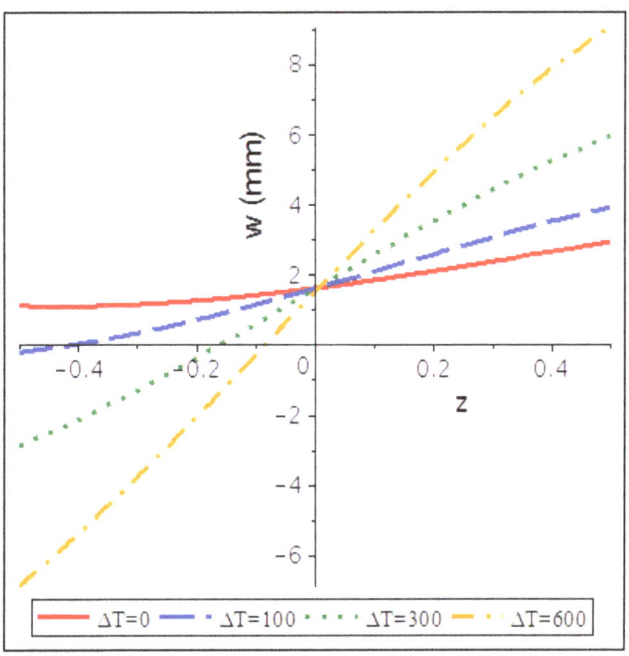

Figure 3.17. The thermal quasi-three-dimensional deflection results for different values of temperature change ΔT.

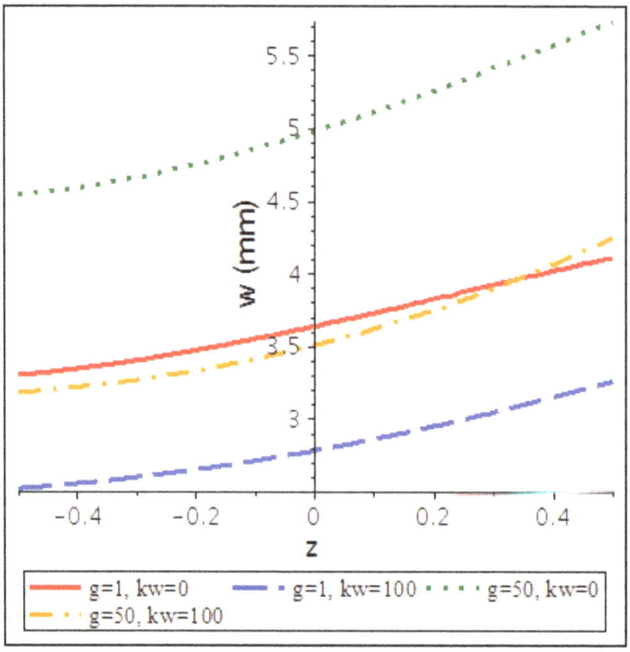

Figure 3.18. The effect of Winkler's elastic foundation on a thick annular/circular sheet for different values of the FGM g parameter.

Figure 3.18 also shows the effect of Winkler's elastic foundation on the quasi-three-dimensional results of FGM structures based on different values of the g coefficient.

Almost similar results can be obtained by assuming some factors, such as the elastic foundation and g factor, in FGM materials. For example, according to figure 3.18, it can be seen that the deformations for different values of $k_w = 0,\ g = 1$ and $k_w = 100$ GPa, $g = 50$ are almost the same. This is a practical conclusion that can be used as a control tool.

3.7 Conclusions and remarks

This chapter investigated the static behavior of structures made of FGMs. For this purpose, the FSDT and quasi-three-dimensional theories were used with von Kàrmàn's strain field and the energy method. The static equations for FGM structures were extracted and solved by the SAPM. The extracted equations are general and can be applied many structures based on their geometric shapes. Environmental factors, including differences in ambient temperature, were also considered in the analysis. The conclusions drawn from this chapter include the following:

1. In the analysis of FGM structures, when the metallic property is higher and the structure is subject to large deformations, nonlinear analysis must be used.

2. The static behavior of FGM structures is unrelated to their geometric shape, and the change process is almost similar for structures with different geometric shapes.
3. In thick structures, three-dimensional elasticity (here, quasi-three-dimensional) should be used to take account of strain changes in the thickness direction.
4. Differences in ambient temperature play a significant role in the deformation of FGM structures. Therefore, in such cases, this issue must be taken into consideration. Also, it must be kept in mind that the simultaneous inclusion of three-dimensional and thermal analysis dramatically increases the accuracy of the results.
5. In order to control some of the results, including the deformations created in the structure, different results can be obtained by changing some factors, including the value of the coefficient g in the FGM and other factors such as the stiffness of the elastic foundation on which the structure is located.

References

[1] Koizumi M 1997 FGM activities in Japan *Composites* B **28** 1–4

[2] Benferhat R, Daouadji T H and Mansour M S 2015 A higher order shear deformation model for bending analysis of functionally graded plates *Trans. Indian Inst. Met.* **68** 7–16

[3] Benyoucef S, Mechab I, Tounsi A, Fekrar A, Atmane H A and Bedia E A A 2010 Bending of thick functionally graded plates resting on Winkler–Pasternak elastic foundations *Mech. Compos. Mater.* **46** 425–34

[4] Brischetto S, Leetsch R, Carrera E, Wallmersperger T and Kroplin B 2008 Thermo-mechanical bending of functionally graded plates *J. Therm. Stresses* **31** 286–308

[5] Daouadji T H, Henni A H, Tounsi A and El Abbes A B 2012 A new hyperbolic shear deformation theory for bending analysis of functionally graded plates *Model. Simul. Eng.* **2012** 2012 159806

[6] Dehsaraji M L, Saidi A R and Mohammadi M 2020 Bending analysis of thick functionally graded piezoelectric rectangular plates using higher-order shear and normal deformable plate theory *Struct. Eng. Mech.* **73** 259–69

[7] Fereidoon A, Seyedmahalle M A and Mohyeddin A 2011 Bending analysis of thin functionally graded plates using generalized differential quadrature method *Arch. Appl. Mech.* **81** 1523–39

[8] Kashtalyan M 2004 Three-dimensional elasticity solution for bending of functionally graded rectangular plates *Eur. J. Mech.* A **23** 853–64

[9] Mantari J L, Oktem A S and Soares C G 2012 Bending response of functionally graded plates by using a new higher order shear deformation theory *Compos. Struct.* **94** 714–23

[10] Mazari A, Attia A, Sekkal M, Kaci A, Tounsi A, Bousahla A A and Mahmoud S R 2018 Bending analysis of functionally graded thick plates with in-plane stiffness variation *Struct. Eng. Mech.* **68** 409–21

[11] Mousavi S M and Tahani M 2012 Analytical solution for bending of moderately thick radially functionally graded sector plates with general boundary conditions using multi-term extended Kantorovich method *Composites* B **43** 1405–16

[12] Naghavi M, Sarrami-Foroushani S and Azhari F 2022 Bending analysis of functionally graded sandwich plates using the refined finite strip method *J. Sandw. Struct. Mater.* **24** 448–83
[13] Orakdogen E, Kucukarslan S, Sofiyev A and Omurtag M H 2010 Finite element analysis of functionally graded plates for coupling effect of extension and bending *Meccanica* **45** 63–72
[14] Panyatong M, Chinnaboon B and Chucheepsakul S 2019 Bending analysis of functionally graded plates with arbitrary shapes and boundary conditions *Struct. Eng. Mech.* **71** 627–41
[15] Phuc P M and Thanh V N 2021 A new sinusoidal shear deformation theory for static bending analysis of functionally graded plates resting on Winkler–Pasternak foundations *Adv. Civ. Eng.* **2021** 6645211
[16] Reddy J N, Wang C M and Kitipornchai S 1999 Axisymmetric bending of functionally graded circular and annular plates *Eur. J. Mech.* A **18** 185–99
[17] Sahraee S 2009 Bending analysis of functionally graded sectorial plates using Levinson plate theory *Compos. Struct.* **88** 548–57
[18] Tounsi A, Houari M S A, Benyoucef S and Bedia E A 2013 A refined trigonometric shear deformation theory for thermoelastic bending of functionally graded sandwich plates *Aerosp. Sci. Technol.* **24** 209–20
[19] Woodward B and Kashtalyan M 2011 Three-dimensional elasticity solution for bending of transversely isotropic functionally graded plates *Eur. J. Mech.* A **30** 705–18
[20] Zenkour A M 2006 Generalized shear deformation theory for bending analysis of functionally graded plates *Appl. Math. Modell.* **30** 67–84
[21] Zenkour A M 2013 A simple four-unknown refined theory for bending analysis of functionally graded plates *Appl. Math. Modell.* **37** 9041–51
[22] Zenkour A M and Alghanmi R A 2018 Bending of functionally graded plates via a refined quasi-3D shear and normal deformation theory *Curved Layer. Struct.* **5** 190–200
[23] Zenkour A M and Hafed Z S 2020 Bending analysis of functionally graded piezoelectric plates via quasi-3D trigonometric theory *Mech. Adv. Mater. Struct.* **27** 1551–62
[24] Zhao X, Lee Y Y and Liew K M 2009 Mechanical and thermal buckling analysis of functionally graded plates *Compos. Struct.* **90** 161–71
[25] Shariat B A S, Javaheri R and Eslami M R 2005 Buckling of imperfect functionally graded plates under in-plane compressive loading *Thin-Walled Struct.* **43** 1020–36
[26] Shariat B A S and Eslami M R 2007 Buckling of thick functionally graded plates under mechanical and thermal loads *Compos. Struct.* **78** 433–9
[27] Najafizadeh M M and Heydari H R 2007 Higher-order theory for buckling of functionally graded circular plates *AIAA J.* **45** 1153–60
[28] Javaheri R and Eslami M R 2002 Thermal buckling of functionally graded plates based on higher order theory *J. Therm. Stresses* **25** 603–25
[29] Feldman E and Aboudi J 1997 Buckling analysis of functionally graded plates subjected to uniaxial loading *Compos. Struct.* **38** 29–36
[30] Chen X L and Liew K M 2004 Buckling of rectangular functionally graded material plates subjected to nonlinearly distributed in-plane edge loads *Smart Mater. Struct.* **13** 1430–7
[31] Zenkour A M and Aljadani M H 2022 Buckling response of functionally graded porous plates due to a quasi-3D refined theory *Mathematics* **10** 565
[32] Thang P T 2016 Analytical solution for thermal buckling analysis of rectangular plates with functionally graded coatings *Aerosp. Sci.Technol.* **55** 465–73

[33] Tebboune W, Merdjah M, Benrahou K H and Tounsi A 2014 Thermoelastic buckling response of thick functionally graded plates *J. Appl. Mech. Tech. Phys.* **55** 857–69
[34] Taj M N A G and Chakrabarti A 2013 Buckling analysis of functionally graded skew plates: an efficient finite element approach *Int. J. Appl. Mech.* **5** 1350041
[35] Naderi A and Saidi A R 2011 Buckling analysis of functionally graded annular sector plates resting on elastic foundations *Proc. Inst. Mech. Eng.* C **225** 312–25
[36] Moita J S, Araujo A L, Correia V F, Soares C M M and Herskovits J 2020 Buckling behavior of composite and functionally graded material plates *Eur. J. Mech.* A **80** 103921
[37] Mohammadi M, Saidi A R and Mohammadi M 2010 Buckling analysis of thin functionally graded rectangular plates resting on elastic foundation *Proc. ASME 10th Biennial Conf. on Engineering Systems Design and Analysis* 2 163–70
[38] Mohammadi M, Saidi A R and Jomehzadeh E 2010 A novel analytical approach for the buckling analysis of moderately thick functionally graded rectangular plates with two simply-supported opposite edges *Proc. Inst. Mech. Eng.* C **224** 1831–41
[39] Ghannadpour S A M, Ovesy H R and Nassirnia M 2012 Buckling analysis of functionally graded plates under thermal loadings using the finite strip method *Comput. Struct.* **108** 93–9
[40] Elmossouess B, Kebdani S, Bouiadjra M B and Tounsi A 2017 A novel and simple HSDT for thermal buckling response of functionally graded sandwich plates *Struct. Eng. Mech.* **62** 401–15
[41] Cho J R 2022 Buckling analysis of functionally graded plates resting on elastic foundation by natural element method *Steel Compos. Struct.* **44** 157–67
[42] Bouazza M, Tounsi A and Adda-Bedia E A 2014 Buckling response of thick functionally graded plates *J. Mater. Eng. Struct.* **1** 137–45 https://www.asjp.cerist.dz/en/article/7335
[43] Zhao X, Lee Y Y and Liew K M 2009 Free vibration analysis of functionally graded plates using the element-free *kp*-Ritz method *J. Sound Vib.* **319** 918–39
[44] Zenkour A M 2005 On vibration of functionally graded plates according to a refined trigonometric plate theory *Int. J. Struct. Stab. Dyn.* **5** 279–97
[45] Sidhoum I A, Boutchicha D, Benyoucef S and Tounsi A 2017 An original HSDT for free vibration analysis of functionally graded plates *Steel Compos. Struct.* **25** 735–45
[46] Reddy J N 2007 Free vibration analysis of functionally graded ceramic-metal plates *Analysis and Design of Plated Structures, Volume 2: Dynamics* ed N E Shanmugam and C M Wang (Sawston: Woodhead Publishing) pp 293–321
[47] Jha D K, Kant T and Singh R K 2013 Free vibration of functionally graded plates with a higher-order shear and normal deformation theory *Int. J. Struct. Stab. Dyn.* **13** 1350004
[48] Fu T, Chen Z B, Yu H Y, Wang Z L and Liu X X 2020 Free vibration of functionally graded sandwich plates based on *n*th-order shear deformation theory via differential quadrature method *J. Sandw. Struct. Mater.* **22** 1660–80
[49] Deshpande G A and Kulkarni S D 2019 Free vibration analysis of functionally graded plates under uniform and linear thermal environment *Acta Mech.* **230** 1347–54
[50] Chen Y T, Xiang S, Zhao W P and Yang K 2017 Free vibration analysis of functionally graded plates based on the fifth-order shear deformation theory *2nd Int. Conf. on Applied Mechanics, Electronics and Mechatronics Engineering (AMEME)* 245–9
[51] Chen W Q and Ding H J 2002 On free vibration of a functionally graded piezoelectric rectangular plate *Acta Mech.* **153** 207–16
[52] Chen C S 2005 Nonlinear vibration of a shear deformable functionally graded plate *Compos. Struct.* **68** 295–302

[53] Bhangale R K and Ganesan N 2006 Free vibration of simply supported functionally graded and layered magneto-electro-elastic plates by finite element method *J. Sound Vib.* **294** 1016–38

[54] Altenbach H and Eremeyev V A 2009 Eigen-vibrations of plates made of functionally graded material *CMC-Comput. Mater. Continua* **9** 153–77

[55] Deepak S A and Shetty R A 2021 Static and free vibration analysis of functionally graded rectangular plates using ANSYS *Mater. Today Proc.* **45** 415–9

[56] Thai C H, Ferreira A J M and Phung-Van P 2020 A nonlocal strain gradient isogeometric model for free vibration and bending analyses of functionally graded plates *Compos. Struct.* **251** 112634

[57] Srividhya S, Raghu P, Rajagopal A and Reddy J N 2018 Nonlocal nonlinear analysis of functionally graded plates using third-order shear deformation theory *Int. J. Eng. Sci.* **125** 1–22

[58] Karami B, Shahsavari D and Janghorban M 2018 Wave propagation analysis in functionally graded (FG) nanoplates under in-plane magnetic field based on nonlocal strain gradient theory and four variable refined plate theory *Mech. Adv. Mater. Struct.* **25** 1047–57

[59] Jung W Y and Han S C 2013 Analysis of sigmoid functionally graded material (S-FGM) nanoscale plates using the nonlocal elasticity theory *Math. Probl. Eng.* **2013** 476131

[60] Dastjerdi S, Malikan M, Dimitri R and Tornabene F 2021 Nonlocal elasticity analysis of moderately thick porous functionally graded plates in a hygro-thermal environment *Compos. Struct.* **255** 112925

[61] Daneshmehr A, Rajabpoor A and Pourdavood M 2014 Stability of size dependent functionally graded nanoplate based on nonlocal elasticity and higher order plate theories and different boundary conditions *Int. J. Eng. Sci.* **82** 84–100

[62] Attar F, Khordad R, Zarifi A and Modabberasl A 2021 Application of nonlocal modified couple stress to study of functionally graded piezoelectric plates *Physica* B **600** 412623

[63] Thai C H, Ferreira A J M, Nguyen-Xuan H and Phung-Van P 2021 A size dependent meshfree model for functionally graded plates based on the nonlocal strain gradient theory *Compos. Struct.* **272** 114169

[64] Yang Z H and He D 2017 Vibration and buckling of orthotropic functionally graded micro-plates on the basis of a re-modified couple stress theory *Results Phys.* **7** 3778–87

[65] Tran T D, Thai C H and Nguyen-Xuan H 2018 A size-dependent functionally graded higher order plate analysis based on modified couple stress theory and moving Kriging meshfree method *CMC-Comput. Mater. Continua* **57** 447–83

[66] Thai H T and Vo T P 2013 A size-dependent functionally graded sinusoidal plate model based on a modified couple stress theory *Compos. Struct.* **96** 376–83

[67] Thai H T and Kim S E 2013 A size-dependent functionally graded Reddy plate model based on a modified couple stress theory *Composites* B **45** 1636–45

[68] Thai H T and Choi D H 2013 Size-dependent functionally graded Kirchhoff and Mindlin plate models based on a modified couple stress theory *Compos. Struct.* **95** 142–53

[69] Reddy J N and Kim J 2012 Modified couple stress theories of functionally graded beams and plates *Proceedings of the IJSSD Symposium 2012 on Progress in Structural Stability and Dynamics (Nanjing, China, 14-16 April 2012)* (Nanjing: Southeast University Press) 1–8

[70] Reddy J N and Kim J 2012 A nonlinear modified couple stress-based third-order theory of functionally graded plates *Compos. Struct.* **94** 1128–43

[71] Mohseni E, Saidi A R and Mohammadi M 2019 Vibration analysis of thick functionally graded micro-plates using HOSNDPT and modified couple stress theory *Iran. J. Sci. Technol. – Trans. Mech. Eng.* **43** 641–51

[72] Kim J, Zur K K and Reddy J N 2019 Bending, free vibration, and buckling of modified couples stress-based functionally graded porous micro-plates *Compos. Struct.* **209** 879–88

[73] Kim J and Reddy J N 2015 A general third-order theory of functionally graded plates with modified couple stress effect and the von Karman nonlinearity: theory and finite element analysis *Acta Mech.* **226** 2973–98

[74] Amar L H H, Kaci A, Yeghnem R and Tounsi A 2018 A new four-unknown refined theory based on modified couple stress theory for size-dependent bending and vibration analysis of functionally graded micro-plate *Steel Compos. Struct.* **26** 89–102

[75] Tang F X, Dong F, Guo Y Z, Shi S N, Jiang J Z and Liu S 2022 Size-dependent buckling and post-buckling analysis of the functionally graded thin plate Al–Cu material based on a modified couple stress theory *Nanomaterials* **12** 3502

[76] Dastjerdi S, Abbasi M and Yazdanparast L 2017 A new modified higher-order shear deformation theory for nonlinear analysis of macro- and nano-annular sector plates using the extended Kantorovich method in conjunction with SAPM *Acta Mech.* **228** 3381–401

[77] Dastjerdi S and Akgöz B 2018 New static and dynamic analyses of macro and nano FGM plates using exact three-dimensional elasticity in thermal environment *Compos. Struct.* **192** 626–41

[78] Dastjerdi S, Tadi Beni Y and Malikan M 2022 A comprehensive study on nonlinear hygro-thermo-mechanical analysis of thick functionally graded porous rotating disk based on two quasi-three-dimensional theories *Mech. Based Des. Struct. Mach.* **50** 3596–625

[79] Dastjerdi S, Akgöz B and Civalek Ö 2020 On the effect of viscoelasticity on behavior of gyroscopes *Int. J. Eng. Sci.* **149** 103236

[80] Dastjerdi S, Alibakhshi A, Akgöz B and Civalek Ö 2023 On a comprehensive analysis for mechanical problems of spherical structures *Int. J. Eng. Sci.* **183** 103796

Chapter 4

The static bending analysis of porous functionally graded sandwich beams

Mehmet Avcar, Lazreg Hadji and Abdelouahed Tounsi

4.1 Introduction

The bending problem of functionally graded sandwich beams (FGSBs) is one of the most interesting subjects for researchers [1–28]. Therefore, as a beginning of the following theme, this chapter presents the static bending analysis of porous/nonporous (P/NP)-FGSBs. In contrast to the approach used in the next chapter, P/NP-FGSBs are assumed to be composed of two porous face sheets constructed of functionally graded materials (FGMs) and a homogenous isotropic ceramic core. To achieve this goal, the material properties of the FGSB are supposed to vary continuously in the thickness direction based on the volume fraction of its constituents, as defined by the modified rule of mixtures, which includes the porosity volume fraction and distinct types of porosity distribution pattern across the cross section. Furthermore, a higher-order shear deformation theory (i.e. hyperbolic shear deformation theory, HypSDT), which does not need shear correction factors, is considered. The governing differential equations are derived using the virtual displacement principle. In addition, a Navier-type analytical solution is presented for the static bending problem of P/NP-FGSBs that have simply supported ends. Detailed comparisons are made with available published data to validate this formulation. Parametric studies are performed to examine the influences of the skin–core–skin thickness ratio, power-law index, span-to-depth ratio, porosity distribution pattern, and porosity volume fraction on the axial and vertical displacements and the axial and shear stresses of FGSBs.

4.2 The basic relations of FGSBs

Figure 4.1 shows the geometry of a P-FGSB that has porous functionally graded (FG) face sheets and a homogeneous isotropic ceramic core, whose axes are defined to lie along the length, width, and height of the FGSB in that order. The FGSB is made up of three elastic layers; from bottom to top, these are called 'Layer 1,'

doi:10.1088/978-0-7503-5301-4ch4

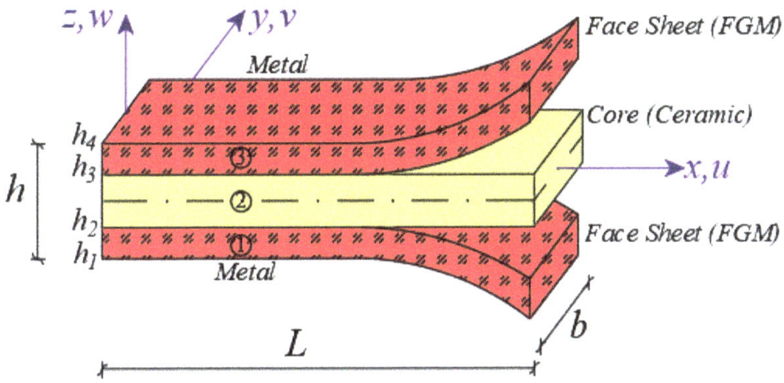

Figure 4.1. The geometry of a P-FGSB.

Table 4.1. The layer schemes of FGSBs.

Layer scheme	h_1	h_2	h_3	h_4
1-1-1	$-h/2$	$-h/6$	$+h/6$	$+h/2$
1-2-1	$-h/2$	$-h/4$	$+h/4$	$+h/2$
2-1-2	$-h/2$	$-h/10$	$+h/10$	$+h/2$
2-1-1	$-h/2$	0	$+h/4$	$+h/2$
2-2-1	$-h/2$	$-h/10$	$+3h/10$	$+h/2$

'Layer 2,' and 'Layer 3.' The two interfaces between the layers and the bottom and top vertical positions are respectively denoted by $h_1 = -h/2,\ h_2,\ h_3,\ h_4 = +h/2$. The skin–core–skin thickness ratio (SCSTR), which defines the thickness of each layer from bottom to top, is denoted by a combination of three numbers, such as 1-8-1, 2-1-2, and so on, due to the necessity for simplicity.

The layer schemes of FGSBs are shown by SCSTR in table 4.1.

This chapter considers four distinct porosity distribution patterns (PDPs), i.e. H-PDP, O-PDP, V-PDP, and X-PDP, as shown in figure 4.2.

It is assumed that the power-law function in the thickness direction governs the volume fraction of FGMs, as follows:

$$V^{(1)}(z) = \left(\frac{z - h_1}{h_2 - h_1}\right)^p \quad z \in [h_1, h_2] \tag{4.1a}$$

$$V^{(2)}(x, z) = 1 \; z \in [h_2, h_3], \tag{4.1b}$$

$$V^{(3)}(x, z) = \left(\frac{z - h_4}{h_3 - h_4}\right)^p, \quad z \in [h_3, h_4]. \tag{4.1c}$$

Here, $V^{(n)}(z)$ is the volume fraction function of the nth layer and can be defined in terms of the volume fraction index p ($0 \leqslant p \leqslant \infty$) in the thickness direction of P-FGSBs.

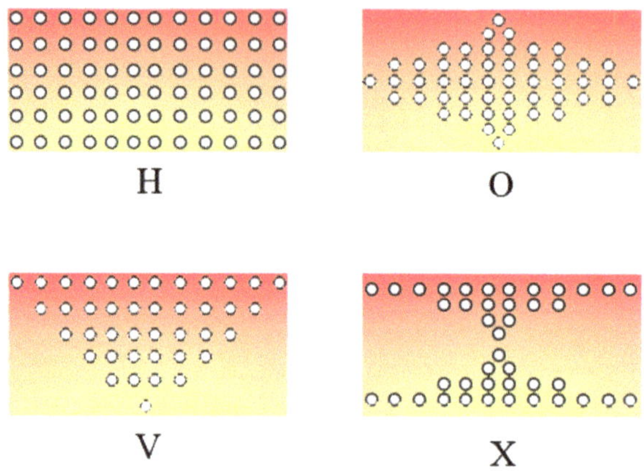

Figure 4.2. PDPs of P-FGSBs.

The following are the material's effective characteristics for each layer:
For H-PDP:

$$\begin{aligned} P^{(1)}(z) &= (P_c - P_m)V^{(1)}(z) + P_m - \frac{\alpha}{2}(P_m + P_c) \\ P^{(2)}(z) &= (P_c - P_m)V^{(2)}(z) + P_m \\ P^{(3)}(z) &= (P_c - P_m)V^{(3)}(z) + P_m - \frac{\alpha}{2}(P_m + P_c) \end{aligned} \tag{4.2}$$

For O-PDP

$$\begin{aligned} P^{(1)}(z) &= (P_c - P_m)V^{(1)}(z) + P_m \\ &\quad - \frac{\alpha}{2}(P_m + P_c)\sqrt{\left(\frac{h_1 - h_2}{2}\right)^2 - \left(z - \left(\frac{h_1 + h_2}{2}\right)\right)^2} \\ P^{(2)}(z) &= (P_c - P_m)V^{(2)}(z) + P_m \\ P^{(3)}(z) &= (P_c - P_m)V^{(3)}(z) + P_m \\ &\quad - \frac{\alpha}{2}(P_m + P_c)\sqrt{\left(\frac{h_3 - h_4}{2}\right)^2 - \left(z - \left(\frac{h_3 + h_4}{2}\right)\right)^2} \end{aligned} \tag{4.3}$$

For V-PDP:

$$\begin{aligned} P^{(1)}(z) &= (P_c - P_m)V^{(1)}(z) + P_m - \frac{\alpha}{2}(P_m + P_c)\left(\frac{z - h_2}{h_1 - h_2}\right) \\ P^{(2)}(z) &= (P_c - P_m)V^{(2)}(z) + P_m \\ P^{(3)}(z) &= (P_c - P_m)V^{(3)}(z) + P_m - \frac{\alpha}{2}(P_m + P_c)\left(\frac{z - h_4}{h_3 - h_4}\right) \end{aligned} \tag{4.4}$$

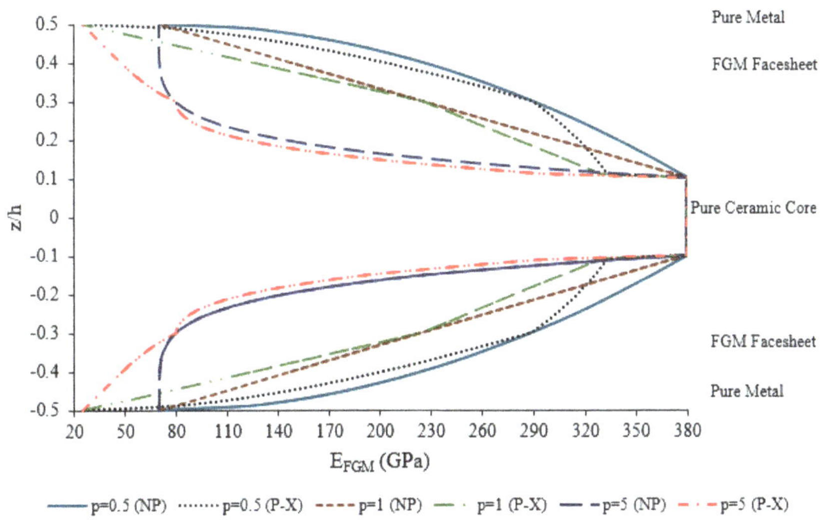

Figure 4.3. The Young's modulus variation of an FGSB that has an SCSTR of 2-1-2 versus the power-law index.

For X-PDP:

$$
\begin{aligned}
P^{(1)}(z) &= (P_c - P_m)V^{(1)}(z) + P_m - \frac{\alpha}{2}(P_m + P_c)\left|\frac{2z - (h_1 + h_2)}{h_1 - h_2}\right| \\
P^{(2)}(z) &= (P_c - P_m)V^{(2)}(z) + P_m \\
P^{(3)}(z) &= (P_c - P_m)V^{(3)}(z) + P_m - \frac{\alpha}{2}(P_m + P_c)\left|\frac{2z - (h_3 + h_4)}{h_3 - h_4}\right|,
\end{aligned}
\tag{4.5}
$$

where $\alpha(\alpha < 1)$ expresses the porosity volume fraction and $P^{(i)}$ $i = 1, 2, 3$ specifies the material features of the layers. In this chapter, Poisson's ratio (ν) is assumed to remain fixed, since its influence is negligible [29, 30]. Unless otherwise specified, the FGSBs are assumed to be made of Al (metal) and Al_2O_3 (ceramic), whose material properties are: $E_m = 70$ GPa, $\nu_m = 0.3$, $\rho_m = 2702$ kgm^{-3}; $E_c = 380$ GPa, $\nu_c = 0.3$, $\rho_c = 3960$ kgm^{-3}. Figure 4.3 illustrates the variation of the Young's modulus of a P-FGSB whose SCSTR is 2-1-2 versus the power-law index for $\alpha = 0.2$ and the X-PDP.

4.3 The basic equations of FGSBs

This section makes the following assumptions:

- The Cartesian coordinate system begins at the middle surface of the FGSB.
- Since the displacements related to the height of the FGSB are insignificant, the stresses involved are also assumed to be negligible.
- The transverse displacements are composed of bending w_b and w_s shear components and are only impacted by x and t:

$$
w(x, z) = w_b(x) + w_s(x). \tag{4.6}
$$

The transverse normal stress, σ_z, is negligible in contrast to the in-plane stress, σ_x. The axial displacement in the x-direction is composed of extension, bending, and shear components, as follows:

$$u = u_0 + u_b + u_s. \tag{4.7}$$

The classical beam theory is used for the displacements of the bending component w_b; thus, u_b is given by

$$u_b = -z\frac{\partial w_b}{\partial x}. \tag{4.8}$$

The shear stress (τ_{xz}) is zero at the top and bottom faces of the FGSB because the shear component (u_s) creates a hyperbolic variation in shear strain (γ_{xz}) when combined with (w_s). Consequently, u_s may be given by

$$u_s = -f(z)\frac{\partial w_s}{\partial x}, \tag{4.9}$$

where $f(z)$ is specified as the function [31]

$$f(z) = z\left[1 + \frac{3\pi}{2}\text{sech}^2\left(\frac{1}{2}\right)\right] - \frac{3\pi}{2}h\tanh\left(\frac{z}{h}\right). \tag{4.10}$$

As a result of the assumptions outlined in the preceding section, equations (4.6)–(4.10) may be used to determine the displacement field of the HypSDT:

$$u(x, z) = u_0(x) - z\frac{\partial w_b}{\partial x} - f(z)\frac{\partial w_s}{\partial x} \tag{4.11a}$$

$$w(x, z, t) = w_b(x, t) + w_s(x, t). \tag{4.11b}$$

The displacements in equations (4.11a) and (4.11b) are connected to the strains shown below:

$$\varepsilon_x = \varepsilon_x^0 + zk_x^b + f(z)k_x^s \tag{4.12a}$$

$$\gamma_{xz} = g(z)\gamma_{xz}^s. \tag{4.12b}$$

Here,

$$\varepsilon_x^0 = \frac{\partial u_0}{\partial x},\ k_x^b = -\frac{\partial^2 w_b}{\partial x^2},\ k_x^s = -\frac{\partial^2 w_s}{\partial x^2},\ \gamma_{xz}^s = \frac{\partial w_s}{\partial x} \tag{4.12c}$$

$$g(z) = 1 - \frac{df(z)}{dz}. \tag{4.12d}$$

If we consider Hooke's law, the stresses become

$$\sigma_x = Q_{11}(z)\varepsilon_x \text{ and } \tau_{xz} = Q_{55}(z)\gamma_{xz}, \tag{4.13a}$$

where

$$Q_{11}(z) = E(z) \text{ and } Q_{55}(z) = \frac{E(z)}{2(1+\nu)}. \tag{4.13b}$$

The virtual displacement principle can be used to generate the governing equations of equilibrium. In this instance, the virtual work concept produces

$$\int_0^L \int_{-h/2}^{h/2} [\sigma_x \delta\varepsilon_x + \tau_{xz}\gamma_{xz}]dzdx - \int_0^L q\delta(w_b + w_s)dx. \tag{4.14}$$

Equation (4.14) may be rewritten by substituting equations (4.12) and (4.13) into it and integrating across the thickness of the FGSB:

$$\int_0^L \left(N_x \delta \frac{\partial u_0}{\partial x} - M_x^b \delta \frac{\partial^2 w_b}{\partial x^2} - M_x^s \delta \frac{\partial^2 w_s}{dx^2} + Q_{xz} \delta \frac{\partial w_s}{dx} \right) dx - \int_0^L q\delta(w_b + w_s)dx, \tag{4.15}$$

where N_x, M_x^b, M_x^s and Q_{xz} are the stress resultants defined by

$$(N_x, M_x^b, M_x^s) = \sum_{n=1}^{3} \int_{h_n}^{h_{n+1}} (1, z, f(z))\sigma_x dz \text{ and } Q_{xz} = \sum_{n=1}^{3} \int_{h_n}^{h_{n+1}} g(z)\tau_{xz} dz. \tag{4.16}$$

Equation (4.15) may be used to obtain the governing equations of equilibrium by integrating the displacement gradients in sections and setting the coefficients zero (i.e. δu_0), δw_b, and δw_s individually. As a result, the equilibrium equations related to the current refined shear deformation theory can be obtained.

$$\delta u_0\text{:} \ \frac{\partial N_x}{\partial x} = 0 \tag{4.17a}$$

$$\delta w_b\text{:} \ \frac{\partial^2 M_x^b}{\partial x^2} + q = 0 \tag{4.17b}$$

$$\delta w_s\text{:} \ \frac{\partial^2 M_x^s}{\partial x^2} + \frac{\partial Q_{xz}}{\partial x} + q = 0 \tag{4.17c}$$

Equations (4.12), (4.13), and (4.16) can be used to write equations (4.17) in terms of the displacements (u_0, w_b, and w_s):

$$A_{11}\frac{\partial^2 u_0}{\partial x^2} - B_{11}\frac{\partial^3 w_b}{\partial x^3} - B_{11}^s\frac{\partial^3 w_s}{\partial x^3} = 0 \tag{4.18a}$$

$$B_{11}\frac{\partial^3 u_0}{\partial x^3} - D_{11}\frac{\partial^4 w_b}{\partial x^4} - D_{11}^s\frac{\partial^4 w_s}{\partial x^4} + q = 0 \tag{4.18b}$$

$$B_{11}^{s}\frac{\partial^3 u_0}{\partial x^3} - D_{11}^{s}\frac{\partial^4 w_b}{\partial x^4} - H_{11}^{s}\frac{\partial^4 w_s}{\partial x^4} + A_{55}^{s}\frac{\partial^2 w_s}{\partial x^2} + q = 0, \tag{4.18c}$$

where A_{11}, D_{11}, etc. are the stiffnesses of the FGSB, specified by

$$(A_{11}, B_{11}, D_{11}, B_{11}^{s}, D_{11}^{s}, H_{11}^{s}) = \sum_{n=1}^{3}\int_{-\frac{h}{2}}^{\frac{h}{2}} Q_{11}(1, z, z^2, f(z), zf(z), f^2(z))dz \tag{4.19a}$$

$$A_{55}^{s} = \sum_{n=1}^{3}\int_{-\frac{h}{2}}^{\frac{h}{2}} Q_{55}[g(z)]^2 dz. \tag{4.19b}$$

4.4 An analytical solution

The basic equations for FGSBs under simply supported end conditions can be solved using Navier's technique by assuming the following variations:

$$\begin{Bmatrix} u_0 \\ w_b \\ w_s \end{Bmatrix} = \sum_{m=1}^{\infty}\begin{Bmatrix} U_m \cos(\lambda x) \\ W_{bm} \sin(\lambda x) \\ W_{sm} \sin(\lambda x) \end{Bmatrix}, \tag{4.20}$$

where U_m, W_{bm}, and W_{sm} are arbitrary parameters to be determined, and $\lambda = m\pi/L$.

In the Fourier series, the transverse load q is extended as follows:

$$q(x) = \sum_{m=1}^{\infty} Q_m \sin(\lambda x), \tag{4.21}$$

where Q_m is the load amplitude and is calculated as follows:

$$Q_m = \frac{2}{L}\int_0^L q(x)\sin(\lambda x)dx. \tag{4.22}$$

In the case of a uniformly distributed load,

$$Q_m = \frac{4q_0}{m\pi}, \quad (m = 1, 3, 5, \ldots). \tag{4.23}$$

If we insert the expansions of u_0, w_b, w_s and q from equations (4.20) and (4.22) into the equilibrium equations of equation (4.18), the following equations can provide an analytical solution:

$$\begin{bmatrix} a_{11} & a_{12} & a_{13} \\ a_{12} & a_{22} & a_{23} \\ a_{13} & a_{23} & a_{33} \end{bmatrix}\begin{Bmatrix} U_m \\ W_{bm} \\ W_{sm} \end{Bmatrix} = \begin{Bmatrix} 0 \\ Q_m \\ Q_m \end{Bmatrix}, \tag{4.24}$$

where

$$\begin{cases} a_{11} = A_{11}\lambda^2 \\ a_{12} = -B_{11}\lambda^3 \\ a_{13} = -B_{11}^s\lambda^3 \\ a_{22} = D_{11}\lambda^4 \\ a_{33} = H_{11}^s\lambda^4 + A_{55}^s\lambda^2 \end{cases}. \tag{4.25}$$

4.5 Numerical results and discussion

This section illustrates the current topic using numerical examples. To prove the correctness of these formulations, we first perform comparison studies. Parametric research is then used to study the impacts of porosity, changes in material properties, and geometrical parameters on the bending analyses of porous (P) and/or nonporous (NP) FGSBs. The uniform pressure is defined as $q_0 = 1 \times 10^6$ N m^{-2}, and the following parameters are used for the numerical results:

$$\text{Nondimensional axial displacement (NDAD): } \bar{u} = 100\frac{E_m h^3}{q_0 L^4}u, \quad \left(x = 0, \quad z = -\frac{h}{2}\right) \tag{4.26}$$

$$\text{Nondimensional vertical displacement (NDVD): } \bar{w} = 100\frac{E_m h^3}{q_0 L^4}w, \quad \left(x = \frac{L}{2}, \quad z = 0\right) \tag{4.27}$$

$$\text{Nondimensional axial stress (NDAS): } \bar{\sigma}_x = \frac{h}{q_0 L}\sigma_x, \quad \left(x = \frac{L}{2}, \quad z = -\frac{h}{2}\right) \tag{4.28}$$

$$\text{Nondimensional shear stress (NDSS): } \bar{\tau}_{xz} = \frac{h}{q_0 L}\tau_{xz}, \quad (x = 0, \quad z = 0) \tag{4.29}$$

Table 4.2 compares the results for the variation of the NDVD of an isotropic homogenous beam subjected to uniform pressure versus the span-to-depth ratio (L/h) with those of [32–34]. As can be seen, the findings are in good agreement with the published ones, especially for long beams.

Table 4.3 compares the results for the variation of the NDVD of a functionally graded beam subjected to uniform pressure versus the porosity volume fraction (α) for $L/h = 5$, $p = 2$, and H-PDP. As can be seen, the results are consistent with those reported.

Table 4.2. The variation of the NDVD of an isotropic beam subjected to uniform pressure versus the span-to-depth ratio.

	Results			
L/*h*	This chapter	Ait Atmane *et al* [32]	Ying *et al* [33]	Chen *et al* [34]
5	1.431 84	1.313 38	1.420 24	1.420 26
15	1.316 52	1.302 26	1.315 27	1.315 28
120	1.3023	1.3009	1.302 29	1.302 29

Table 4.3. The variation of the NDVD of a P functionally graded beam subjected to uniform pressure versus the porosity volume fraction.

	Results		
α	This chapter	Zghal *et al* [4]	Ait Atmane *et al* [32]
0	5.198	5.20	5.35
0.1	5.828	5.82	6.22
0.2	6.650	6.63	7.38

Tables 4.4–4.6 compare the results for the variation of the NDVD, NDAS, and NDSS of NP-FGSBs versus the power-law index (p) with those of [35, 36]. As can be seen, the findings are in good agreement with the published ones.

Note that the small discrepancies between the comparison studies given in tables 4.2–4.6 arise from the beam theories utilized and/or inclusion of the thickness stretching effect.

Table 4.7 presents the variations of the nondimensional (ND) displacements and stresses of NP/P-FGSBs versus the power-law index for $L/h = 5$ and H-PDP. It can be observed that the values of the NDAD, NDAV, and NDSS rise as p increases, while the NDAS reduces with increasing p. The variation of p has the greatest influence on the values of the NDAD, whereas it has the lowest influence on those of the NDSS. The effect of p on these values changes significantly, depending on the SCSTR. In addition, taking account of the porosity increases the impact of the variation in p on the values of the ND displacements and stresses of NP/P-FGSBs.

Table 4.8 shows the variations of the ND displacements and stresses of NP/P-FGSBs versus the span-to-depth ratio, L/h, for $p = 5$ and H-PDP. It shows that the values of the NDAD, NDAV, and NDSS increase as L/h rises, whereas the NDAS reduces with the growth of L/h. The values of the NDAD are the most impacted by the fluctuation of p, whereas the NDSS values are the least impacted. Depending on the SCSTR, L/h has a considerable impact on the values of the ND displacements

Table 4.4. The variation of the NDVD of NP-FGSBs versus the power-law index.

		$L/h = 5$					$L/h = 20$				
p	Results	1-1-1	1-2-1	2-1-2	2-1-1	1-1-1	1-2-1	1-2-1	2-1-2	1-1-1	1-2-1
0	Vo *et al* [35]	3.1397	3.1397	—	3.1397	3.1397	2.8947	2.8947	—	2.8947	2.8947
	Sayyad and Avhad [36]	3.1241	3.1241	3.1241	—	—	2.8585	2.8585	2.8585	—	—
	This chapter	3.1651	3.1651	3.1651	3.1651	3.1651	2.8962	2.8962	2.8962	2.8962	2.8962
1	Vo *et al* [35]	6.2098	5.3612	—	6.4719	5.7777	5.9364	5.0975	—	6.1810	5.5040
	Sayyad and Avhad [36]	6.3011	5.0341	6.8424	—	—	5.9561	5.3415	6.4978	—	—
	This chapter	6.2686	5.4125	6.9312	6.5443	5.8401	5.9400	5.1006	6.5836	6.1977	5.5161
5	Vo *et al* [35]	11.1175	8.4276	—	11.6384	9.6459	10.8309	8.1589	—	11.2886	9.3498
	Sayyad and Avhad [36]	11.0708	8.0576	13.0323	—	—	10.6766	8.4045	12.5898	—	—
	This chapter	11.2233	8.5133	13.1959	11.8209	9.7900	10.8374	8.1642	12.7558	11.3784	9.4122
10	Vo *et al* [35]	12.4453	9.3099	—	12.8026	10.6769	12.1519	9.0413	—	12.4206	10.3715
	Sayyad and Avhad [36]	12.3910	8.9290	14.4346	—	—	11.9795	9.2824	13.9531	—	—
	This chapter	12.5599	9.4039	14.6127	13.0087	10.8458	12.1589	9.0470	14.6127	12.5283	10.4528

Table 4.5. The variation of the NDAS of NP-FGSBs versus the power-law index.

		$L/h = 5$					$L/h = 20$				
p	Results	1-1-1	1-2-1	2-1-2	2-1-1	2-2-1	1-1-1	1-2-1	2-1-2	2-1-1	2-2-1
0	Vo *et al* [35]	3.8005	3.8005	—	3.8005	3.8005	15.0125	15.0125	—	15.0125	15.0125
	Sayyad and Avhad [36]	3.8025	3.8025	3.8025	—	—	15.0135	15.0136	15.0136	—	—
	This chapter	3.8042	3.8042	3.8042	3.8042	3.8042	15.0134	15.0134	15.0134	15.0134	15.0134
1	Vo *et al* [35]	1.433	1.2315	—	1.3866	1.2459	5.6845	4.8797	—	5.4955	4.936
	Sayyad and Avhad [36]	1.4614	1.2331	1.5900	—	—	5.7370	4.8802	6.3020	—	—
	This chapter	1.4355	1.2336	1.5904	1.3891	1.2481	5.6851	4.8802	6.3019	5.4962	4.9365
5	Vo *et al* [35]	2.6079	1.9672	—	2.403	2.016	10.3824	7.8185	—	9.5498	8.01
	Sayyad and Avhad [36]	2.6101	1.9707	3.0733	—	—	10.2712	7.8196	12.2223	—	—
	This chapter	2.6130	1.9713	3.0736	2.4078	2.0202	10.3837	7.8196	12.2222	9.5510	8.0110
10	Vo *et al* [35]	2.9245	2.1788	—	2.6254	2.2161	11.65	8.6655	—	10.4346	8.8094
	Sayyad and Avhad [36]	2.9268	2.1829	3.4047	—	—	11.5237	8.6667	13.5459	—	—
	This chapter	2.9300	2.1834	3.4048	2.6305	2.2207	11.6514	8.6667	13.5455	10.4359	8.8106

Table 4.6. The variation of the NDSS of NP-FGSBs versus the power-law index.

		$L/h = 5$					$L/h = 20$				
p	Results	1-1-1	1-2-1	2-1-2	2-1-1	2-2-1	1-1-1	1-2-1	2-1-2	2-1-1	2-2-1
0	Vo *et al* [35]	0.7233	0.7233	—	0.7233	0.7233	0.7432	0.7432	—	0.7432	0.7432
	Sayyad and Avhad [36]	0.7285	0.7285	0.7285	—	—	0.7355	0.7355	0.7355	—	—
	This chapter	0.7486	0.7486	0.7486	0.7486	0.7486	0.7606	0.7606	0.7606	0.7606	0.7606
1	Vo *et al* [35]	0.8444	0.7993	—	0.8940	0.8342	0.8657	0.8193	—	0.9166	0.8552
	Sayyad and Avhad [36]	0.8767	0.8056	0.905	—	—	0.8726	0.8106	0.9107	—	—
	This chapter	0.8723	0.8279	0.9251	0.9236	0.8628	0.8816	0.8369	0.9348	0.9336	0.8722
5	Vo *et al* [35]	0.9931	0.8763	—	1.1532	0.9683	1.0194	0.8986	—	1.1826	0.9927
	Sayyad and Avhad [36]	1.0048	0.8851	1.1766	—	—	1.0101	0.8897	1.1835	—	—
	This chapter	1.0193	0.9056	1.1870	1.1854	0.9973	1.0290	0.9141	1.1982	1.1968	1.0066
10	Vo *et al* [35]	1.0458	0.898	—	1.2646	1.0148	1.0736	0.9214	—	1.2969	1.0405
	Sayyad and Avhad [36]	1.0586	0.9083	1.2982	—	—	1.0642	0.9128	1.3055	—	—
	This chapter	1.0700	0.9267	1.3040	1.2984	1.0435	1.0801	0.9355	1.3160	1.3109	1.0531

Table 4.7. ND displacements and stresses of NP/P FGBs versus the power-law index.

		$\alpha = 0$				$\alpha = 0.1$			
p	SCSTR	$\bar{u}$	$\bar{w}$	$\bar{\sigma}_x$	$\bar{\tau}_{xz}$	$\bar{u}$	$\bar{w}$	$\bar{\sigma}_x$	$\bar{\tau}_{xz}$
0	1-1-1	0.9406	3.1651	3.8042	0.7486	0.9968	3.3465	3.7937	0.7674
	1-2-1	0.9406	3.1651	3.8042	0.7486	0.9913	3.3266	3.7729	0.7586
	2-1-2	0.9406	3.1651	3.8042	0.7486	0.9988	3.3557	3.8011	0.7774
	2-1-1	0.9406	3.1651	3.8042	0.7486	0.9993	3.3454	3.7744	0.7767
	2-2-1	0.9406	3.1651	3.8042	0.7486	0.9970	3.3321	3.7586	0.7661
1	1-1-1	1.9237	6.2686	1.4355	0.8723	2.1774	7.0697	1.1029	0.9037
	1-2-1	1.6539	5.4125	1.2336	0.8279	1.8191	5.9331	0.9209	0.8451
	2-1-2	2.1307	6.9312	1.5904	0.9251	2.4581	7.9680	1.2453	0.9751
	2-1-1	2.1516	6.5443	1.3891	0.9236	2.4803	7.4451	1.0639	0.9716
	2-2-1	1.9011	5.8401	1.2481	0.8628	2.1422	6.4992	0.9361	0.8908
5	1-1-1	3.4957	11.2233	2.6130	1.0193	4.4435	14.1983	2.2547	1.0716
	1-2-1	2.6397	8.5133	1.9713	0.9056	3.0907	9.9291	1.5666	0.9346
	2-1-2	4.1104	13.1959	3.0736	1.1870	5.5447	17.7021	2.8146	1.2761
	2-1-1	4.1175	11.8209	2.4078	1.1854	5.5100	15.4442	2.0970	1.2693
	2-2-1	3.3728	9.7900	2.0202	0.9973	4.2050	11.9794	1.6562	1.0415
10	1-1-1	3.9183	12.5599	2.9300	1.0700	5.1516	16.4283	2.6150	1.1268
	1-2-1	2.9227	9.4039	2.1834	0.9267	3.4868	11.1739	1.7680	0.9586
	2-1-2	4.5522	14.6127	3.4048	1.3040	6.3817	20.3582	3.2406	1.4062
	2-1-1	4.5577	13.0087	2.6305	1.2984	6.3335	17.5895	2.3636	1.3956
	2-2-1	3.7717	10.8458	2.2207	1.0435	4.8435	13.6449	1.8685	1.0915

and stresses of NP/P-FGSBs. Furthermore, the presence of porosity increases the impact of the variation in L/h on the values of the ND displacements and stresses of NP/P-FGSBs.

Table 4.9 illustrates the variations of the ND displacements and stresses of NP/P-FGSBs versus several PDPs for $p = 5$, $L/h = 5$. It is clear that depending on the PDP, the ND displacements and stresses of NP/P-FGSBs undergo substantial changes, and this effect becomes more pronounced as the porosity volume fraction increases. The impact of PDPs on the ND displacements and stresses of NP/P-FGSBs varies depending on the SCSTR. However, among the PDPs, the H-PDP appears to have the greatest impact, whilst the O-PDP appears to have the least. In addition, a change of PDP has the greatest impact on the NDAD values but the least impact on the NDSS values.

4.6 Conclusions

This chapter presented the static bending analysis of P/NP-FGSBs composed of two porous face sheets constructed of FGMs and a homogenous isotropic ceramic core.

Table 4.8. ND displacements and stresses of NP/P-FGSBs versus the span-to-depth ratio.

		$\alpha = 0$				$\alpha = 0.1$			
L/h	SCSTR	$\bar{u}$	$\bar{w}$	$\bar{\sigma}_x$	$\bar{\tau}_{xz}$	$\bar{u}$	$\bar{w}$	$\bar{\sigma}_x$	$\bar{\tau}_{xz}$
5	1-1-1	1.9237	6.2686	1.4355	0.8723	2.1774	7.0697	1.1029	0.9037
	1-2-1	1.6539	5.4125	1.2336	0.8279	1.8191	5.9331	0.9209	0.8451
	2-1-2	2.1307	6.9312	1.5904	0.9251	2.4581	7.9680	1.2453	0.9751
	2-1-1	2.1516	6.5443	1.3891	0.9236	2.4803	7.4451	1.0639	0.9716
	2-2-1	1.9011	5.8401	1.2481	0.8628	2.1422	6.4992	0.9361	0.8908
10	1-1-1	0.9507	6.0058	2.8482	0.8791	1.0767	6.7955	2.1891	0.9102
	1-2-1	0.8163	5.1630	2.4455	0.8345	0.8983	5.6766	1.8262	0.8514
	2-1-2	1.0536	6.6531	3.1569	0.9322	1.2162	7.6729	2.4730	0.9821
	2-1-1	1.0644	6.2671	2.7541	0.9309	1.2277	7.1514	2.1099	0.9787
	2-2-1	0.9397	5.5809	2.4738	0.8697	1.0596	6.2298	1.8561	0.8974
20	1-1-1	0.4739	5.9400	5.6851	0.8816	0.5368	6.7269	4.3699	0.9125
	1-2-1	0.4068	5.1006	4.8802	0.8369	0.4478	5.6124	3.6447	0.8537
	2-1-2	0.5253	6.5836	6.3019	0.9348	0.6065	7.5991	4.9371	0.9845
	2-1-1	0.5307	6.1977	5.4962	0.9336	0.6123	7.0780	4.2110	0.9813
	2-2-1	0.4685	5.5161	4.9365	0.8722	0.5283	6.1625	3.7041	0.8998
50	1-1-1	0.1894	5.9216	14.2049	0.8824	0.2146	6.7077	10.9191	0.9133
	1-2-1	0.1626	5.0832	12.1930	0.8377	0.1789	5.5945	9.1064	0.8544
	2-1-2	0.2100	6.5641	15.7465	0.9357	0.2424	7.5784	12.3367	0.9853
	2-1-1	0.2121	6.1783	13.7320	0.9345	0.2447	7.0574	10.5213	0.9821
	2-2-1	0.1872	5.4979	12.3336	0.8730	0.2112	6.1436	9.2546	0.9006

HypSDT (which does not need shear correction factors) was employed. The governing differential equations were derived using the virtual work principle. In addition, a Navier-type analytical solution was presented for the static bending problem of P/NP-FGSBs that have simply supported ends. Detailed comparisons were made with available published data to validate the presented formulation. Parametric studies were performed to examine the influences of the skin–core–skin thickness ratio, power-law index, span-to-depth ratio, porosity distribution pattern, and porosity volume fraction on the axial and vertical displacements and the axial and shear stresses of FGSBs.

This chapter showed that the parameters under consideration significantly affect the static bending of FGSBs. In addition to handling the given scenario successfully and producing positive outcomes, the proposed HypSDT also made the issue simpler to address. The mechanical behaviors of other types of structure composed of various materials at the macroscale and the microscale will be examined in future research using the solution approach described here.

Table 4.9. ND displacements and stresses of NP/P-FGSBs for different PDPs (p = 5; L/h = 5).

			$\alpha = 0.1$				$\alpha = 0.2$			
	SCSTR	$\alpha = 0$	H	O	V	X	H	O	V	X
$\bar{u}$	1-1-1	3.4957	4.4435	3.5149	3.9296	3.9317	6.1015	3.5343	4.4847	4.4924
	1-2-1	2.6397	3.0907	2.6472	2.8529	2.8517	3.7288	2.6547	3.1033	3.1009
	2-1-2	4.1104	5.5447	4.1429	4.7569	4.7679	8.5280	4.1759	5.6405	5.6766
	2-1-1	4.1175	5.5100	4.1530	4.8971	4.7595	8.3063	4.1891	6.0397	5.6351
	2-2-1	3.3728	4.2050	3.3912	3.8268	3.7588	5.5734	3.4098	4.4204	4.2429
$\bar{w}$	1-1-1	11.2233	14.1983	11.2836	12.5344	12.5932	19.3959	11.3446	14.2060	14.3538
	1-2-1	8.5133	9.9291	8.5369	9.1656	9.1793	11.9309	8.5606	9.9301	9.9618
	2-1-2	13.1959	17.7021	13.2981	15.1195	15.2642	27.056	13.4019	17.733	18.1195
	2-1-1	11.8209	15.4442	11.9090	13.8032	13.5057	22.6254	11.9986	16.688 48	15.7891
	2-2-1	9.7900	11.9794	9.8363	10.9664	10.8109	15.5501	9.8831	12.4980	12.0863
$\bar{\sigma}_x$	1-1-1	2.6130	2.2547	2.6274	2.9134	1.9945	1.6300	2.6419	3.2956	1.1996
	1-2-1	1.9713	1.5666	1.9770	2.1226	1.4452	0.9950	1.9826	2.2998	0.8272
	2-1-2	3.0736	2.8146	3.0979	3.5060	2.4195	2.2795	3.1227	4.0909	1.5163
	2-1-1	2.4078	2.0970	2.4231	2.7699	1.8512	1.5802	2.4387	3.2924	1.1284
	2-2-1	2.0202	1.6562	2.0284	2.2417	1.5045	1.1144	2.0368	2.5286	0.8794
$\bar{\tau}_{xz}$	1-1-1	1.0193	1.0716	1.0204	1.0439	1.0499	1.1430	1.0214	1.0719	1.0856
	1-2-1	0.9056	0.9346	0.9061	0.9193	0.9216	0.9718	0.9065	0.9346	0.9397
	2-1-2	1.1870	1.2761	1.1892	1.2291	1.2399	1.3975	1.1914	1.2768	1.3030
	2-1-1	1.1854	1.2693	1.1877	1.2184	1.2334	1.3781	1.1901	1.2567	1.2895
	2-2-1	0.9973	1.0415	0.9983	1.0166	1.0228	1.0982	0.9993	1.0391	1.0518

References

[1] Larbi L O, Kaci A, Houari M S A and Tounsi A 2013 An efficient shear deformation beam theory based on neutral surface position for bending and free vibration of functionally graded beams(#) *Mech. Based Des. Struct. Mach.* **41** 421–33

[2] Zouatnia N, Hadji L and Kassoul A 2017 An analytical solution for bending and vibration responses of functionally graded beams with porosities *Wind Struct.* **25** 329–42

[3] Sayyad A S, Avhad P V and Hadji L 2022 On the static deformation and frequency analysis of functionally graded porous circular beams *Forces Mech.* **7** 100093

[4] Zghal S, Ataoui D and Dammak F 2020 Static bending analysis of beams made of functionally graded porous materials *Mech. Based Des. Struct. Mach.* **50** 1012–29

[5] Garg A, Chalak H D, Belarbi M O and Zenkour A M 2021 Hygro-thermo-mechanical based bending analysis of symmetric and unsymmetric power-law, exponential and sigmoidal FG sandwich beams *Mech. Adv. Mater. Struct.* **29** 4523–45

[6] Nguyen N D, Vo T P and Nguyen T K 2020 An improved shear deformable theory for bending and buckling response of thin-walled FG sandwich I-beams resting on the elastic foundation *Compos. Struct.* **254** 112823

[7] Patil R, Joladarashi S and Kadoli R 2022 Bending and vibration studies of FG porous sandwich beam with viscoelastic boundary conditions: FE approach *Mech. Adv. Mater. Struct.* 1–20
[8] Zenkour A M, Allam M N M and Sobhy M 2010 Bending analysis of FG viscoelastic sandwich beams with elastic cores resting on Pasternak's elastic foundations *Acta Mech.* **212** 233–52
[9] Belarbi M O, Khechai A, Bessaim A, Houari M S A, Garg A, Hirane H and Chalak H D 2021 Finite element bending analysis of symmetric and non-symmetric functionally graded sandwich beams using a novel parabolic shear deformation theory *Proc. Inst. Mech. Eng.* L **235** 2482–504
[10] Draiche K, Bousahla A A, Tounsi A and Hussain M 2021 An integral shear and normal deformation theory for bending analysis of functionally graded sandwich curved beams *Arch. Appl. Mech.* **91** 4669–91
[11] Fereidoon A, Andalib M and Hemmatian H 2015 Bending analysis of curved sandwich beams with functionally graded core *Mech. Adv. Mater. Struct.* **22** 564–77
[12] Hadji L and Safa A 2020 Bending analysis of softcore and hardcore functionally graded sandwich beams *Earthq. Struct.* **18** 481–92
[13] Karamanli A 2017 Bending behaviour of two directional functionally graded sandwich beams by using a quasi-3D shear deformation theory *Compos. Struct.* **174** 70–86
[14] Koutoati K, Mohri F and Daya E 2021 Finite element approach of axial bending coupling on static and vibration behaviors of functionally graded material sandwich beams *Mech. Adv. Mater. Struct.* **28** 1537–53
[15] Li C, Shen H S and Wang H 2019 Nonlinear bending of sandwich beams with functionally graded negative Poisson's ratio honeycomb core *Compos. Struct.* **212** 317–25
[16] Liu J, Hao C K, Ye W B and Zang Q S 2021 Application of a new semi-analytic method in bending behavior of functionally graded material sandwich beams *Mech. Based Des. Struct. Mach.* **51** 2130–53
[17] Nguyen N D, Nguyen T K, Vo T P and Nguyen L B 2020 Bending, buckling and free vibration behaviors of thin-walled functionally graded sandwich and composite channel-section beams *Mech. Based Des. Struct. Mach.* **51** 932–60
[18] Sayyad A S and Ghugal Y M 2017 A unified shear deformation theory for the bending of isotropic, functionally graded, laminated and sandwich beams and plates *Int. J. Appl. Mech.* **9** 1750007
[19] Sayyad A S and Ghugal Y M 2021 A unified five-degree-of-freedom theory for the bending analysis of softcore and hardcore functionally graded sandwich beams and plates *J. Sandw. Struct. Mater.* **23** 473–506
[20] Srikarun B, Songsuwan W and Wattanasakulpong N 2021 Linear and nonlinear static bending of sandwich beams with functionally graded porous core under different distributed loads *Compos. Struct.* **276** 114538
[21] Tran T T, Nguyen N H, Do T V, Minh P V and Duc N D 2021 Bending and thermal buckling of unsymmetric functionally graded sandwich beams in high-temperature environment based on a new third-order shear deformation theory *J. Sandw. Struct. Mater.* **23** 906–30
[22] Berkia A, Benguediab S, Menasria A, Bouhadra A, Bourada F, Mamen B, Tounsi A, Benrahou K H, Benguediab M and Hussain M 2022 Static buckling analysis of bi-directional functionally graded sandwich (BFGSW) beams with two different boundary conditions *Steel Compos. Struct.* **44** 489–503

[23] Chinh T H, Tu T M, Duc D M and Hung T Q 2021 Static flexural analysis of sandwich beam with functionally graded face sheets and porous core via point interpolation meshfree method based on polynomial basic function *Arch. Appl. Mech.* **91** 933–47
[24] Hamed M A, Abo-bakr R M, Mohamed S A and Eltaher M A 2020 Influence of axial load function and optimization on static stability of sandwich functionally graded beams with porous core *Eng. Comput.* **36** 1929–46
[25] Kouami K, Foudil M, El Mostafa D and Erasmo C 2021 A finite element approach for the static and vibration analyses of functionally graded material viscoelastic sandwich beams with nonlinear material behavior *Compos. Struct.* **274** 114315
[26] Nguyen T K and Nguyen B D 2015 A new higher-order shear deformation theory for static, buckling and free vibration analysis of functionally graded sandwich beams *J. Sandw. Struct. Mater.* **17** 613–31
[27] Rezaiee-Pajand M, Masoodi A R and Mokhtari M 2018 Static analysis of functionally graded non-prismatic sandwich beams *Adv. Comput. Des.* **3** 165–90
[28] Yang L, Peng J S, Fang F and Yang J 2019 Static pull-in instability and free vibration of functionally graded graphene nanoplatelet reinforced micro-sandwich beams under thermo-electrical actuation *Microsyst. Technol.* **25** 3599–608
[29] Vo T P, Thai H T, Nguyen T K, Inam F and Lee J H 2015 A quasi-3D theory for vibration and buckling of functionally graded sandwich beams *Compos. Struct.* **119** 1–12
[30] Delale F and Erdogan F 1983 The crack problem for a nonhomogeneous plane *J. Appl. Mech.* **50** 609–14
[31] Avcar M, Hadji L and Civalek O 2021 Natural frequency analysis of sigmoid functionally graded sandwich beams in the framework of high order shear deformation theory *Compos. Struct.* **276** 114564
[32] Ait Atmane H, Tounsi A and Bernard F 2017 Effect of thickness stretching and porosity on mechanical response of a functionally graded beams resting on elastic foundations *Int. J. Mech. Mater. Des.* **13** 71–84
[33] Ying J, Lu C F and Chen W Q 2008 Two-dimensional elasticity solutions for functionally graded beams resting on elastic foundations *Compos. Struct.* **84** 209–19
[34] Chen W Q, Lu C F and Bian Z G 2004 A mixed method for bending and free vibration of beams resting on a Pasternak elastic foundation *Appl. Math. Model.* **28** 877–90
[35] Vo T P, Thai H T, Nguyen T K, Inam F and Lee J 2015 Static behaviour of functionally graded sandwich beams using a quasi-3D theory *Composites* B **68** 59–74
[36] Sayyad A S and Avhad P V 2019 On static bending, elastic buckling and free vibration analysis of symmetric functionally graded sandwich beams *J. Solid Mech.* **11** 166–80

Chapter 5

Dynamic analysis of porous functionally graded shell structures using the generalized differential quadrature method

Francesco Tornabene, Matteo Viscoti and Rossana Dimitri

Functionally graded materials (FGMs) are becoming popular in structural engineering. In many applications, a continuous variation of material properties is required, and the structural vibration can be oriented according to the design requirements. In this chapter, we investigate the dynamic behavior of doubly curved shell structures using the equivalent single layer (ESL) approach and higher-order theories. The shell geometry is described using differential geometry principles, and an efficient mapping is adopted for arbitrarily shaped domains. A smooth variation of the constituent materials is characterized by a generalized five-parameter expression, and the elastic constants are determined through a homogenization procedure. Furthermore, linear, trigonometric, and hyperbolic functions are adopted to define the porosity dispersion. The governing equations for doubly curved laminated structures with FGM layers are derived from Hamilton's principle. Furthermore, the generalized differential quadrature (GDQ) method is adopted for the numerical implementation. The outcome of this formulation is compared to refined three-dimensional finite element solutions in a preliminary validation analysis, which highlights the computational efficiency of this model. An extensive parametric investigation is then performed, showing the influence of the FGM and porosity distribution parameters on the structural responses of doubly curved structures.

5.1 Introduction

Recent advances in many engineering applications very frequently require smart innovative materials characterized by enhanced mechanical properties [1, 2]. Such materials can be used to produce components that have complicated shapes and unconventional structural behaviors due to the anisotropy of the constitutive material. Laminated structures are usually adopted in this process, since a

doi:10.1088/978-0-7503-5301-4ch5

suitable layer sequence design can significantly orient the overall structural response [3, 4]. As a particular case of anisotropic materials, classic composite laminae belong to the class of orthotropic media [5–7]. They are produced by embedding long fibers in isotropic polymeric matrixes. The equivalent homogenized materials show enhanced mechanical properties in one direction of the lamina. On the other hand, layers characterized by a transverse isotropic response can be obtained if a pattern of honeycomb cells is adopted. Lattice pantographic laminated structures turn out to be very soft in the core region of the panel, whereas the external part turns out to be more rigid [8–10].

Another important class of enhanced materials is that of the granular composites, known as FGMs [11–13]. They are obtained from a coating process in which two different isotropic phases, namely a ceramic (C) and a metal (M), are mixed in the same layer, creating a volume fraction distribution. It has been shown that the employment of FGMs leads to an improvement in the global vibrational responses of structures in both lower and higher modes [14–17]. Generally speaking, the homogenized material properties of FGMs are obtained from the well-known rule of mixture, taking into account the mechanical properties of the isotropic constituent materials. In this way, it is possible to obtain an inhomogeneous layer that has a smooth variation in its material properties. Depending on the particular application, the material variation acts in the direction of the structure's thickness or along the main geometric dimensions [18, 19].

When very complicated materials are considered in a model, three-dimensional formulations are usually adopted because these materials have unusual structural behaviors [20, 21]. On the other hand, a high computational cost usually results because of the elevated number of degrees of freedom (DOFs) involved in the problem. In contrast, two-dimensional approaches have been defined; such approaches start from the homogenization of the through-the-thickness material properties [22–25]. As shown in some previous works, the accuracy of such formulations is based on axiomatic assumptions about the unknown field variable in the direction of the thickness of the structure. In the research literature, thin plates and shells are usually modeled using classical approaches such as the classical plate theory (CPT) [26] and the first-order shear deformation theory (FSDT) [27]. However, when the lamination scheme includes a soft region within the shell thickness, refined formulations such as the third-order shear deformation theory (TSDT) should be adopted [28, 29]. On the other hand, in the latest applications, stretching effects and shear deformations play an important role in the global deflection mechanism. Therefore, a higher-order expansion of the field variable should be considered for both in-plane and out-of-plane displacement field components. A milestone in this research topic was the so-called unified formulation [30–34], in which the governing relations of the structural problem are derived by means of generalized thickness functions. Furthermore, when laminated structures are investigated, zigzag effects should be embedded in the model [35–37]. In [38], an efficient 2D formulation was used for laminated doubly curved shells made of generally anisotropic materials, together with a generalized Murakami's function, whereas in references [39, 40], a refined zigzag theory was proposed for orthotropic lamination schemes. It has been shown that when higher-order thickness functions are selected, the three-dimensional responses of structures made of complicated lattice and honeycomb cells can be derived even though a two-dimensional approach is adopted [41]. On the other hand, when the lamination scheme accounts for very complicated deformations of the structure, the two-dimensional layer-wise (LW) strategy [42–44] is the best alternative to three-dimensional models for the prediction of both the static and vibrational responses.

As far as the numerical implementation of the fundamental relations is concerned, the classical applications adopt domain decomposition methods [45, 46] such as the well-known finite element method (FEM). It should be said that FEM-based formulations lead to solutions with a lower differentiability order since they are based on a local interpolation of the solution within a single element. The real (i.e. smooth) structural response can be only approximated with a good level of accuracy if the number of DOFs is very high. In contrast, spectral collocation formulations [47, 48] account for a global interpolation of the solution over the entire physical domain, thus allowing the solution to be smooth even though a reduced number of DOFs is considered. One member of this class is the well-known GDQ method [49–52], which has been demonstrated to be a valid methodology capable of overcoming several computational drawbacks that affect other numerical techniques. Based on the Weierstrass interpolation theorem, this method approximates the derivatives of the unknown function as a linear combination of the values assumed by the function itself in a predetermined discrete grid selected from the definition domain. In particular, it has been shown that the GDQ rule embeds several numerical approaches that belong to the weighted residual class. Moreover, when it is applied to nonuniform grids, very fast convergence of the solution can be seen compared to those ones from other numerical approaches [53, 54].

In this chapter, an ESL model is developed for the dynamic analysis of doubly curved shell structures made of generally anisotropic layers. The geometry of the shell is described in terms of a reference surface parametrized using principal coordinates located in the middle thickness of the structure, taking into account the main outcomes of the differential geometry. A generalized approach is adopted for the assessment of the field variable. In particular, a through-the-thickness assumption is made for each component of the three-dimensional displacement field vector. Higher-order theories are considered for the description of the field variable, whereas the interlaminar zigzag effects are accounted for by Murakami's function for both in-plane and out-of-plane components. As far as the constitutive relationship is concerned, each layer of the stacking sequence is modeled as a three-dimensional generally anisotropic material. Therefore, all the possible coupling effects between stress and strain components are considered in the model. The FGM layers are modeled from a homogenization based on the rule of mixture, accounting for the presence of a ceramic phase and a metallic phase. Moreover, the porosity effect is considered in the computation of the equivalent elastic properties according to different approaches, whereas the volume fraction dispersion is modeled according to a five-parameter power law. The fundamental governing equations are directly derived from the Hamilton's principle in the strong form by computing the elastic strain energy and the kinetic energy of the shell by means of a higher-order ESL approach. Furthermore, the natural boundary conditions are outlined using the same energy computation. The numerical implementation adopts the GDQ method starting from a nonuniform two-dimensional computational grid taken from the physical domain.

The accuracy of the formulation is demonstrated by some validation examples, in which the mode frequencies and shapes of structures characterized by different curvatures and material properties derived from this model are compared to those produced by refined three-dimensional simulations based on FEM. The effect of the lamination scheme, material properties, and volume fraction distributions on the dynamic response of these structures are shown and used to tune the governing dispersion parameters. A free vibration analysis is also performed for porous shells, accounting for both polynomial and trigonometric through-the-thickness variations of the material properties. Unlike other previous studies on FGM structures, this formulation provides an efficient methodology for the dynamic analysis of anisotropic laminated shells with a double curvature made of porous FGM granular composites. Furthermore, the adoption of a higher-order model allows one to obtain the three-dimensional vibration response of the shell from a two-dimensional model, thus requiring a reduced number of DOFs. This formulation has been implemented in the research software DiQuMASPAB [55] based on the GDQ numerical technique, which allows one to perform the static and free vibration analyses of laminated anisotropic shell structures. The graphical user interface of the code allows one to develop the geometries of very complicated structures using differential geometry tools, whereas arbitrarily shaped structures are modeled by means of an efficient physical domain mapping. Furthermore, a postprocessing tool is adopted to determine the three-dimensional vibrational response of the shell.

5.2 A geometrical description of the shell

According to the ESL approach, the geometry of an arbitrary doubly curved shell can be described once a parametrization of the reference middle surface $\mathbf{r}(\alpha_1, \alpha_2)$ has been provided. For this purpose, a coordinate system $O'\alpha_1\zeta$ in principal coordinates is defined starting from the principal parametric directions α_1 and α_2 of the surface at issue, while the through-the-thickness direction is denoted by ζ (figure 5.1). In this way, the generic position vector of the structure can be written as follows [20]:

$$\mathbf{R}(\alpha_1, \alpha_2, \zeta) = \mathbf{r}(\alpha_1, \alpha_2) + \frac{h(\alpha_1, \alpha_2)}{2} z\, \mathbf{n}(\alpha_1, \alpha_2). \tag{5.1}$$

where the dimensionless coordinate $z = 2\zeta/h(\alpha_1, \alpha_2) \in [-1,1]$ and $h(\alpha_1, \alpha_2)$ is the total thickness of the laminated shell obtained from the superimposition of l layers of thickness $h_k(\alpha_1, \alpha_2) = \zeta_{k+1}(\alpha_1, \alpha_2) - \zeta_k(\alpha_1, \alpha_2)$:

$$h(\alpha_1, \alpha_2) = \sum_{k=1}^{l} h_k(\alpha_1, \alpha_2). \tag{5.2}$$

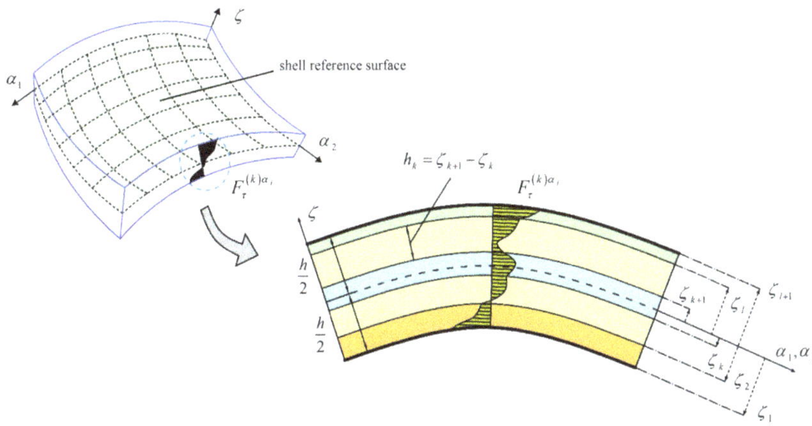

Figure 5.1. A geometrical definition of a doubly curved shell structure according to the ESL approach. The geometry of the structure is described using the shell reference surface located in the middle thickness. The higher-order thickness functions provide a quasi-3D assessment of the kinematic field variable.

In order to give a physical meaning to equation (5.1), a rectangular physical domain $[\alpha_1^0, \alpha_1^1] \times [\alpha_2^0, \alpha_2^1]$ is introduced, where α_i^0, α_i^1 with $i = 1,2$ are the extremes of the definition interval of each principal coordinate. Furthermore, the normal unit vector $\mathbf{n}(\alpha_1, \alpha_2)$ is computed for each point of the physical domain by taking into account the partial derivatives of $\mathbf{r}$ with respect to the α_1, α_2 principal directions [20], identified as $\mathbf{r}_{,1}$ and $\mathbf{r}_{,2}$:

$$\mathbf{n}(\alpha_1, \alpha_2) = \frac{\mathbf{r}_{,1} \times \mathbf{r}_{,2}}{A_1 A_2}. \tag{5.3}$$

In the previous relation, the well-known Lamé parameters $A_1(\alpha_1, \alpha_2)$ and $A_2(\alpha_1, \alpha_2)$ have been used. They can be computed at each point of the physical domain from the following relation:

$$A_i(\alpha_1, \alpha_2) = \sqrt{\mathbf{r}_{,i} \cdot \mathbf{r}_{,i}}, \quad i = 1,2. \tag{5.4}$$

On the other hand, the principal radii of curvature $R_1(\alpha_1, \alpha_2)$ and $R_2(\alpha_1, \alpha_2)$ are computed at each point of the physical domain, setting $\mathbf{r}_{,11}$, $\mathbf{r}_{,22}$, the second-order derivatives of $\mathbf{r}(\alpha_1, \alpha_2)$ with respect to α_1 and α_2, respectively [20]:

$$R_i(\alpha_1, \alpha_2) = -\frac{\mathbf{r}_{,i} \cdot \mathbf{r}_{,i}}{\mathbf{r}_{,ii} \cdot \mathbf{n}}, \quad i = 1,2 \tag{5.5}$$

The quantities introduced in the previous relation are adopted in the analysis by means of the through-the-thickness parameter $H_i(\alpha_1, \alpha_2, \zeta)$, which is defined for each principal direction $\alpha_i = \alpha_1, \alpha_2$, so that $H_i = 1 + \zeta/R_i$.

5.3 Kinematic relations

In this section, the displacement field vector $\mathbf{U}^{(k)}(\alpha_1, \alpha_2, \zeta, t) = [U_1^{(k)} U_2^{(k)} U_3^{(k)}]^T$ associated with each point of a doubly curved three-dimensional shell solid is assessed within a two-dimensional ESL model according to higher-order theories and a generalized formulation. For this purpose, a set of thickness functions $F^{(k\tau)\alpha_1}$, $F^{(k\tau)\alpha_2}$, $F^{(k\tau)\alpha_3}$ is defined in each kth layer, for each τth kinematic expansion order, and for each $\alpha_i = \alpha_1, \alpha_2, \alpha_3$, setting $\tau = 0, \ldots, N+1$ (figure 5.1). In a compact matrix notation, this gives [20]:

$$\mathbf{U}^{(k)}(\alpha_1, \alpha_2, \zeta, t) = \sum_{\tau=0}^{N+1} \mathbf{F}^{(k\tau)}(\zeta)\mathbf{u}^{(\tau)}(\alpha_1, \alpha_2, t) \Leftrightarrow \begin{bmatrix} U_1^{(k)} \\ U_2^{(k)} \\ U_3^{(k)} \end{bmatrix} = \sum_{\tau=0}^{N+1} \begin{bmatrix} F^{(k\tau)\alpha_1} & 0 & 0 \\ 0 & F^{(k\tau)\alpha_2} & 0 \\ 0 & 0 & F^{(k\tau)\alpha_3} \end{bmatrix} \begin{bmatrix} u_1^{(\tau)} \\ u_2^{(\tau)} \\ u_3^{(\tau)} \end{bmatrix}. \tag{5.6}$$

In particular, a power expansion has been adopted for each generalized thickness function together with a Murakami's function for the last kinematic expansion order, so that zigzag effects are embedded in the model:

$$F^{(k\tau)\alpha_i}(\zeta) = \begin{cases} \zeta^\tau & \text{for } \tau = 0, \ldots, N \\ (-1)^k z_k = (-1)^k \dfrac{2}{\zeta_{k+1} - \zeta_k}\zeta - \dfrac{\zeta_{k+1} + \zeta_k}{\zeta_{k+1} - \zeta_k} & \text{for } \tau = N+1. \end{cases} \tag{5.7}$$

For a rapid identification of the selected displacement field in equation (5.7), the nomenclature EDZ–N is introduced. In particular, 'E' means that the formulation is based on the ESL approach, while 'D' denotes that the displacement field vector is assumed as unknown field variable of the problem. The letter 'Z' is adopted when Murakami's function is embedded within the thickness functions set, according to equation (5.7). When zigzag effects are not considered in the model, the axiomatic assumption is expanded up to the Nth order.

The kinematic relations for this ESL formulation are obtained from those of the three-dimensional elasticity problem, which are shown below for the sake of completeness [20]:

$$\boldsymbol{\varepsilon}^{(k)}(\alpha_1, \alpha_2, \zeta, t) = \mathbf{D}\mathbf{U}^{(k)} = \mathbf{D}_\zeta\left(\sum_{i=1}^{3}\mathbf{D}_\Omega^{\alpha_i}\right)\mathbf{U}^{(k)} \tag{5.8}$$

where $\boldsymbol{\varepsilon}^{(k)}(\alpha_1, \alpha_2, \zeta, t) = \left[\varepsilon_1^{(k)} \ \varepsilon_2^{(k)} \ \gamma_{12}^{(k)} \ \gamma_{13}^{(k)} \ \gamma_{23}^{(k)} \ \varepsilon_3^{(k)}\right]^T$ denotes the three-dimensional strain vector that refers to $O'\alpha_1\zeta$. The through-the-thickness kinematic operator $\mathbf{D}_\zeta$, which accounts for the derivatives with respect to the thickness coordinate ζ, reads as follows:

$$\mathbf{D}_{\zeta} = \begin{bmatrix} \frac{1}{H_1} & 0 & 0 & 0 & 0 & 0 & 0 & 0 & 0 \\ 0 & \frac{1}{H_2} & 0 & 0 & 0 & 0 & 0 & 0 & 0 \\ 0 & 0 & \frac{1}{H_1} & \frac{1}{H_2} & 0 & 0 & 0 & 0 & 0 \\ 0 & 0 & 0 & 0 & \frac{1}{H_1} & 0 & \frac{\partial}{\partial \zeta} & 0 & 0 \\ 0 & 0 & 0 & 0 & 0 & \frac{1}{H_2} & 0 & \frac{\partial}{\partial \zeta} & 0 \\ 0 & 0 & 0 & 0 & 0 & 0 & 0 & 0 & \frac{\partial}{\partial \zeta} \end{bmatrix} \tag{5.9}$$

whereas the kinematic operators $\mathbf{D}_{\Omega}^{\alpha_1}$, $\mathbf{D}_{\Omega}^{\alpha_2}$, $\mathbf{D}_{\Omega}^{\alpha_3}$ containing the derivatives with respect to α_1, α_2 are reported below:

$$\mathbf{D}_{\Omega}^{\alpha_1} = \begin{bmatrix} \frac{1}{A_1}\frac{\partial}{\partial \alpha_1} & 0\,0 \\ \frac{1}{A_1 A_2}\frac{\partial A_2}{\partial \alpha_1} & 0\,0 \\ -\frac{1}{A_1 A_2}\frac{\partial A_1}{\partial \alpha_2} & 0\,0 \\ \frac{1}{A_2}\frac{\partial}{\partial \alpha_2} & 0\,0 \\ -\frac{1}{R_1} & 0\,0 \\ 0 & 0\,0 \\ 1 & 0\,0 \\ 0 & 0\,0 \\ 0 & 0\,0 \end{bmatrix} \quad \mathbf{D}_{\Omega}^{\alpha_2} = \begin{bmatrix} 0 & \frac{1}{A_1 A_2}\frac{\partial A_1}{\partial \alpha_2} & 0 \\ 0 & \frac{1}{A_2}\frac{\partial}{\partial \alpha_2} & 0 \\ 0 & \frac{1}{A_1}\frac{\partial}{\partial \alpha_1} & 0 \\ 0 & -\frac{1}{A_1 A_2}\frac{\partial A_2}{\partial \alpha_1} & 0 \\ 0 & 0 & 0 \\ 0 & -\frac{1}{R_2} & 0 \\ 0 & 0 & 0 \\ 0 & 1 & 0 \\ 0 & 0 & 0 \end{bmatrix} \quad \mathbf{D}_{\Omega}^{\alpha_3} = \begin{bmatrix} 0\,0 & \frac{1}{R_1} \\ 0\,0 & \frac{1}{R_2} \\ 0\,0 & 0 \\ 0\,0 & 0 \\ 0\,0 & \frac{1}{A_1}\frac{\partial}{\partial \alpha_1} \\ 0\,0 & \frac{1}{A_2}\frac{\partial}{\partial \alpha_2} \\ 0\,0 & 0 \\ 0\,0 & 0 \\ 0\,0 & 1 \end{bmatrix}. \tag{5.10}$$

By introducing the unified expression for the displacement field variable of equation (5.6) into the kinematic relations of equation (5.8), which refer to a three-dimensional solid, we can derive the ESL kinematic equations a doubly curved shell structure that embeds the higher-order model of interest:

$$\begin{aligned} \boldsymbol{\varepsilon}^{(k)} &= \mathbf{D}_{\zeta}\left(\sum_{i=1}^{3}\mathbf{D}_{\Omega}^{\alpha_i}\right)\mathbf{U}^{(k)} = \sum_{\tau=0}^{N+1}\sum_{i=1}^{3}\mathbf{D}_{\zeta}\mathbf{D}_{\Omega}^{\alpha_i}\mathbf{F}^{(k\tau)}\mathbf{u}^{(\tau)} \\ &= \sum_{\tau=0}^{N+1}\sum_{i=1}^{3}\mathbf{Z}^{(k\tau)\alpha_i}\mathbf{D}_{\Omega}^{\alpha_i}\mathbf{u}^{(\tau)} = \sum_{\tau=0}^{N+1}\sum_{i=1}^{3}\mathbf{Z}^{(k\tau)\alpha_i}\boldsymbol{\varepsilon}^{(\tau)\alpha_i}. \end{aligned} \tag{5.11}$$

The complete derivation of equation (5.11) can be found in reference [20]. In the previous relation, the three-dimensional strain vector $\boldsymbol{\varepsilon}^{(k)}$ is expanded by means of the so-called generalized strain vector $\boldsymbol{\varepsilon}^{(\tau)\alpha_i}(\alpha_1, \alpha_2, t) = \left[\varepsilon_1^{(\tau)\alpha_i}\ \varepsilon_2^{(\tau)\alpha_i}\ \gamma_1^{(\tau)\alpha_i}\ \gamma_2^{(\tau)\alpha_i}\ \gamma_{13}^{(\tau)\alpha_i}\ \gamma_{23}^{(\tau)\alpha_i}\ \omega_{13}^{(\tau)\alpha_i}\ \omega_{23}^{(\tau)\alpha_i}\ \varepsilon_3^{(\tau)\alpha_i}\right]^T$, defined for each τth kinematic expansion order, setting $\alpha_i = \alpha_1, \alpha_2, \alpha_3$. The generalized kinematic operator $\mathbf{Z}^{(k\tau)\alpha_i}$ of equation (5.11) reads as follows:

$$\mathbf{Z}^{(k\tau)\alpha_i} = \begin{bmatrix} \frac{F^{(k\tau)\alpha_i}}{H_1} & 0 & 0 & 0 & 0 & 0 & 0 & 0 & 0 \\ 0 & \frac{F^{(k\tau)\alpha_i}}{H_2} & 0 & 0 & 0 & 0 & 0 & 0 & 0 \\ 0 & 0 & \frac{F^{(k\tau)\alpha_i}}{H_1} & \frac{F^{(k\tau)\alpha_i}}{H_2} & 0 & 0 & 0 & 0 & 0 \\ 0 & 0 & 0 & 0 & \frac{F^{(k\tau)\alpha_i}}{H_1} & 0 & \frac{\partial F^{(k\tau)\alpha_i}}{\partial \zeta} & 0 & 0 \\ 0 & 0 & 0 & 0 & 0 & \frac{F^{(k\tau)\alpha_i}}{H_2} & 0 & \frac{\partial F^{(k\tau)\alpha_i}}{\partial \zeta} & 0 \\ 0 & 0 & 0 & 0 & 0 & 0 & 0 & 0 & \frac{\partial F^{(k\tau)\alpha_i}}{\partial \zeta} \end{bmatrix}. \tag{5.12}$$

5.4 The higher-order constitutive relation

In this chapter, we consider a doubly curved shell structure whose layers are intended to be generally anisotropic and to have a linear elastic behavior. As a consequence, the following constitutive relation is assumed for each kth layer of the stacking sequence, referred to the curvilinear reference system $O'\alpha_1\alpha_2\zeta$ introduced in the previous sections, according to the notation in reference [20]:

$$\boldsymbol{\sigma}^{(k)} = \overline{\mathbf{E}}^{(k)}\boldsymbol{\varepsilon}^{(k)} \begin{bmatrix} \sigma_1^{(k)} \\ \sigma_2^{(k)} \\ \tau_{12}^{(k)} \\ \tau_{13}^{(k)} \\ \tau_{23}^{(k)} \\ \sigma_3^{(k)} \end{bmatrix} = \begin{bmatrix} \overline{E}_{11}^{(k)} & \overline{E}_{12}^{(k)} & \overline{E}_{16}^{(k)} & \overline{E}_{14}^{(k)} & \overline{E}_{15}^{(k)} & \overline{E}_{13}^{(k)} \\ \overline{E}_{12}^{(k)} & \overline{E}_{22}^{(k)} & \overline{E}_{26}^{(k)} & \overline{E}_{24}^{(k)} & \overline{E}_{25}^{(k)} & \overline{E}_{23}^{(k)} \\ \overline{E}_{16}^{(k)} & \overline{E}_{26}^{(k)} & \overline{E}_{66}^{(k)} & \overline{E}_{46}^{(k)} & \overline{E}_{56}^{(k)} & \overline{E}_{36}^{(k)} \\ \overline{E}_{14}^{(k)} & \overline{E}_{24}^{(k)} & \overline{E}_{46}^{(k)} & \overline{E}_{44}^{(k)} & \overline{E}_{45}^{(k)} & \overline{E}_{34}^{(k)} \\ \overline{E}_{15}^{(k)} & \overline{E}_{25}^{(k)} & \overline{E}_{56}^{(k)} & \overline{E}_{45}^{(k)} & \overline{E}_{55}^{(k)} & \overline{E}_{35}^{(k)} \\ \overline{E}_{13}^{(k)} & \overline{E}_{23}^{(k)} & \overline{E}_{36}^{(k)} & \overline{E}_{34}^{(k)} & \overline{E}_{35}^{(k)} & \overline{E}_{33}^{(k)} \end{bmatrix} \begin{bmatrix} \varepsilon_1^{(k)} \\ \varepsilon_2^{(k)} \\ \gamma_{12}^{(k)} \\ \gamma_{13}^{(k)} \\ \gamma_{23}^{(k)} \\ \varepsilon_3^{(k)} \end{bmatrix} \quad \text{for} \quad k = 1, \ldots, l \tag{5.13}$$

where $\overline{E}_{ij}^{(k)}$ for $i, j = 1,\ldots,6$, are the generally anisotropic elastic coefficients related to the ith element of the three-dimensional stress vector $\boldsymbol{\sigma}^{(k)}$ and the jth element of the strain vector $\boldsymbol{\varepsilon}^{(k)}$. The three-dimensional stiffness matrix $\overline{\mathbf{E}}^{(k)}$ of a generic kth layer of the structure refers to the geometric reference system $O'\alpha_1\alpha_2\zeta$. The elastic properties of the kth lamina are expressed with respect to the material symmetry axes by means of the elastic stiffness matrix $\mathbf{E}^{(k)}$:

$$\mathbf{E}^{(k)} = \begin{bmatrix} E_{11}^{(k)} & E_{12}^{(k)} & E_{16}^{(k)} & E_{14}^{(k)} & E_{15}^{(k)} & E_{13}^{(k)} \\ E_{12}^{(k)} & E_{22}^{(k)} & E_{26}^{(k)} & E_{24}^{(k)} & E_{25}^{(k)} & E_{23}^{(k)} \\ E_{16}^{(k)} & E_{26}^{(k)} & E_{66}^{(k)} & E_{46}^{(k)} & E_{56}^{(k)} & E_{36}^{(k)} \\ E_{14}^{(k)} & E_{24}^{(k)} & E_{46}^{(k)} & E_{44}^{(k)} & E_{45}^{(k)} & E_{34}^{(k)} \\ E_{15}^{(k)} & E_{25}^{(k)} & E_{56}^{(k)} & E_{45}^{(k)} & E_{55}^{(k)} & E_{35}^{(k)} \\ E_{13}^{(k)} & E_{23}^{(k)} & E_{36}^{(k)} & E_{34}^{(k)} & E_{35}^{(k)} & E_{33}^{(k)} \end{bmatrix} \quad \text{for} \quad k = 1, \ldots, l. \tag{5.14}$$

As can be seen, the three-dimensional stiffness matrix of the previous equation can be specialized for both orthotropic and isotropic materials. In [20], the interested reader can find further useful insights for the computation of such engineering constants. For an orthotropic medium, three elastic moduli $E_1^{(k)}$, $E_2^{(k)}$, $E_3^{(k)}$, three shear moduli $G_{12}^{(k)}$, $G_{13}^{(k)}$, $G_{23}^{(k)}$, and three Poisson's coefficients $\nu_{12}^{(k)}$, $\nu_{13}^{(k)}$, $\nu_{23}^{(k)}$ are introduced, whereas the quantities $E^{(k)}$, $\nu^{(k)}$ refer to an isotropic layer. The density of the material is denoted by $\rho^{(k)}$.

As a matter of fact, the matrices $\overline{\mathbf{E}}^{(k)}$ and $\mathbf{E}^{(k)}$ are related to each other by means of the material orientation angle $\vartheta^{(k)}$, which is obtained from the transformation matrix $\mathbf{T}^{(k)} = \mathbf{T}^{(k)}(\vartheta^{(k)})$ and embedded in the model. More details can be found in reference [20]. The arbitrary stiffness coefficient $E_{ij}^{(k)}$ of the matrix reported in equation (5.14) is usually intended to be the three-dimensional elastic constant of the anisotropic material, namely $E_{ij}^{(k)} = C_{ij}^{(k)}$. On the other hand, for plane stress assumptions, the reduced elastic coefficients $E_{ij}^{(k)} = Q_{ij}^{(k)}$ are adopted in the following form:

$$Q_{ij}^{(k)} = C_{ij}^{(k)} - \frac{C_{i3}^{(k)} C_{j3}^{(k)}}{C_{33}^{(k)}} \quad \text{for} \quad \begin{array}{c} i, j = 1, \ldots, 6, \\ k = 1, \ldots, l. \end{array} \tag{5.15}$$

For an arbitrary kth layer of a laminate, it is possible to assess the equivalent properties of a generic FGM region by applying the well-known rule of mixture. If $V_C^{(k)}(\zeta)$ and $V_M^{(k)}(\zeta)$ denote the volume fraction of the constituent ceramic and metal, respectively, the following relation should be considered:

$$V_C^{(k)}(\zeta) + V_M^{(k)}(\zeta) + \beta^{(k)}(\zeta) = 1 \tag{5.16}$$

where $\beta^{(k)}$ denotes the porosity of the lamina. In the case of a dense material, the condition $\beta^{(k)} = 0$ is assumed.

The evaluation of the volume fraction within each layer of the structure is performed by means of a five-parameter power law. If the ceramic material is located at the top of the layer ($\zeta = \zeta_{k+1}$), the following relation should be used [20]:

$$V_C^{(k)}(\zeta) = \left(d^{(k)} - a^{(k)}\left(\frac{\zeta_{k+1} - \zeta}{h_k}\right) + b^{(k)}\left(\frac{\zeta_{k+1} - \zeta}{h_k}\right)^{c^{(k)}}\right)^{p^{(k)}}. \tag{5.17}$$

On the other hand, if the metal phase is concentrated at the external side, equation (5.17) should be written as:

$$V_C^{(k)}(\zeta) = \left(d^{(k)} - a^{(k)}\left(\frac{\zeta - \zeta_k}{h_k}\right) + b^{(k)}\left(\frac{\zeta - \zeta_k}{h_k}\right)^{c^{(k)}}\right)^{p^{(k)}}. \tag{5.18}$$

The actual distribution profile depends on the calibration of the distribution parameters $a^{(k)}$, $b^{(k)}$, $c^{(k)}$, $d^{(k)}$, $p^{(k)}$. In particular, when $p^{(k)} = 0$, a uniform distribution of the ceramic material is obtained, whereas an isotropic metal layer is present, when $p^{(k)} = +\infty$.

Based on equations (5.17) and (5.18), the equivalent elastic properties $\overline{P}^{(k)}(\zeta) = \overline{\rho}^{(k)}, \overline{E}^{(k)}, \overline{\nu}^{(k)}$ of the FGM material can be derived by setting $\overline{\rho}^{(k)}(\zeta)$ the equivalent density, $\overline{E}^{(k)}(\zeta)$ the equivalent elastic modulus and $\overline{\nu}^{(k)}(\zeta)$ the Poisson's coefficient. In the case of dense material, the application of the rule of mixture leads to the following relation, setting $P_C^{(k)} = \rho_C^{(k)}, E_C^{(k)}, \nu_C^{(k)}$ and $P_M^{(k)} = \rho_M^{(k)}, E_M^{(k)}, \nu_M^{(k)}$ as the corresponding material properties of the ceramic and metal phases, respectively [20]:

$$\overline{P}^{(k)}(\zeta) = P_C^{(k)} V_C^{(k)}(\zeta) + P_M^{(k)} V_M(\zeta). \tag{5.19}$$

For porous layers, the quantities outlined in equation (5.19) should be reduced by means of the porosity distribution $\beta^{(k)}(\zeta)$ introduced in equation (5.16). To this end, three approaches have been adopted, leading to the relations reported in the following:

$$\begin{aligned}
&\text{Approach 1(A1)} \rightarrow P^{(k)}(\zeta) = \overline{P}^{(k)}(\zeta)(1 - \beta^{(k)}(\zeta)) \\
&\text{Approach 2(A2)} \rightarrow P^{(k)}(\zeta) = \overline{P}^{(k)}(\zeta) - \frac{P_M^{(k)} + P_C^{(k)}}{2}\beta^{(k)}(\zeta) \\
&\text{Approach 3(A3)} \rightarrow P^{(k)}(\zeta) = \overline{P}^{(k)}(\zeta) - \min\left(P_M^{(k)}, P_C^{(k)}\right)\beta^{(k)}(\zeta).
\end{aligned} \tag{5.20}$$

It should be noted that the relations A1, A2, and A3 lead to the same equivalent properties in the case of an isotropic material. Furthermore, the evaluation of the equivalent elastic properties depends on the selection of the porosity distribution, which is defined in terms of scaling parameter $\beta_0^{(k)}$ and dimensionless coordinate $\xi_k = z_k/2$ assessed from the quantities already introduced in equation (5.7). Table 5.1 lists all the porosity profiles embedded in the model, accounting for a power and trigonometric analytical expression with both symmetric and asymmetric through-the-thickness profiles.

By introducing the higher-order kinematic relation (5.11) into the three-dimensional constitutive relationship (5.13), we obtain the following equation [20]:

Table 5.1. Through-the-thickness distributions of porosity for FGM layers embedded in the model, with both symmetric and asymmetric profiles.

Dimensionless coordinate transformation		
$\beta^{(k)}(\zeta) = \beta^{(k)}(\xi_k)$	$\xi_k = \frac{z_k}{2} = \frac{1}{\zeta_{k+1} - \zeta_k}\zeta - \frac{1}{2}\frac{\zeta_{k+1} + \zeta_k}{\zeta_{k+1} - \zeta_k}$	
Designation	Symbol	Analytical expression
Uniform	U^{β_0}	$\beta^{(k)}(\xi_k) = \beta_0^{(k)}$
Linear I	$L\text{–}I^{\beta_0}$	$\beta^{(k)}(\xi_k) = \beta_0^{(k)}(1-2 \mid \xi_k \mid)$
Linear II	$L\text{–}II^{\beta_0}$	$\beta^{(k)}(\xi_k) = 2\beta_0^{(k)} \mid \xi_k \mid$
Linear III	$L\text{–}III^{\beta_0}$	$\beta^{(k)}(\xi_k) = \frac{\beta_0^{(k)}}{2}(1+2\xi_k)$
Linear IV	$L\text{–}IV^{\beta_0}$	$\beta^{(k)}(\xi_k) = \frac{\beta_0^{(k)}}{2}(1-2\xi_k)$
Hyperbolic sine I	$HS\text{–}I^{\beta_0}$	$\beta^{(k)}(\xi_k) = \beta_0^{(k)}\left(1 - \left\lvert \frac{\sinh \xi_k}{\sinh\left(\frac{1}{2}\right)} \right\rvert^3\right)$
Hyperbolic sine II	$HS\text{–}II^{\beta_0}$	$\beta^{(k)}(\xi_k) = \beta_0^{(k)} \left\lvert \frac{\sinh \xi_k}{\sinh\left(\frac{1}{2}\right)} \right\rvert^3$
Symmetric cosine I	$SC\text{–}I^{\beta_0}$	$\beta^{(k)}(\xi_k) = \beta_0^{(k)}(1 - \cos(\pi\xi_k))$
Symmetric cosine II	$SC\text{–}II^{\beta_0}$	$\beta^{(k)}(\xi_k) = \beta_0^{(k)} \cos(\pi\xi_k)$
Asymmetric cosine I	$UC\text{–}I^{\beta_0}$	$\beta^{(k)}(\xi_k) = \beta_0^{(k)}\left(1 - \cos\left(\frac{\pi}{2}\xi_k + \frac{\pi}{4}\right)\right)$
Asymmetric cosine II	$UC\text{–}II^{\beta_0}$	$\beta^{(k)}(\xi_k) = \beta_0^{(k)} \cos\left(\frac{\pi}{2}\xi_k + \frac{\pi}{4}\right)$
Asymmetric cosine III	$UC\text{–}III^{\beta_0}$	$\beta^{(k)}(\xi_k) = \beta_0^{(k)}\left(1 - \cos\left(\frac{\pi}{2}\xi_k - \frac{\pi}{4}\right)\right)$
Asymmetric cosine IV	$UC\text{–}IV^{\beta_0}$	$\beta^{(k)}(\xi_k) = \beta_0^{(k)} \cos\left(\frac{\pi}{2}\xi_k - \frac{\pi}{4}\right)$
Hyperbolic tangent I	$HT\text{–}I^{\beta_0}$	$\beta^{(k)}(\xi_k) = \beta_0^{(k)}\left(1 - \left\lvert \frac{\tanh \xi_k}{\tanh\left(\frac{1}{2}\right)} \right\rvert^3\right)$
Hyperbolic tangent II	$HT\text{–}II^{\beta_0}$	$\beta^{(k)}(\xi_k) = \beta_0^{(k)} \left\lvert \frac{\tanh \xi_k}{\tanh\left(\frac{1}{2}\right)} \right\rvert^3$

$$\boldsymbol{\sigma}^{(k)} = \sum_{\eta=0}^{N+1}\sum_{j=1}^{3}\overline{\mathbf{E}}^{(k)}\mathbf{Z}^{(k\eta)\alpha_j}\boldsymbol{\varepsilon}^{(\eta)\alpha_j} \quad \text{for} \quad k = 1, \ldots, l. \tag{5.21}$$

The relation above is now adopted to compute the virtual variation $\delta\Phi$ of the elastic strain energy of the doubly curved laminated structure. In this way, the stress resultant vector $\mathbf{S}^{(\tau)\alpha_i}(\alpha_1, \alpha_2) = [N_1^{(\tau)\alpha_i} N_2^{(\tau)\alpha_i} N_{12}^{(\tau)\alpha_i} N_{21}^{(\tau)\alpha_i} T_1^{(\tau)\alpha_i} T_2^{(\tau)\alpha_i} P_1^{(\tau)\alpha_i} P_2^{(\tau)\alpha_i} S_3^{(\tau)\alpha_i}]^T$ is associated with each kinematic expansion order, setting $\tau = 0, \ldots, N+1$ and $i = 1,2,3$ [20]:

$$\begin{aligned}
\delta\Phi &= \sum_{k=1}^{l}\int_{\alpha_1}\int_{\alpha_2}\int_{\zeta_k}^{\zeta_{k+1}}(\delta\boldsymbol{\varepsilon}^{(k)})^T\boldsymbol{\sigma}^{(k)}A_1A_2H_1H_2d\alpha_1d\alpha_2d\zeta \\
&= \sum_{k=1}^{l}\int_{\alpha_1}\int_{\alpha_2}\int_{\zeta_k}^{\zeta_{k+1}}\left(\sum_{\tau=0}^{N+1}\sum_{i=1}^{3}\mathbf{Z}^{(k\tau)\alpha_i}\delta\boldsymbol{\varepsilon}^{(\tau)\alpha_i}\right)^T\boldsymbol{\sigma}^{(k)}A_1A_2H_1H_2d\alpha_1d\alpha_2d\zeta \\
&= \sum_{\tau=0}^{N+1}\sum_{i=1}^{3}\int_{\alpha_1}\int_{\alpha_2}(\delta\boldsymbol{\varepsilon}^{(\tau)\alpha_i})^T\sum_{k=1}^{l}\int_{\zeta_k}^{\zeta_{k+1}}(\mathbf{Z}^{(k\tau)\alpha_i})^T\boldsymbol{\sigma}^{(k)}A_1A_2H_1H_2d\alpha_1d\alpha_2d\zeta \\
&= \sum_{\tau=0}^{N+1}\sum_{i=1}^{3}\int_{\alpha_1}\int_{\alpha_2}(\delta\boldsymbol{\varepsilon}^{(\tau)\alpha_i})^T\mathbf{S}^{(\tau)\alpha_i}A_1A_2d\alpha_1d\alpha_2.
\end{aligned} \tag{5.22}$$

A higher-order generalized elastic constitutive relationship is thus derived between the generalized stress resultant vector and the generalized strain vector:

$$\begin{aligned}
\mathbf{S}^{(\tau)\alpha_i} &= \sum_{\eta=0}^{N+1}\sum_{j=1}^{3}\left(\sum_{k=1}^{l}\int_{\zeta_k}^{\zeta_{k+1}}(\mathbf{Z}^{(k\tau)\alpha_i})^T\overline{\mathbf{E}}^{(k)}\mathbf{Z}^{(k\eta)\alpha_j}H_1H_2d\zeta\right)\boldsymbol{\varepsilon}^{(\eta)\alpha_j} \\
&= \sum_{\eta=0}^{N+1}\sum_{j=1}^{3}\mathbf{A}^{(\tau\eta)\alpha_i\alpha_j}\boldsymbol{\varepsilon}^{(\eta)\alpha_j} \quad \text{for} \quad \begin{matrix}\tau = 0, \ldots, N+1, \\ i = 1,2,3\end{matrix}
\end{aligned} \tag{5.23}$$

where $\mathbf{A}^{(\tau\eta)\alpha_i\alpha_j}$ denotes the so-called generalized constitutive matrix, accounting for the material properties, the curvature of the reference surface, and the selected thickness function. An extended form of the matrix in question is shown below [20]:

$$\mathbf{A}^{(\tau\eta)\alpha_i\alpha_j} = \left[\begin{array}{cccccc|ccc}
A_{11(20)}^{(\tau\eta)[00]\alpha_i\alpha_j} & A_{12(11)}^{(\tau\eta)[00]\alpha_i\alpha_j} & A_{16(20)}^{(\tau\eta)[00]\alpha_i\alpha_j} & A_{16(11)}^{(\tau\eta)[00]\alpha_i\alpha_j} & A_{14(20)}^{(\tau\eta)[00]\alpha_i\alpha_j} & A_{15(11)}^{(\tau\eta)[00]\alpha_i\alpha_j} & A_{14(10)}^{(\tau\eta)[01]\alpha_i\alpha_j} & A_{15(10)}^{(\tau\eta)[01]\alpha_i\alpha_j} & A_{13(10)}^{(\tau\eta)[01]\alpha_i\alpha_j} \\
A_{12(11)}^{(\tau\eta)[00]\alpha_i\alpha_j} & A_{22(02)}^{(\tau\eta)[00]\alpha_i\alpha_j} & A_{26(11)}^{(\tau\eta)[00]\alpha_i\alpha_j} & A_{26(02)}^{(\tau\eta)[00]\alpha_i\alpha_j} & A_{24(11)}^{(\tau\eta)[00]\alpha_i\alpha_j} & A_{25(02)}^{(\tau\eta)[00]\alpha_i\alpha_j} & A_{24(01)}^{(\tau\eta)[01]\alpha_i\alpha_j} & A_{25(01)}^{(\tau\eta)[01]\alpha_i\alpha_j} & A_{23(01)}^{(\tau\eta)[01]\alpha_i\alpha_j} \\
A_{16(20)}^{(\tau\eta)[00]\alpha_i\alpha_j} & A_{26(11)}^{(\tau\eta)[00]\alpha_i\alpha_j} & A_{66(20)}^{(\tau\eta)[00]\alpha_i\alpha_j} & A_{66(11)}^{(\tau\eta)[00]\alpha_i\alpha_j} & A_{46(20)}^{(\tau\eta)[00]\alpha_i\alpha_j} & A_{56(11)}^{(\tau\eta)[00]\alpha_i\alpha_j} & A_{46(10)}^{(\tau\eta)[01]\alpha_i\alpha_j} & A_{56(10)}^{(\tau\eta)[01]\alpha_i\alpha_j} & A_{36(10)}^{(\tau\eta)[01]\alpha_i\alpha_j} \\
A_{16(11)}^{(\tau\eta)[00]\alpha_i\alpha_j} & A_{26(02)}^{(\tau\eta)[00]\alpha_i\alpha_j} & A_{66(11)}^{(\tau\eta)[00]\alpha_i\alpha_j} & A_{66(02)}^{(\tau\eta)[00]\alpha_i\alpha_j} & A_{46(11)}^{(\tau\eta)[00]\alpha_i\alpha_j} & A_{56(02)}^{(\tau\eta)[00]\alpha_i\alpha_j} & A_{46(01)}^{(\tau\eta)[01]\alpha_i\alpha_j} & A_{56(01)}^{(\tau\eta)[01]\alpha_i\alpha_j} & A_{36(01)}^{(\tau\eta)[01]\alpha_i\alpha_j} \\
A_{14(20)}^{(\tau\eta)[00]\alpha_i\alpha_j} & A_{24(11)}^{(\tau\eta)[00]\alpha_i\alpha_j} & A_{46(20)}^{(\tau\eta)[00]\alpha_i\alpha_j} & A_{46(11)}^{(\tau\eta)[00]\alpha_i\alpha_j} & A_{44(20)}^{(\tau\eta)[00]\alpha_i\alpha_j} & A_{45(11)}^{(\tau\eta)[00]\alpha_i\alpha_j} & A_{44(10)}^{(\tau\eta)[01]\alpha_i\alpha_j} & A_{45(10)}^{(\tau\eta)[01]\alpha_i\alpha_j} & A_{34(10)}^{(\tau\eta)[01]\alpha_i\alpha_j} \\
A_{15(11)}^{(\tau\eta)[00]\alpha_i\alpha_j} & A_{25(02)}^{(\tau\eta)[00]\alpha_i\alpha_j} & A_{56(11)}^{(\tau\eta)[00]\alpha_i\alpha_j} & A_{56(02)}^{(\tau\eta)[00]\alpha_i\alpha_j} & A_{45(11)}^{(\tau\eta)[00]\alpha_i\alpha_j} & A_{55(02)}^{(\tau\eta)[00]\alpha_i\alpha_j} & A_{45(01)}^{(\tau\eta)[01]\alpha_i\alpha_j} & A_{55(01)}^{(\tau\eta)[01]\alpha_i\alpha_j} & A_{35(01)}^{(\tau\eta)[01]\alpha_i\alpha_j} \\
\hline
A_{14(10)}^{(\tau\eta)[10]\alpha_i\alpha_j} & A_{24(01)}^{(\tau\eta)[10]\alpha_i\alpha_j} & A_{46(10)}^{(\tau\eta)[10]\alpha_i\alpha_j} & A_{46(01)}^{(\tau\eta)[10]\alpha_i\alpha_j} & A_{44(10)}^{(\tau\eta)[10]\alpha_i\alpha_j} & A_{45(01)}^{(\tau\eta)[10]\alpha_i\alpha_j} & A_{44(00)}^{(\tau\eta)[11]\alpha_i\alpha_j} & A_{45(00)}^{(\tau\eta)[11]\alpha_i\alpha_j} & A_{34(00)}^{(\tau\eta)[11]\alpha_i\alpha_j} \\
A_{15(10)}^{(\tau\eta)[10]\alpha_i\alpha_j} & A_{25(01)}^{(\tau\eta)[10]\alpha_i\alpha_j} & A_{56(10)}^{(\tau\eta)[10]\alpha_i\alpha_j} & A_{56(01)}^{(\tau\eta)[10]\alpha_i\alpha_j} & A_{45(10)}^{(\tau\eta)[10]\alpha_i\alpha_j} & A_{55(01)}^{(\tau\eta)[10]\alpha_i\alpha_j} & A_{45(00)}^{(\tau\eta)[11]\alpha_i\alpha_j} & A_{55(00)}^{(\tau\eta)[11]\alpha_i\alpha_j} & A_{35(00)}^{(\tau\eta)[11]\alpha_i\alpha_j} \\
A_{13(10)}^{(\tau\eta)[10]\alpha_i\alpha_j} & A_{23(01)}^{(\tau\eta)[10]\alpha_i\alpha_j} & A_{36(10)}^{(\tau\eta)[10]\alpha_i\alpha_j} & A_{36(01)}^{(\tau\eta)[10]\alpha_i\alpha_j} & A_{34(10)}^{(\tau\eta)[10]\alpha_i\alpha_j} & A_{35(01)}^{(\tau\eta)[10]\alpha_i\alpha_j} & A_{34(00)}^{(\tau\eta)[11]\alpha_i\alpha_j} & A_{35(00)}^{(\tau\eta)[11]\alpha_i\alpha_j} & A_{33(00)}^{(\tau\eta)[11]\alpha_i\alpha_j}
\end{array}\right] \tag{5.24}$$

The arbitrary term of equation (5.24) is computed by means of the following condensed expression, setting $\partial^0 F^{(k\tau)\alpha_i}/\partial\zeta^0 = F^{(k\tau)\alpha_i}$ and $\partial^0 F^{(k\eta)\alpha_j}/\partial\zeta^0 = F^{(k\eta)\alpha_j}$ for the sake of convenience [20]:

$$A_{nm(pq)}^{(\tau\eta)[fg]\alpha_i\alpha_j} = \sum_{k=1}^{l} \int_{\zeta_k}^{\zeta_{k+1}} \overline{B}_{nm}^{(k)} \frac{\partial^f F^{(k\eta)\alpha_j}}{\partial\zeta^f} \frac{\partial^g F^{(k\tau)\alpha_i}}{\partial\zeta^g} \frac{H_1 H_2}{H_1^p H_2^q} d\zeta$$

$$\text{for} \quad \begin{array}{c} \tau, \eta = 0, \ldots, N+1, \\ n, m = 1, \ldots, 6, \\ p, q = 0,1,2, \\ \alpha_i, \alpha_j = \alpha_1, \alpha_2, \alpha_3, \\ f, g = 0,1 \end{array} . \tag{5.25}$$

The elastic properties of the material of the generic kth layer are computed in the model by means of the quantities $\overline{B}_{nm}^{(k)}$ with $n, m = 1, \ldots, 6$, according to the following expression:

$$\overline{B}_{nm}^{(k)} = \begin{cases} \overline{E}_{nm}^{(k)} \\ \kappa(\zeta)\overline{E}_{nm}^{(k)} \end{cases} \text{for} \quad \begin{array}{l} n, m = 1,2,3,6, \\ n, m = 4,5 \end{array} \tag{5.26}$$

setting $\kappa(\zeta)$ the shear correction factor, which is assumed equal to 5/6 for a linear dispersion of the in-plane displacement components in equation (5.6). By introducing the definition of the generalized strain component given in equation (5.11) into the higher-order ESL constitutive relationship of equation (5.23), we can derive the expression for the generalized stress resultant vector $\mathbf{S}^{(\tau)\alpha_i}$ in terms of the higher-order displacement field vector:

$$\mathbf{S}^{(\tau)\alpha_i} = \sum_{\eta=0}^{N+1} \sum_{j=1}^{3} \mathbf{A}^{(\tau\eta)\alpha_i\alpha_j} \mathbf{D}_\Omega^{\alpha_j} \mathbf{u}^{(\eta)} \quad \text{for} \quad \begin{array}{l} \tau = 0, \ldots, N+1, \\ i = 1,2,3 \end{array} . \tag{5.27}$$

5.5 Fundamental governing equations

In this section, the fundamental relations for the dynamic analysis of laminated doubly curved shells are presented. They are derived from the virtual variation of the elastic strain energy, Φ, and the kinetic energy, T, of the system. More specifically, the Hamilton's principle is adopted, where δ is the virtual variation of the corresponding physical quantities. In the arbitrary time interval $[t_1, t_2]$ with extremes t_1, t_2, we obtain a stationary energy configuration of the system as follows [20]:

$$\delta \int_{t_1}^{t_2} (T - \Phi) dt = 0 \rightarrow \int_{t_1}^{t_2} \delta T dt - \int_{t_1}^{t_2} \delta\Phi dt = 0. \tag{5.28}$$

The evaluation of the virtual variation of the kinetic energy of the system is performed by means of the following time integral, accounting for the three-dimensional displacement field vector $\mathbf{U}^{(k)}(\alpha_1, \alpha_2, \zeta, t)$ introduced in equation (5.6), together with its second-order time derivative $\ddot{\mathbf{U}}^{(k)}$ of components $\ddot{U}_i^{(k)} = \partial^2 U_i^{(k)}/\partial t^2$ for $i = 1,2,3$ [20]:

$$\int_{t_1}^{t_2} \delta T dt = -\sum_{k=1}^{l} \int_{t_1}^{t_2} \int_{\zeta_k}^{\zeta_{k+1}} \int_{\alpha_1} \int_{\alpha_2} \rho^{(k)} (\delta \mathbf{U}^{(k)})^T \ddot{\mathbf{U}}^{(k)} H_1 H_2 A_1 A_2 d\alpha_1 d\alpha_2 d\zeta dt. \quad (5.29)$$

The application of the integration by parts rule to the expression for the virtual variation of the elastic strain energy (5.22) provides a relation that outlines the virtual variation of the displacement field $\delta \mathbf{u}^{(\tau)}$. It gives:

$$\begin{aligned}
\int_{t_1}^{t_2} \delta\Phi dt = & \sum_{\tau=0}^{N+1} \int_{t_1}^{t_2} \int_{\alpha_1} \int_{\alpha_2} \Bigg(\Bigg(-\left(\frac{\partial\left(N_1^{(\tau)\alpha_1} A_2\right)}{\partial\alpha_1} + \frac{\partial\left(N_{21}^{(\tau)\alpha_1} A_1\right)}{\partial\alpha_2} + N_{12}^{(\tau)\alpha_1} \frac{\partial A_1}{\partial\alpha_2} - N_2^{(\tau)\alpha_1} \frac{\partial A_2}{\partial\alpha_1} + \left(\frac{T_1^{(\tau)\alpha_1}}{R_1} - P_1^{(\tau)\alpha_1} \right) A_1 A_2 \right) \delta u_1^{(\tau)} + \\
& -\left(\frac{\partial\left(N_2^{(\tau)\alpha_2} A_1\right)}{\partial\alpha_2} + \frac{\partial\left(N_{12}^{(\tau)\alpha_2} A_2\right)}{\partial\alpha_1} + N_{21}^{(\tau)\alpha_2} \frac{\partial A_2}{\partial\alpha_1} - N_1^{(\tau)\alpha_2} \frac{\partial A_1}{\partial\alpha_2} + \left(\frac{T_2^{(\tau)\alpha_2}}{R_2} \quad P_2^{(\tau)\alpha_2} \right) A_1 A_2 \right) \delta u_2^{(\tau)} + \\
& -\left(\frac{\partial\left(T_1^{(\tau)\alpha_3} A_2\right)}{\partial\alpha_1} + \frac{\partial\left(T_2^{(\tau)\alpha_3} A_1\right)}{\partial\alpha_2} - \left(\frac{N_1^{(\tau)\alpha_3}}{R_1} + \frac{N_2^{(\tau)\alpha_3}}{R_2} \right) A_1 A_2 - S_3^{(\tau)\alpha_3} A_1 A_2 \right) \delta u_3^{(\tau)} \Bigg) d\alpha_1 d\alpha_2 dt + \\
& + \sum_{\tau=0}^{N+1} \int_{t_1}^{t_2} \oint_{\alpha_1} \left(N_{21}^{(\tau)\alpha_1} \delta u_1^{(\tau)} + N_2^{(\tau)\alpha_2} \delta u_2^{(\tau)} + T_2^{(\tau)\alpha_3} \delta u_3^{(\tau)} \right) A_1 d\alpha_1 dt + \\
& + \sum_{\tau=0}^{N+1} \int_{t_1}^{t_2} \oint_{\alpha_2} \left(N_1^{(\tau)\alpha_1} \delta u_1^{(\tau)} + N_{12}^{(\tau)\alpha_2} \delta u_2^{(\tau)} + T_1^{(\tau)\alpha_3} \delta u_3^{(\tau)} \right) A_2 d\alpha_2 dt
\end{aligned} \quad , \quad (5.30)$$

where the last two integrals are performed along the edges of the physical domain characterized by the constants α_2 and α_1, respectively.

By introducing the unified formulation of the displacement field variable of equation (5.6) in the previous relation, we can express the time integral of the kinetic energy in terms of the second-order time derivative of the generalized displacement field components $\ddot{u}_1^{(\tau)}$, $\ddot{u}_2^{(\tau)}$, $\ddot{u}_3^{(\tau)}$ with $\tau = 0,\ldots, N + 1$, as follows [20]:

$$\begin{aligned}
\int_{t_1}^{t_2} \delta T dt = & -\sum_{\tau=0}^{N+1} \int_{t_1}^{t_2} \int_{\alpha_1} \int_{\alpha_2} \left(\left(\sum_{\eta=0}^{N+1} I_0^{(\tau\eta)\alpha_1\alpha_1} \ddot{u}_1^{(\eta)} \right) \delta u_1^{(\tau)} + \left(\sum_{\eta=0}^{N+1} I_0^{(\tau\eta)\alpha_2\alpha_2} \ddot{u}_2^{(\eta)} \right) \delta u_2^{(\tau)} + \right. \\
& \left. + \left(\sum_{\eta=0}^{N+1} I_0^{(\tau\eta)\alpha_3\alpha_3} \ddot{u}_3^{(\eta)} \right) \delta u_3^{(\tau)} \right) A_1 A_2 d\alpha_1 d\alpha_2 dt
\end{aligned} \quad , \quad (5.31)$$

where $I_0^{(\tau\eta)\alpha_i\alpha_j}$ for $i, j = 1,2,3$ and $\tau, \eta = 0,\ldots, N + 1$ are the higher-order inertial coefficients. They are evaluated at each point of the physical domain using the following condensed relation:

$$I_0^{(\tau\eta)\alpha_i\alpha_j} = \sum_{k=1}^{l} \int_{\zeta_k}^{\zeta_{k+1}} \rho^{(k)} F^{(k\tau)\alpha_i} F^{(k\eta)\alpha_j} H_1 H_2 d\zeta \ \text{ for } \ \begin{array}{l} \alpha_i, \alpha_j = \alpha_1, \alpha_2, \alpha_3, \\ \tau, \eta = 0, \ldots, N+1 \end{array} . \quad (5.32)$$

Such terms are collected for each $\tau, \eta = 0,\ldots, N + 1$ kinematic expansion orders in the generalized inertial matrix $\mathbf{M}^{(\tau\eta)}$, which reads:

$$\mathbf{M}^{(\tau\eta)} = \begin{bmatrix} I_0^{(\tau\eta)\alpha_1\alpha_1} & 0 & 0 \\ 0 & I_0^{(\tau\eta)\alpha_2\alpha_2} & 0 \\ 0 & 0 & I_0^{(\tau\eta)\alpha_3\alpha_3} \end{bmatrix} \quad \text{for } \tau, \eta = 0,\ldots, N+1. \tag{5.33}$$

Using the Hamilton's principle (5.28), the equilibrium equations are obtained for the higher-order model, which introduces the second-order time derivative of the generalized displacement field vector $\ddot{\mathbf{u}}^{(\eta)} = [\ddot{u}_1^{(\eta)}\ddot{u}_2^{(\eta)}\ddot{u}_3^{(\eta)}]^T$ [20]:

$$\sum_{i=1}^{3} \mathbf{D}_{\Omega}^{*\alpha_i}\mathbf{S}^{(\tau)\alpha_i} - \sum_{\eta=0}^{N+1} \mathbf{M}^{(\tau\eta)}\ddot{\mathbf{u}}^{(\eta)} = \mathbf{0} \quad \text{for } \tau = 0, \ldots, N+1, \tag{5.34}$$

where $\mathbf{D}_{\Omega}^{*\alpha_i}$ for $i = 1,2,3$ are the equilibrium operators, which are reported below in an expanded form:

$$\mathbf{D}_{\Omega}^{*\alpha_1} = \begin{bmatrix} \frac{1}{A_1}\frac{\partial}{\partial\alpha_1} + \frac{1}{A_1A_2}\frac{\partial A_2}{\partial\alpha_1} & -\frac{1}{A_1A_2}\frac{\partial A_2}{\partial\alpha_1} & \frac{1}{A_1A_2}\frac{\partial A_1}{\partial\alpha_2} & \frac{1}{A_2}\frac{\partial}{\partial\alpha_2} + \frac{1}{A_1A_2}\frac{\partial A_1}{\partial\alpha_2} & \frac{1}{R_1} & 0 & -1 & 0 & 0 \\ 0 & 0 & 0 & 0 & 0 & 0 & 0 & 0 & 0 \\ 0 & 0 & 0 & 0 & 0 & 0 & 0 & 0 & 0 \end{bmatrix} \tag{5.35}$$

$$\mathbf{D}_{\Omega}^{*\alpha_2} = \begin{bmatrix} 0 & 0 & 0 & 0 & 0 & 0 & 0 & 0 & 0 \\ -\frac{1}{A_1A_2}\frac{\partial A_1}{\partial\alpha_2} & \frac{1}{A_2}\frac{\partial}{\partial\alpha_2} + \frac{1}{A_1A_2}\frac{\partial A_1}{\partial\alpha_2} & \frac{1}{A_1}\frac{\partial}{\partial\alpha_1} + \frac{1}{A_1A_2}\frac{\partial A_2}{\partial\alpha_1} & \frac{1}{A_1A_2}\frac{\partial A_2}{\partial\alpha_1} & 0 & \frac{1}{R_2} & 0 & -1 & 0 \\ 0 & 0 & 0 & 0 & 0 & 0 & 0 & 0 & 0 \end{bmatrix} \tag{5.36}$$

$$\mathbf{D}_{\Omega}^{*\alpha_3} = \begin{bmatrix} 0 & 0 & 0 & 0 & 0 & 0 & 0 & 0 & 0 \\ 0 & 0 & 0 & 0 & 0 & 0 & 0 & 0 & 0 \\ -\frac{1}{R_1} & -\frac{1}{R_2} & 0 & 0 & \frac{1}{A_1}\frac{\partial}{\partial\alpha_1} + \frac{1}{A_1A_2}\frac{\partial A_2}{\partial\alpha_1} & \frac{1}{A_1}\frac{\partial}{\partial\alpha_1} + \frac{1}{A_1A_2}\frac{\partial A_2}{\partial\alpha_1} & 0 & 0 & -1 \end{bmatrix}. \tag{5.37}$$

Starting from equation (5.34), the final form of the fundamental governing equations for the dynamic analysis of anisotropic doubly curved shell structures can be obtained by using (5.27) for the generalized stress resultants $\mathbf{S}^{(\tau)\alpha_i}$ in terms of $\mathbf{u}^{(\tau)}$:

$$\sum_{\eta=0}^{N+1} \mathbf{L}^{(\tau\eta)}\mathbf{u}^{(\eta)} = \sum_{\eta=0}^{N+1} \mathbf{M}^{(\tau\eta)}\ddot{\mathbf{u}}^{(\eta)} \quad \text{for } \tau = 0, \ldots, N+1, \tag{5.38}$$

where $\mathbf{L}^{(\tau\eta)}$ is the fundamental matrix that refers to each $\tau, \eta = 0,\ldots, N+1$:

$$\mathbf{L}^{(\tau\eta)} = \sum_{i=1}^{3}\sum_{j=1}^{3} \mathbf{D}_{\Omega}^{*\alpha_i}\mathbf{A}^{(\tau\eta)\alpha_i\alpha_j}\mathbf{D}_{\Omega}^{\alpha_j} = \begin{bmatrix} L_{11}^{(\tau\eta)\alpha_1\alpha_1} & L_{12}^{(\tau\eta)\alpha_1\alpha_2} & L_{13}^{(\tau\eta)\alpha_1\alpha_3} \\ L_{21}^{(\tau\eta)\alpha_2\alpha_1} & L_{22}^{(\tau\eta)\alpha_2\alpha_2} & L_{23}^{(\tau\eta)\alpha_2\alpha_3} \\ L_{31}^{(\tau\eta)\alpha_3\alpha_1} & L_{32}^{(\tau\eta)\alpha_3\alpha_2} & L_{33}^{(\tau\eta)\alpha_3\alpha_3} \end{bmatrix}. \tag{5.39}$$

The complete expression for the generalized fundamental coefficients $L_{ij}^{(\tau\eta)\alpha_i\alpha_j}$ for $i, j = 1,2,3$ can be found in reference [20]. The application of the integration by parts rule to the Hamiltonian principle (5.28) also provides the boundary conditions for the differential problem. At the boundary edges of the physical domain characterized by $\alpha_2 \in [\alpha_2^0, \alpha_2^1]$, the following relations are obtained:

$$\begin{aligned} N_1^{(\tau)\alpha_1} &= \overline{N}_1^{(\tau)\alpha_1} \quad \text{or} \quad u_1^{(\tau)} = \bar{u}_1^{(\tau)} \\ N_{12}^{(\tau)\alpha_1} &= \overline{N}_{12}^{(\tau)\alpha_1} \quad \text{or} \quad u_2^{(\tau)} = \bar{u}_2^{(\tau)} \\ T_1^{(\tau)\alpha_1} &= \overline{T}_1^{(\tau)\alpha_1} \quad \text{or} \quad u_3^{(\tau)} = \bar{u}_3^{(\tau)}, \end{aligned} \tag{5.40}$$

where $\overline{N}_1^{(\tau)\alpha_1}$, $\overline{N}_{12}^{(\tau)\alpha_1}$, $\overline{T}_1^{(\tau)\alpha_1}$ and $\bar{u}_1^{(\tau)}$, $\bar{u}_2^{(\tau)}$, $\bar{u}_3^{(\tau)}$ are the prescribed values of the generalized stress and displacement field components, respectively. In the same way, for edges located at $\alpha_2 = \alpha_2^0$ or at $\alpha_2 = \alpha_2^1$, the application of the integration by parts rule gives:

$$\begin{aligned} N_{21}^{(\tau)\alpha_1} &= \overline{N}_{21}^{(\tau)\alpha_1} \quad \text{or} \quad u_1^{(\tau)} = \bar{u}_1^{(\tau)} \\ N_2^{(\tau)\alpha_1} &= \overline{N}_2^{(\tau)\alpha_1} \quad \text{or} \quad u_2^{(\tau)} = \bar{u}_2^{(\tau)} \\ T_2^{(\tau)\alpha_1} &= \overline{T}_2^{(\tau)\alpha_1} \quad \text{or} \quad u_3^{(\tau)} = \bar{u}_3^{(\tau)}, \end{aligned} \tag{5.41}$$

where $\overline{N}_2^{(\tau)\alpha_1}$, $\overline{N}_{21}^{(\tau)\alpha_1}$, $\overline{T}_2^{(\tau)\alpha_1}$ are the assigned higher-order generalized stress components defined for each $\tau = 0,\ldots, N+1$. As a particular case of equations (5.40) and (5.41), the clamped (C) and free (F) boundary conditions read as follows [20]:

$$\begin{aligned} &\text{Clamped(C)} \quad u_1^{(\tau)} = u_2^{(\tau)} = u_3^{(\tau)} = 0 \quad \text{at} \ \alpha_1 = \alpha_1^0 \ \text{or} \ \alpha_1 = \alpha_1^1 \\ &\qquad\qquad\quad\ u_1^{(\tau)} = u_2^{(\tau)} = u_3^{(\tau)} = 0 \quad \text{at} \ \alpha_2 = \alpha_2^0 \ \text{or} \ \alpha_2 = \alpha_2^1 \\ &\text{Free(F)} \quad N_1^{(\tau)\alpha_1} = N_{12}^{(\tau)\alpha_1} = T_1^{(\tau)\alpha_1} = 0 \quad \text{at} \ \alpha_1 = \alpha_1^0 \ \text{or} \ \alpha_1 = \alpha_1^1 \\ &\qquad\qquad N_{21}^{(\tau)\alpha_1} = N_2^{(\tau)\alpha_1} = T_2^{(\tau)\alpha_1} = 0 \quad \text{at} \ \alpha_2 = \alpha_2^0 \ \text{or} \ \alpha_2 = \alpha_2^1. \end{aligned} \tag{5.42}$$

5.6 Numerical implementation

In this section, the numerical solution is found for the fundamental governing relations reported in equation (5.38). In particular, the GDQ method is adopted to approximate derivatives of arbitrary order. Let us consider a univariate smooth function $f = f(x)$ defined in a closed interval of extremes x_0, x_1 so that $x \in [x_0, x_1]$. According to this numerical technique, the nth-order derivative of f at a generic point $x_i \in [x_0, x_1]$ with $i = 1,\ldots, I_Q$ can be expressed as follows [20]:

$$\left.\frac{d^n f(x)}{dx^n}\right|_{x=x_i} \cong \sum_{j=1}^{I_Q} \varsigma_{ij}^{(n)} f(x_j) \qquad i = 1, 2,\ldots, I_Q \tag{5.43}$$

where $f(x_j)$ for $i = 1,\ldots, I_Q$ are the values assumed by the function in a predetermined discrete grid made of I_Q points belonging to the definition domain, while $\varsigma_{ij}^{(n)}$ with $i, j = 1,\ldots, I_Q$ are the GDQ weighting coefficients for the nth-order derivative. They are computed using a recursive procedure [20] based on the properties of the Lagrange polynomials $\mathcal{L}$ and their first-order derivatives, denoted by $\mathcal{L}^{(1)}$:

$$\begin{aligned} &\varsigma_{ij}^{(1)} = \frac{\mathcal{L}^{(1)}(x_i)}{(x_i - x_j)\mathcal{L}^{(1)}(x_j)}, \quad \varsigma_{ij}^{(n)} = n\left(\varsigma_{ij}^{(1)}\varsigma_{ii}^{(n-1)} - \frac{\varsigma_{ij}^{(n-1)}}{x_i - x_j}\right) && i \neq j \\ &\varsigma_{ii}^{(n)} = -\sum_{j=1 j\neq i}^{N} \varsigma_{ij}^{(n)} && i = j \end{aligned} \tag{5.44}$$

In this chapter, equation (5.38) is numerically solved in the computational domain starting from a two-dimensional nonuniform discrete grid obtained from the Chebyshev–Gauss–Lobatto (CGL) harmonic distribution, since it has been demonstrated to yield stable and accurate results. The arbitrary element $\bar{x}_i$ within a dimensionless interval $[-1,1]$ reads as follows:

$$\bar{x}_i = -\cos\left(\frac{i-1}{I_Q - 1}\pi\right), \; i = 1, \ldots, I_Q. \tag{5.45}$$

The mode frequencies and the corresponding mode shapes of a doubly curved shell are calculated from equation (5.38) based on a harmonic solution. To this end, the generalized displacement field vector $\mathbf{u}^{(\tau)}(\alpha_1, \alpha_2, t)$ is expressed as follows for each τth order of the kinematic expansion:

$$\mathbf{u}^{(\tau)}(\alpha_1, \alpha_2, t) = \mathbf{U}^{(\tau)}(\alpha_1, \alpha_2)e^{i\omega t}. \tag{5.46}$$

In the previous relation, $\mathbf{U}^{(\tau)} = \left[U_1^{(\tau)}(\alpha_1, \alpha_2) \;\; U_2^{(\tau)}(\alpha_1, \alpha_2) \;\; U_3^{(\tau)}(\alpha_1, \alpha_2)\right]^T$ denotes the modal shape vector that refers to each $\tau = 0,\ldots, N + 1$, while $\omega = 2\pi f$ is the circular frequency and f is the natural frequency. In this way, the fundamental relations that refer to the τth kinematic expansion order of equation (5.38) are expressed as:

$$\sum_{\eta=0}^{N+1} \mathbf{L}^{(\tau\eta)}\mathbf{U}^{(\eta)} + \omega^2 \sum_{\eta=0}^{N+1} \mathbf{M}^{(\tau\eta)}\mathbf{U}^{(\eta)} = \mathbf{0} \;\; \text{for} \quad \tau = 0, \ldots, N + 1. \tag{5.47}$$

The assembled form of equation (5.47) that refers to a complete expansion of the displacement field variable of equation (5.6) is as follows:

$$\begin{bmatrix} \mathbf{L}^{(00)} & \mathbf{L}^{(01)} & \cdots & \mathbf{L}^{(0(N+1))} \\ \mathbf{L}^{(10)} & \mathbf{L}^{(11)} & \cdots & \mathbf{L}^{(1(N+1))} \\ \vdots & \vdots & \ddots & \vdots \\ \mathbf{L}^{((N+1)0)} & \mathbf{L}^{((N+1)1)} & \cdots & \mathbf{L}^{((N+1)(N+1))} \end{bmatrix} \begin{bmatrix} \mathbf{U}^{(0)} \\ \mathbf{U}^{(1)} \\ \vdots \\ \mathbf{U}^{(N+1)} \end{bmatrix} + \omega^2 \begin{bmatrix} \mathbf{M}^{(00)} & \mathbf{M}^{(01)} & \cdots & \mathbf{M}^{(0(N+1))} \\ \mathbf{M}^{(10)} & \mathbf{M}^{(11)} & \cdots & \mathbf{M}^{(1(N+1))} \\ \vdots & \vdots & \ddots & \vdots \\ \mathbf{M}^{((N+1)0)} & \mathbf{M}^{((N+1)1)} & \cdots & \mathbf{M}^{((N+1)(N+1))} \end{bmatrix} \begin{bmatrix} \mathbf{U}^{(0)} \\ \mathbf{U}^{(1)} \\ \vdots \\ \mathbf{U}^{(N+1)} \end{bmatrix} = \begin{bmatrix} \mathbf{0} \\ \mathbf{0} \\ \vdots \\ \mathbf{0} \end{bmatrix}. \tag{5.48}$$

The assembled differential equations, together with the boundary conditions (5.40) and (5.41), are numerically computed via the GDQ method in a two-dimensional computational grid of $I_N \times I_M$ discrete points developed from the CGL distribution of equation (5.45). In this way, the discrete stiffness matrix $\mathbf{K}$ and the discrete inertial matrix $\mathbf{M}$ are obtained, while the DOF vector is denoted by $\boldsymbol{\delta} = [\boldsymbol{\delta}_b\, \boldsymbol{\delta}_d]^T$. A separation of variables is then performed so that $\boldsymbol{\delta}_b$ collects the unknowns related to the boundaries of the computational domain, whereas $\boldsymbol{\delta}_d$ is related to the others. As a consequence, the discrete form of equation (5.48) is arranged as follows [20]:

$$\mathbf{K}\boldsymbol{\delta} = \omega^2\mathbf{M}\boldsymbol{\delta} \Leftrightarrow \begin{bmatrix} \mathbf{K}_{bb} & \mathbf{K}_{bd} \\ \mathbf{K}_{db} & \mathbf{K}_{dd} \end{bmatrix} \begin{bmatrix} \boldsymbol{\delta}_b \\ \boldsymbol{\delta}_d \end{bmatrix} = \omega^2 \begin{bmatrix} \mathbf{0} & \mathbf{0} \\ \mathbf{0} & \mathbf{M}_{dd} \end{bmatrix} \begin{bmatrix} \boldsymbol{\delta}_b \\ \boldsymbol{\delta}_d \end{bmatrix}. \tag{5.49}$$

A more efficient numerical system can be obtained from a static condensation of equation (5.49):

$$(\mathbf{M}_{dd}^{-1}(\mathbf{K}_{dd} - \mathbf{K}_{db}\mathbf{K}_{bb}^{-1}\mathbf{K}_{bd}) - \omega^2\mathbf{I})\boldsymbol{\delta}_d = \mathbf{0}, \tag{5.50}$$

$\mathbf{I}$ being the identity matrix.

The integrations that occur in equation (5.25) for the evaluation of the generalized stiffness constants are performed using the generalized integral quadrature (GIQ) method [20]. Starting from the GDQ rule of equation (5.43), it provides the integral of an arbitrary smooth function $f(x)$ restricted to a closed interval $[x_i, x_j]$ as a weighted sum with coefficients $\tilde{w}_k^{ij}$ of the values $f(x_k)$ with $k = 1, \ldots, I_Q$ assumed by the function itself:

$$\int_{x_i}^{x_j} f(x)dx = \sum_{k=1}^{I_Q} \tilde{w}_k^{ij} f(x_k) = \sum_{k=1}^{I_Q} (w_{jk} - w_{ik}) f(x_k) \tag{5.51}$$

The coefficients w_{ik} and w_{jk} with $i, j, k = 1, \ldots, I_Q$ are derived from the GDQ coefficients (5.44) that refer to $n = 1$. In particular, the latter are adopted for the derivation of the GDQ shifted coefficients denoted by $\bar{\varsigma}_{ij}^{(1)}$ for $i, j = 1, \ldots, I_Q$, collected in the matrix $\bar{\boldsymbol{\varsigma}}^{(1)}$ of dimensions $I_Q \times I_Q$. They are computed by means of the following expression:

$$\begin{aligned} \bar{\varsigma}_{ij}^{(1)} &= \frac{x_i - \varepsilon}{x_j - \varepsilon} \varsigma_{ij}^{(1)} \quad \text{for } i \neq j \\ \bar{\varsigma}_{ij}^{(1)} &= \varsigma_{ii}^{(1)} + \frac{1}{x_i - \varepsilon} \quad \text{for } i = j, \end{aligned} \tag{5.52}$$

where the parameter ε has been calibrated to 1×10^{-10}. The GIQ coefficients w_{ik}, w_{jk} are the elements of the inverse matrix of $\bar{\boldsymbol{\varsigma}}^{(1)}$, denoted by $\mathbf{W}$:

$$\mathbf{W} = (\bar{\boldsymbol{\varsigma}}^{(1)})^{-1}. \tag{5.53}$$

5.7 Application and results

We now provide some examples stemming from the formulation presented in this chapter to evaluate the mode frequencies and shapes of laminated structures of different curvatures. More specifically, a reference solution is provided in each case by a refined three-dimensional finite element model, as implemented in a commercial software. The assumption (5.7) for the field variable is then calibrated to highlight the three-dimensional capability of this ESL formulation. The accuracy of the solution is checked for structures with different curvatures and lamination schemes. Moreover, it is shown that the introduction of a soft region within the stacking sequence due to the presence of porosity does not affect the reliability of the formulation for the prediction of the three-dimensional response.

In this chapter, the FGM layers are obtained by combining two phases of isotropic materials. The ceramic is characterized by a density equal to $\rho_{\mathrm{C}}^{(k)} = 5700\mathrm{kg\ m^{-3}}$, whereas the metal density has been set to $\rho_{\mathrm{M}}^{(k)} = 2707\mathrm{kg\ m^{-3}}$. The elastic material properties of the two phases under consideration are:

$$\begin{aligned} &\text{Ceramic} \rightarrow E_{\mathrm{C}}^{(k)} = 168\mathrm{GPa}, \quad \nu_{\mathrm{C}}^{(k)} = 0.3 \\ &\text{Metal} \ \ \ \rightarrow E_{\mathrm{M}}^{(k)} = 70\mathrm{GPa}, \quad \ \ \nu_{\mathrm{M}}^{(k)} = 0.3. \end{aligned} \tag{5.54}$$

On the other hand, the porosity of the material has been modeled by means of the three approaches outlined in equation (5.20), while taking into account of various through-the-thickness distributions with polynomial and trigonometric analytical expressions. To this end, the interested reader can find some useful relations $\beta^{(k)}(\zeta) = \beta^{(k)}(\xi_k)$ in table 5.1 involving the porosity effects within each kth layer of the stacking sequence. Such quantities are defined in terms of dimensionless coordinate ξ_k within the layer, independently of its effective thickness $h_k = \zeta_{k+1} - \zeta_k$. As a matter of fact, the scaling parameter of the porosity distribution associated with the kth lamina is denoted by $\beta_0^{(k)}$. Both symmetric and unsymmetric profiles have been considered for the dispersion of porosity within the structure. Furthermore, different distributions have been selected so that the porosity effect can be concentrated in the core region or in the external skins of each layer.

The first example consists of an elliptic cylinder made of a single FGM lamina. The granular composite under consideration has been obtained by combining the two isotropic constituent materials of equation (5.54). A two-dimensional ESL model has been developed from the present formulation. It accounts for a CGL grid (5.45) characterized by $I_N = I_M = 31$, and the first ten natural frequencies have been evaluated through the use of higher-order theories, taking into account the porosity effects due to linear, cosine, and hyperbolic tangent through-the-thickness dispersions. In each case, a reference solution has been calculated using a three-dimensional finite element model. The results have been collected in table 5.2. Figure 5.2 presents all the information related about the lamination scheme, boundary conditions, and shell geometry. It has been shown that the outcomes of finite element simulations can be obtained with a high level of accuracy with an ESL approach if the kinematic expansion order $N = 4$ is selected. Classical displacement field assumptions such as FSDT and TSDT provide higher values of the mode frequencies, since the stretching effect is not considered in the kinematic assumptions. Furthermore, finite element models account for a discrete variation of the material properties along the thickness of the cylinder. Therefore, very complicated

Table 5.2. A free vibration analysis of an elliptic FGM cylinder made of a single porous layer. The analysis was performed using higher-order theories.

Elliptic cylinder (FCFC)							
Mode f (Hz) DOFs	3D FEM 1804 563	FSDT 5046	TSDT 10 092	ED1 5046	ED2 7569	ED3 10 092	ED4 12 615
No porosity—$FGM^{C(\beta_0=0)}_{M(a=1/b=0/c=1/d=1/p=0.5)}$							
1	136.911	139.034	139.068	138.991	139.387	139.513	139.493
2	229.808	231.678	231.700	230.944	231.891	231.976	231.969
3	281.765	285.767	285.883	285.272	286.380	286.812	286.768
4	285.357	289.203	289.333	288.537	289.753	290.185	290.144
5	388.055	390.808	390.843	384.059	390.725	390.867	390.860
6	405.420	410.486	410.615	407.857	410.553	411.062	411.030
7	493.127	494.816	494.853	490.947	495.091	495.238	495.223
8	497.765	504.499	504.804	504.143	505.502	506.689	506.600
9	523.404	529.622	529.859	523.109	530.193	530.958	530.913
10	530.710	535.543	535.677	532.211	535.573	536.257	536.213
Linear I porosity—$FGM^{C(A1-L-I/\beta_0=0.7)}_{M(a=1/b=0/c=1/d=1/p=0.5)}$							
1	151.741	153.681	153.263	153.617	154.260	153.956	153.548
2	241.394	247.361	247.033	246.928	247.791	247.455	247.196
3	308.791	313.573	312.277	313.119	314.623	313.701	312.897
4	310.512	316.810	315.503	316.223	317.709	316.717	316.047
5	401.919	407.458	406.959	403.413	408.156	407.242	406.977
6	435.405	444.869	443.233	442.940	445.541	444.332	443.583
7	499.876	512.661	512.257	510.357	513.357	512.704	512.464
8	544.073	554.587	551.171	550.778	556.247	554.111	552.528
9	548.886	556.657	555.767	554.211	557.501	556.198	555.770
10	572.099	577.888	577.446	570.479	579.711	578.014	577.812

Hyperbolic tangent II porosity—$FGM^{C(A1-HT-II/\beta_0=0.7)}_{M(a=1/b=0/c=1/d=1/p=0.5)}$							
1	117.332	121.621	121.704	121.606	122.294	122.399	121.910
2	215.134	221.134	221.198	220.471	221.614	221.647	221.337
3	243.727	252.190	252.474	251.961	253.503	253.866	252.916
4	248.488	258.538	258.859	258.226	259.691	260.089	259.271
5	363.656	375.252	375.371	370.021	376.119	375.754	375.354
6	368.912	375.607	375.948	373.648	376.711	376.952	376.070
7	430.665	445.028	445.775	444.873	447.299	448.332	446.660
8	456.255	473.234	474.017	473.190	475.353	476.437	474.872
9	479.558	487.083	487.264	483.584	487.977	487.894	487.426
10	503.740	510.731	510.899	503.104	511.869	511.309	510.720
Asymmetric cosine II porosity—$FGM^{C(A1-UC-II/\beta_0=0.7)}_{M(a=1/b=0/c=1/d=1/p=0.5)}$							
1	130.547	125.970	125.936	125.644	126.340	126.093	126.043
2	224.474	227.204	227.171	226.620	227.369	227.235	227.231
3	269.707	260.602	260.499	259.849	261.187	260.845	260.736
4	273.922	267.551	267.434	266.824	268.003	267.761	267.693
5	381.008	386.015	385.959	381.544	386.182	385.921	385.889
6	391.060	392.320	392.170	389.626	392.617	392.281	392.189
7	476.443	460.435	460.169	459.215	461.208	460.861	460.640
8	490.202	490.542	490.240	489.477	491.130	490.887	490.713
9	502.224	497.103	497.058	494.145	497.356	497.200	497.159
10	520.424	529.454	529.368	522.269	529.807	529.311	529.277

Geometric inputs: $a = 1.50m$, $b = 2.00\text{m}$, $h_1 = 0.10\text{m}$
Material properties: first layer: $FGM^{C(A1/\beta_0=0.7)}_{M(a=1/b=0/c=1/d=1/p=0.5)}$
Computational grid: CGL distribution with $I_N \times I_M = 31\times31$

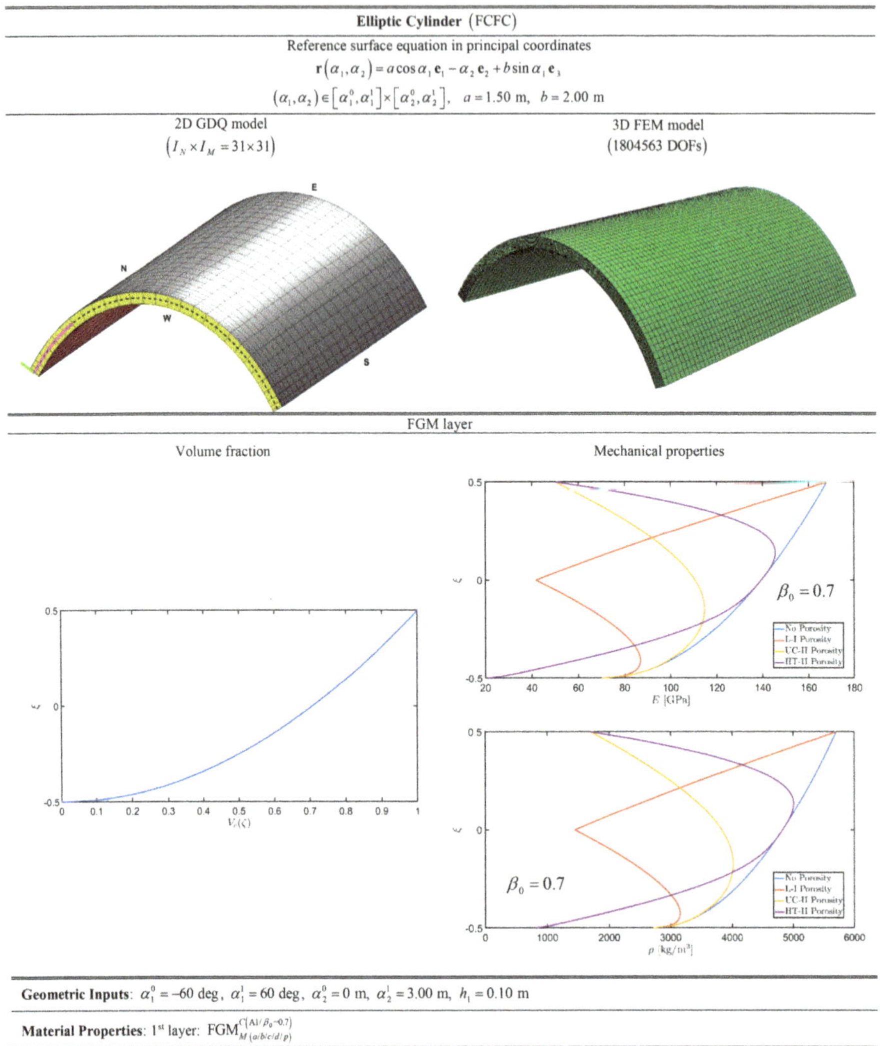

Figure 5.2. A representation of an elliptic cylinder made of a porous FGM layer. The reference surface of the structure is described according to the ESL approach, taking account of the curvilinear principal coordinates.

porosity profiles such as the trigonometric ones are only approximated within the model, thus requiring a significative increase in the number of elements so that stable results can be provided. On the other hand, the computation of the generalized stiffness constants of equation (5.25) using the GIQ rule provides a more accurate description of the distribution, thus reducing the computational cost with respect to the 3D FEM. In figure 5.3, the first nine mode shapes of the cylinder are shown. They have been calculated from the two-dimensional model using the ED4 displacement field assumption, taking into account a hyperbolic tangent

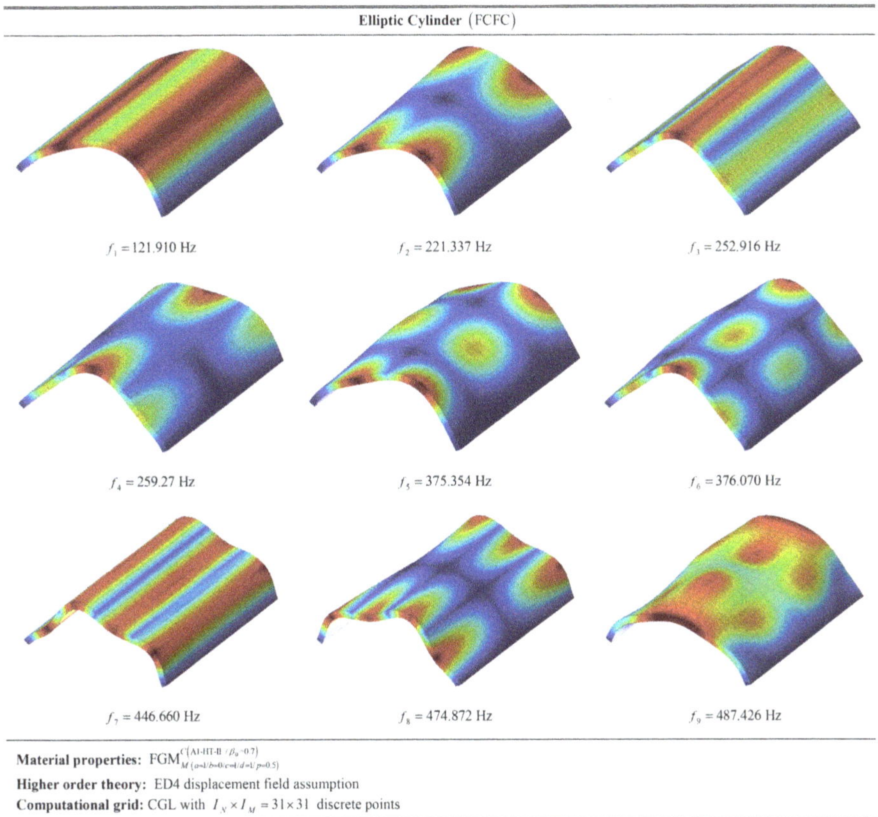

Figure 5.3. The first nine mode shapes of an elliptic cylinder made of a porous FGM layer calculated using the ED4 higher-order theory. The through-the-thickness distribution of the constituent material is modeled by means of a trigonometric analytical expression.

distribution of porosity in the direction of the shell thickness. It can be seen that the deformation effects act in both longitudinal and circumferential principal directions, with some complicating effects that are especially visible in higher modes.

The higher-order model has been validated for a revolution shell with a parabolic meridian characterized by different distributions of the material porosity. As can be seen from figure 5.4, two different configurations have been considered for the FGM layer, since two different values have been assigned to the parameter $p^{(k)}$ of the distribution of equation (5.17). In each case, mode frequencies and shapes have been evaluated for both a dense material and a porous material while taking account of linear through-the-thickness porosity distributions with symmetric and asymmetric configurations. As can be seen from figure 5.4, the through-the-thickness distributions of the material properties with a nonuniform dispersion of porosity fall within a dense configuration and uniform configuration. In table 5.3, a validation of the higher-order model has been performed in the absence of porosity within the model through the use of a finite element three-dimensional numerical prediction as

Revolution Surface with Parabolic Meridian (FCFF)

Reference surface equation in principal coordinates

$$\mathbf{r}(\varphi,\vartheta)=R_0(\varphi)\cos\vartheta\,\mathbf{e}_1-R_0(\varphi)\sin\vartheta\,\mathbf{e}_2+\frac{\left(R_0(\varphi)-R_b\right)^2}{\kappa}\mathbf{e}_3,\quad R_0(\varphi)=\frac{\kappa\tan\varphi}{2}+R_b,\quad \kappa=\frac{a^2-d^2}{b}$$

$$(\varphi,\vartheta)\in\left[\varphi_0,\vartheta_0\right]\times\left[\varphi_1,\vartheta_1\right],\quad a=3.00\text{ m},\ b=4.00\text{ m},\ d=1.00\text{ m}$$

2D GDQ model $\left(I_N\times I_M=31\times31\right)$

3D FEM model (1608963 DOFs)

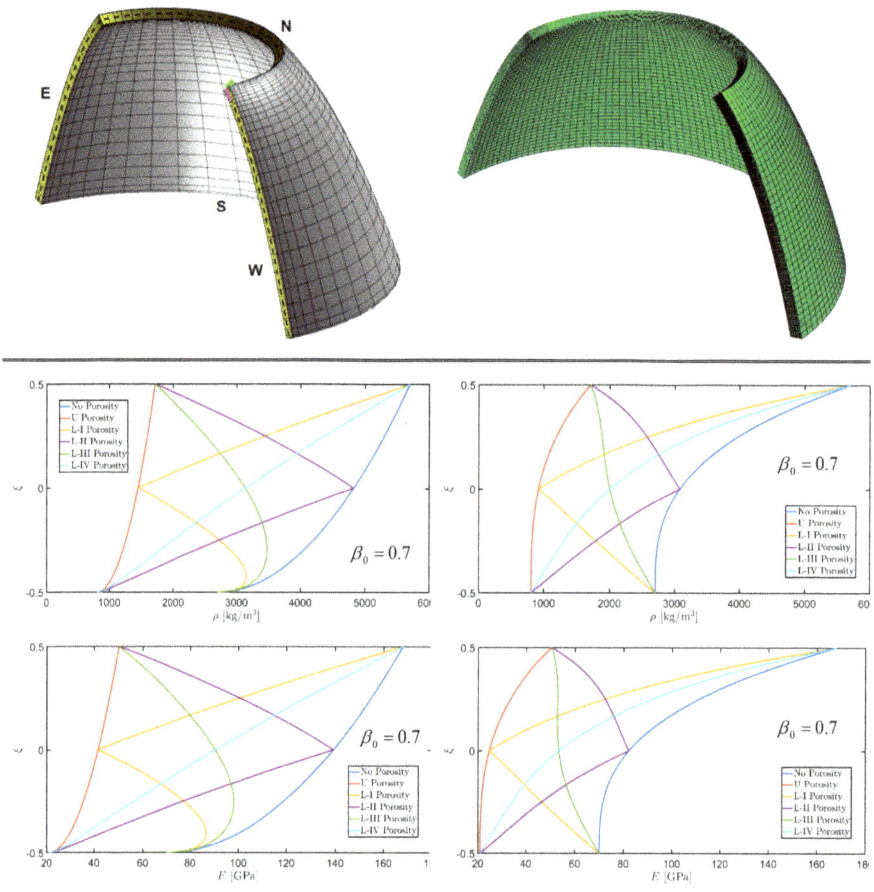

$\mathrm{FGM}_{M\,(a=1/b=0/c=1/d=1/p=0.5)}^{C(\mathrm{Al}/\beta_0=0.7)}$ $\mathrm{FGM}_{M\,(a=1/b=0/c=1/d=1/p=3)}^{C(\mathrm{Al}/\beta_0=0.7)}$

Geometric Inputs: $\varphi_0=\arctan\left(\frac{2c}{\kappa}\right)$, $\varphi_1=\arctan\left(\frac{2a}{\kappa}\right)$, $\vartheta_0=0$, $\vartheta_1=\pi$, $h_1=0.20$ m

Material Properties: 1st layer: $\mathrm{FGM}_{M\,(a/b/c/d/p)}^{C(\mathrm{Al}/\beta_0=0.7)}$

Figure 5.4. A representation of a revolution surface of a parabolic meridian made of a porous FGM layer. The reference surface of the structure is described according to the ESL approach while taking account of the curvilinear principal coordinates.

reference solution. The analysis has been performed for a homogeneous isotropic ceramic material ($p = 0$) and a metal ($p = +\infty$). Furthermore, the FGM configurations characterized by $p = 0.5$ and $p = 3$ have been also investigated. A fixed

Table 5.3. A free vibration analysis of an FGM revolution surface of a parabolic meridian made of a single porous layer; the analysis is performed using higher-order theories.

Revolution surface with parabolic meridian (FCFF)									
	No porosity				$A1-\beta_0=0.7$				
					U	L–I	L–II	L–III	L–IV
Mode f (Hz)	3D FEM	FSDT	TSDT	ED4	ED4	ED4	ED4	ED4	ED4
DOFs	1608 963	5046	10 092	12 615	12 615	12 615	12 615	12 615	12 615
				$FGM_{M(a=1/b=0/c=1/d=1/p=0)}^{C(A1/\beta_0)}$					
1	41.292	41.258	41.285	41.111	43.292	45.908	37.074	39.532	39.100
2	50.169	50.140	50.152	50.028	52.192	53.610	45.187	48.701	47.734
3	66.555	66.504	66.530	66.577	69.460	74.952	59.647	63.306	63.137
4	87.128	87.033	87.000	87.286	90.708	99.257	72.347	81.078	79.657
5	110.421	110.290	110.252	110.690	115.395	124.511	94.910	104.701	102.399
6	120.674	120.614	120.637	120.666	125.476	127.076	109.242	117.326	116.677
7	130.953	130.898	130.898	131.088	134.924	142.825	125.075	127.866	126.986
8	159.579	159.661	159.693	159.516	157.341	177.394	134.750	143.772	148.074
9	182.386	182.317	182.310	182.461	184.900	197.101	157.321	168.216	171.004
10	185.711	185.681	185.676	185.792	188.288	203.732	161.514	176.058	177.019
				$FGM_{M(a=1/b=0/c=1/d=1/p=0.5)}^{C(A1/\beta_0)}$					
1	39.652	40.046	40.072	39.912	42.024	44.327	36.270	39.073	37.736
2	48.521	48.904	48.916	48.792	50.893	52.304	44.117	48.106	46.058
3	63.758	64.441	64.468	64.548	67.376	72.141	58.416	62.560	60.997
4	82.686	83.850	83.833	84.129	87.392	95.468	70.251	80.291	75.634
5	105.298	106.575	106.556	107.013	111.575	119.883	92.505	103.599	97.906
6	117.465	118.144	118.168	118.211	122.539	124.449	106.537	115.837	112.500
7	126.449	127.426	127.437	127.674	132.029	138.377	123.330	126.062	124.570
8	152.425	154.402	154.448	154.337	152.294	171.146	131.176	142.047	142.028
9	174.456	176.491	176.514	176.669	179.007	192.513	153.336	166.283	164.214
10	179.799	181.200	181.224	181.322	183.889	196.438	156.811	174.262	169.671

(Continued)

Table 5.3. *(Continued)*

Revolution surface with parabolic meridian (FCFF)									
	No porosity				$A1-\beta_0=0.7$				
					U	L–I	L–II	L–III	L–IV
Mode f (Hz)	3D FEM	FSDT	TSDT	ED4	ED4	ED4	ED4	ED4	ED4
DOFs	1608 963	5046	10 092	12 615	12 615	12 615	12 615	12 615	12 615
$FGM_{M(a=1/b=0/c=1/d=1/p=3)}^{C(A1/\beta_0)}$									
1	40.024	39.936	39.955	39.836	41.895	43.792	36.033	39.210	36.876
2	48.562	48.453	48.454	48.362	50.430	51.437	44.028	47.986	44.975
3	64.501	64.375	64.374	64.468	67.198	71.311	57.845	62.742	59.579
4	84.547	84.404	84.284	84.561	87.891	94.284	70.920	81.890	73.752
5	107.040	106.853	106.696	107.122	111.686	118.249	92.751	104.896	95.508
6	116.593	116.317	116.320	116.392	121.047	122.073	106.239	114.994	109.717
7	126.643	126.428	126.354	126.582	130.140	136.350	120.371	124.077	121.716
8	154.663	154.537	154.459	154.415	152.190	169.249	130.436	142.930	138.479
9	176.778	176.601	176.392	176.576	178.938	189.221	152.663	167.443	160.212
10	179.553	179.293	179.171	179.281	181.686	193.884	156.908	173.113	165.478
$FGM_{M(a=1/b=0/c=1/d=1/p=+\infty)}^{C(A1/\beta_0)}$									
1	38.677	38.664	38.690	38.508	40.522	43.000	34.726	37.177	36.601
2	46.991	46.982	46.993	46.860	48.870	50.215	42.326	45.756	44.697
3	62.340	62.326	62.350	62.361	65.047	70.205	55.870	59.271	59.014
4	81.611	81.574	81.544	81.758	84.998	92.971	67.765	75.844	74.488
5	103.429	103.367	103.331	103.681	108.138	116.626	88.900	97.788	95.781
6	113.032	113.004	113.026	113.025	117.523	119.029	102.324	109.896	109.221
7	122.660	122.659	122.659	122.787	126.389	133.780	117.155	119.606	118.871
8	149.473	149.634	149.664	149.415	147.375	166.160	126.217	134.773	138.728
9	170.836	170.866	170.860	170.907	173.220	184.619	147.359	157.469	160.194
10	173.950	173.983	173.978	174.026	176.380	190.830	151.286	164.973	165.808

Geometric inputs: $a=1.50m$, $b=2.00\text{m}$, $h_1=0.20\text{m}$
Material properties: first layer: $FGM_{M(a=1/b=0/c=1/d=1/p)}^{C(A1/\beta_0)}$
Computational grid: CGL distribution with $I_N \times I_M = 31\times31$

value of $\beta_0 = 0.7$ of the porosity distribution scaling parameter was then selected, and the variation of the natural frequencies of the structure was obtained using the ED4 higher-order theory. The results show that different vibrational responses can be obtained for symmetric and asymmetric profile assumptions. Moreover, different results are obtained when the porosity is concentrated in the middle thickness of the structure or in the extrados and intrados, regardless of the configuration of the dense FGM material. As can be seen from the mode shapes reported in figure 5.5 where a L–IV porosity layup has been selected, the variation of the elastic stiffness of the homogenized material leads to some local deformation effects in the three-dimensional shell that classical two-dimensional models are not able to predict. Once the higher-order model was validated, a sensitivity analysis was performed for the same structure, which showed the influence of porosity on the vibrational response of the shell. In table 5.4, the first three mode frequencies are evaluated for different distributions of the volume fractions of the ceramic material. Nonuniform

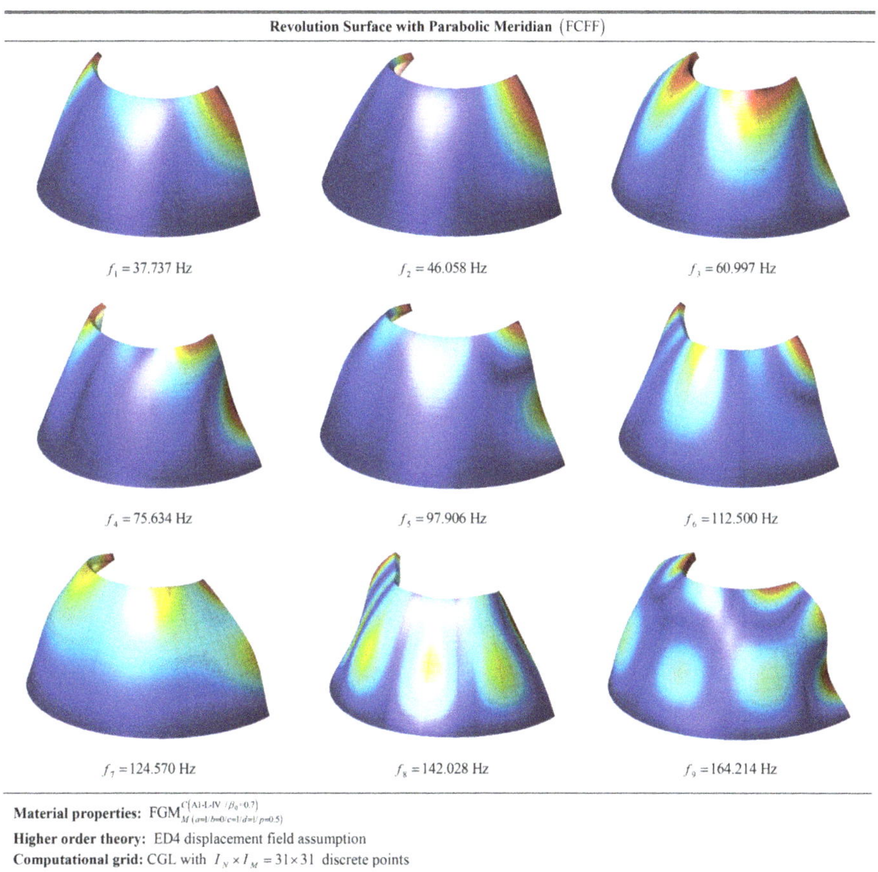

Figure 5.5. The first nine mode shapes of a revolution surface of a parabolic meridian made of a porous FGM layer calculated using the ED4 higher-order theory. The through-the-thickness distribution of the constituent material is modeled by means of a linear analytical expression.

Table 5.4. A modal analysis of a revolution surface with a parabolic meridian made of a single porous FGM layer. The first three mode frequencies of the structure are calculated while taking account of different porosity distributions and different FGM volume fraction parameters.

Revolution surface with parabolic meridian (FCFF)									
β_0	Linear I porosityFGM$_{M(a=1/b=0/c=1/d=1/p)}^{C(A1-L-I/\beta_0=0.7)}$			Asymmetric cosine II porosity FGM$_{M(a=1/b=0/c=1/d=1/p)}^{C(A1-UC-II/\beta_0=0.7)}$			Hyperbolic tangent II porosityFGM$_{M(a=1/b=0/c=1/d=1/p)}^{C(A1-HT-II/\beta_0=0.7)}$		
Mode f(Hz)	f_1	f_2	f_3	f_1	f_2	f_3	f_1	f_2	f_3
$p = 0$									
0.1	41.111	50.028	66.577	41.111	50.028	66.577	41.111	50.028	66.577
0.2	41.471	50.345	67.171	41.308	50.187	66.822	40.934	49.872	66.128
0.3	41.860	50.679	67.825	41.467	50.312	67.033	40.722	49.679	65.652
0.4	42.284	51.032	68.545	41.579	50.395	67.188	40.475	49.446	65.144
0.5	42.746	51.405	69.340	41.627	50.422	67.253	40.192	49.166	64.601
0.6	43.253	51.800	70.223	41.581	50.363	67.170	39.869	48.832	64.015
0.7	43.813	52.217	71.206	41.389	50.164	66.848	39.500	48.434	63.381
0.8	44.436	52.657	72.308	40.951	49.724	66.127	39.080	47.959	62.690
0.9	45.129	53.123	73.549	40.083	48.807	64.719	38.604	47.385	61.932
$p = 0.5$									
0.1	39.912	48.792	64.548	39.912	48.792	64.548	39.912	48.792	64.548
0.2	40.249	49.104	65.092	40.044	48.894	64.688	39.752	48.635	64.150
0.3	40.611	49.433	65.690	40.134	48.957	64.784	39.563	48.445	63.729
0.4	41.004	49.779	66.346	40.171	48.971	64.815	39.343	48.218	63.283
0.5	41.431	50.146	67.071	40.140	48.917	64.749	39.091	47.948	62.807
0.6	41.899	50.534	67.873	40.012	48.765	64.538	38.803	47.627	62.294
0.7	42.414	50.943	68.764	39.742	48.458	64.106	38.474	47.248	61.740
0.8	42.984	51.375	69.761	39.247	47.891	63.329	38.100	46.796	61.136
0.9	43.618	51.830	70.880	38.384	46.837	62.004	37.671	46.257	60.474

0.1	39.547	48.349	63.937	39.547	48.349	63.937	39.547	48.349	63.937
0.2	39.870	48.650	64.460	39.647	48.423	64.027	39.392	48.196	63.550
0.3	40.217	48.967	65.031	39.703	48.454	64.067	39.208	48.011	63.141
0.4	40.591	49.302	65.656	39.701	48.432	64.036	38.993	47.788	62.704
0.5	40.998	49.653	66.343	39.629	48.337	63.903	38.746	47.522	62.237
0.6	41.440	50.025	67.099	39.457	48.137	63.622	38.461	47.206	61.732
0.7	41.925	50.417	67.935	39.142	47.772	63.122	38.136	46.830	61.183
0.8	42.459	50.829	68.862	38.605	47.134	62.291	37.763	46.380	60.583
0.9	43.048	51.262	69.895	37.713	45.997	60.946	37.334	45.839	59.920
					$p = 3$				
0.1	39.836	48.362	64.468	39.836	48.362	64.468	39.836	48.362	64.468
0.2	40.152	48.639	64.993	39.913	48.424	64.513	39.675	48.227	64.052
0.3	40.491	48.929	65.563	39.937	48.442	64.497	39.483	48.058	63.609
0.4	40.854	49.235	66.183	39.900	48.403	64.396	39.256	47.853	63.133
0.5	41.247	49.557	66.858	39.781	48.289	64.174	38.993	47.604	62.617
0.6	41.673	49.897	67.594	39.551	48.062	63.780	38.688	47.304	62.055
0.7	42.135	50.256	68.400	39.161	47.662	63.137	38.335	46.940	61.438
0.8	42.639	50.632	69.283	38.527	46.970	62.124	37.927	46.498	60.756
0.9	43.190	51.027	70.252	37.506	45.739	60.553	37.453	45.958	59.996
					$p = 5$				
0.1	40.072	48.427	64.924	40.072	48.427	64.924	40.072	48.427	64.924
0.2	40.395	48.699	65.468	40.169	48.510	65.001	39.903	48.296	64.484
0.3	40.740	48.985	66.057	40.213	48.552	65.015	39.701	48.133	64.014
0.4	41.112	49.287	66.696	40.195	48.539	64.941	39.463	47.933	63.507
0.5	41.512	49.605	67.391	40.094	48.455	64.741	39.187	47.689	62.957
0.6	41.946	49.941	68.148	39.879	48.263	64.360	38.866	47.393	62.356
0.7	42.417	50.294	68.975	39.498	47.905	63.712	38.494	47.032	61.694
0.8	42.930	50.666	69.879	38.861	47.264	62.664	38.063	46.591	60.960
0.9	43.490	51.056	70.869	37.813	46.088	61.003	37.561	46.047	60.139

Geometric inputs: $\varphi_0 = \arctan\left(\frac{2c}{\kappa}\right)$, $\varphi_1 = \arctan\left(\frac{2a}{\kappa}\right)$, $\vartheta_0 = 0$, $\vartheta_1 = \pi$, $h_1 = 0.20\text{m}$

Material properties: first layer: $\text{FGM}^{C(\text{A1}/\beta_0=0.7)}_{M(a=1/b=0/c=1/d=1/p)}$

Computational grid: CGL distribution with $I_N \times I_M = 31\times31$

dispersions of porosity are considered by setting different values of the distribution scaling parameter β_0. The equivalent elastic properties of the homogenized porous material have been computed using the A1 approach (5.20). The numerical results of the parametric investigation under discussion are plotted in figure 5.6. It can be seen that the employment of the L-I porosity increases natural frequencies in case of dense material ($\beta_0 = 0$), regardless of the nature of the constituent phases. For both isotropic materials and FGM granular composites, the introduction of a symmetric porosity distribution leads to an increase in the natural frequencies. On the other hand, for unsymmetric dispersions such as the UC-II and HT-II distributions, a decreased frequency can be observed for all considered FGM configurations. However, in case of UC-II, a slight increase in the first three mode frequencies is observed for lower values of β_0.

The effect of porosity has been, also, checked for a laminated doubly curved shell structure with negative curvature, namely a revolution hyperbolic hyperboloid. The analytical expression of the reference surface in principal coordinates can be

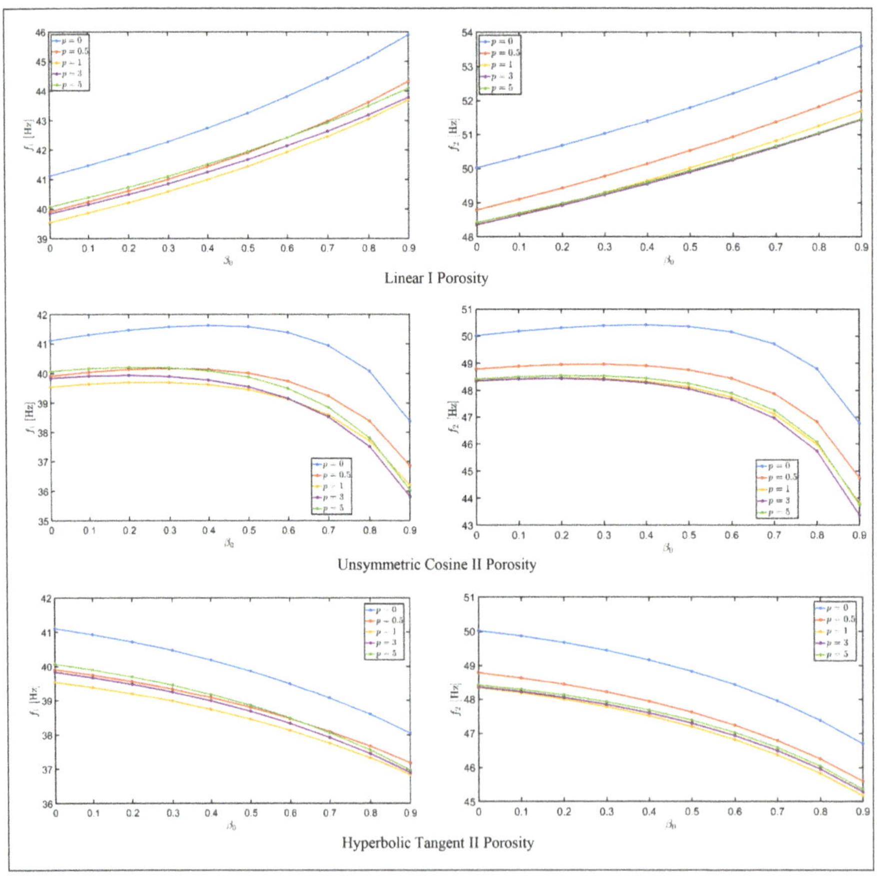

Figure 5.6. A parametric investigation of the natural frequencies of a revolution surface with a parabolic meridian made of an FGM layer with different through-the-thickness porosity distributions.

found in figure 5.7, together with a representation of the three-dimensional mesh employed for finite element numerical predictions, as well as the GDQ model. The lamination scheme of the structure consists of three layers. The central core is made of an FGM produced using the constitutive materials reported in equation (5.54), while the external sheets are made of an orthotropic graphite epoxy and a generally anisotropic triclinic material [55]. The three-dimensional stiffness matrix of the triclinic material ($\rho^{(k)} = 7750 \text{kg m}^{-3}$) expressed in the material reference system is shown below, setting $C_{nm}^{(k)} = E_{nm}^{(k)}$ for $n, m = 1, \ldots, 6$ according to equation (5.14):

Revolution Hyperbolic Hyperboloid (FCFC)

Reference surface equation in principal coordinates

$$\mathbf{r}(\alpha_1, \alpha_2) = a \cosh \alpha_1 \cos \alpha_2 \, \mathbf{e}_1 - a \cosh \alpha_1 \sin \alpha_2 \, \mathbf{e}_2 + c \sinh \alpha_1 \, \mathbf{e}_3$$

$$(\alpha_1, \alpha_2) \in [\alpha_1^0, \alpha_1^1] \times [\alpha_2^0, \alpha_2^1], \quad a = 0.50 \text{ m}, \quad c = 0.70 \text{ m}$$

2D GDQ model $(I_N \times I_M = 31 \times 31)$ | 3D FEM model (1924851 DOFs)

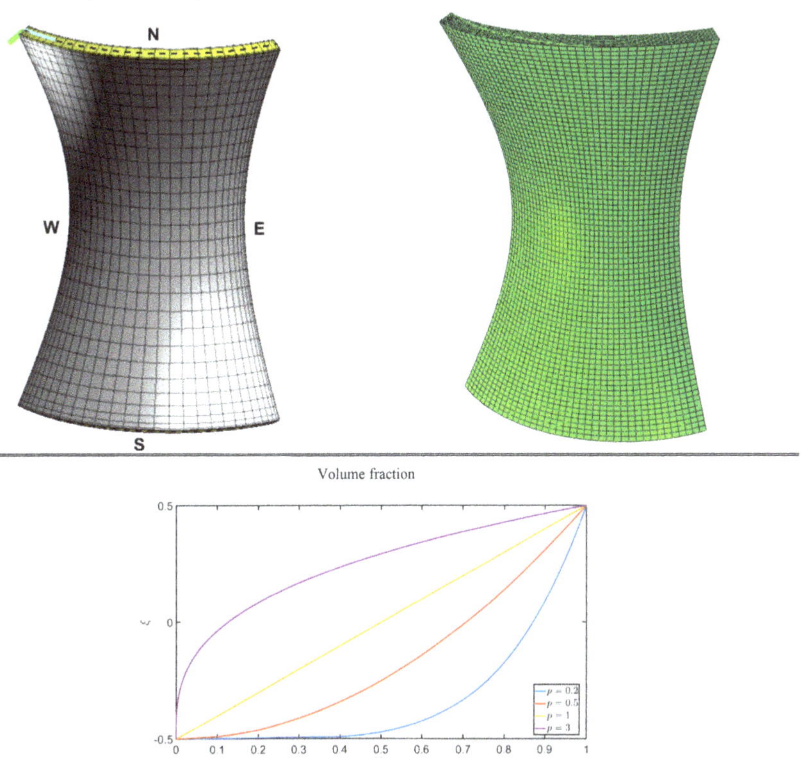

Geometric Inputs: $\alpha_1^0 = -1$, $\alpha_1^1 = 1$, $\alpha_2^0 = 0$, $\alpha_2^1 = \pi/2$, $h_1 = h_3 = 0.005$ m, $h_2 = 0.05$ m

Material Properties: $\left(70 / \text{FGM}_{M(a=1/b=0/c=1/d=1/p)}^{C(A1/\beta_0=0.8)} / 30\right)$, 1st layer: triclinic material, 2nd layer: FGM material, 3rd layer: graphite-epoxy

Figure 5.7. A representation of a revolution hyperbolic hyperboloid with a porous FGM layer within the lamination scheme. The reference surface of the structure is described according to the ESL approach, taking account of the curvilinear principal coordinates.

$$\mathbf{E}^{(k)} = \begin{bmatrix} C_{11}^{(k)} & C_{12}^{(k)} & C_{16}^{(k)} & C_{14}^{(k)} & C_{15}^{(k)} & C_{13}^{(k)} \\ C_{12}^{(k)} & C_{22}^{(k)} & C_{26}^{(k)} & C_{24}^{(k)} & C_{25}^{(k)} & C_{23}^{(k)} \\ C_{16}^{(k)} & C_{26}^{(k)} & C_{66}^{(k)} & C_{46}^{(k)} & C_{56}^{(k)} & C_{36}^{(k)} \\ C_{14}^{(k)} & C_{24}^{(k)} & C_{46}^{(k)} & C_{44}^{(k)} & C_{45}^{(k)} & C_{34}^{(k)} \\ C_{15}^{(k)} & C_{25}^{(k)} & C_{56}^{(k)} & C_{45}^{(k)} & C_{55}^{(k)} & C_{35}^{(k)} \\ C_{13}^{(k)} & C_{23}^{(k)} & C_{36}^{(k)} & C_{34}^{(k)} & C_{35}^{(k)} & C_{33}^{(k)} \end{bmatrix} = \begin{bmatrix} 98.84 & 53.92 & 0.03 & 1.05 & -0.1 & 50.78 \\ 53.92 & 99.19 & 0.03 & 0.55 & -0.18 & 50.87 \\ 0.03 & 0.03 & 22.55 & -0.04 & 0.25 & 0.02 \\ 1.05 & 0.55 & -0.04 & 21.1 & 0.07 & 1.03 \\ -0.1 & -0.18 & 0.25 & 0.07 & 21.14 & -0.18 \\ 50.78 & 50.87 & 0.02 & 1.03 & -0.18 & 87.23 \end{bmatrix} \text{GPa}. \tag{5.55}$$

The orthotropic graphite epoxy is described by well-known engineering constants, namely the elastic moduli $E_1^{(k)}$, $E_2^{(k)}$, $E_3^{(k)}$, the shear moduli $G_{12}^{(k)}$, $G_{13}^{(k)}$, $G_{23}^{(k)}$ and the Poisson's coefficients $\nu_{12}^{(k)}$, $\nu_{13}^{(k)}$, $\nu_{23}^{(k)}$:

$$\begin{array}{lll} E_1^{(k)} = 137.90\text{GPa} & G_{23}^{(k)} = 6.21\text{GPa} & \nu_{23}^{(k)} = 0.49 \\ E_2^{(k)} = E_3^{(k)} = 8.96\text{GPa} & G_{12}^{(k)} = G_{13}^{(k)} = 7.10\text{GPa} & \nu_{12}^{(k)} = \nu_{13}^{(k)} = 0.30. \end{array} \tag{5.56}$$

In [20], the interested reader can find the complete expression for $C_{nm}^{(k)}$ in terms of engineering constants.

Table 5.5 presents a validation of the higher-order displacement field assumption for the structure here proposed. The first ten mode frequencies have been calculated using the GDQ model for different distributions of porosity in the direction of the shell thickness. In each case, a 3D FEM solution has been calculated using a model based on parabolic brick elements. As far as the ESL model is concerned, the higher-order displacement field assumption also accounts for the Murakami's zigzag function introduced in equation (5.7), to model suitably the interlaminar phenomena of the lamination scheme. As can be seen, the introduction of different distributions of material porosity requires the employment of a higher-order displacement field assumption, especially for higher modes in which the coupling effects are more pronounced. Moreover, the accuracy of the ESL model is not affected by different configurations of the material porosity, which shows that the model is able to predict accurately both the conventional deflections and the effects caused by the presence of a soft core.

After such accuracy check for both dense and porous lamination schemes, the sensitivity of the vibrational response of the hyperbolic hyperboloid under discussion is evaluated, as shown in tables 5.6 and 5.7, for both symmetric and unsymmetric reductions of the material properties, respectively. In figure 5.8 we report the first nine mode shapes of the structure for SC-II porosity distributions, showing the three-dimensional vibrational behavior of the porous shell object under analysis.

Table 5.5. A free vibration analysis of a revolution hyperbolic hyperboloid with a porous FGM layer within the lamination scheme, employing higher-order theories.

Revolution hyperbolic hyperboloid (FCFC)											
Mode f(Hz) DOFs	3D FEM 1804 563	FSDT 5046	TSDT 10 092	ED1 5046	EDZ1 7569	ED2 7569	EDZ2 10 092	ED3 10 092	EDZ3 12 615	ED4 12 615	EDZ4 15 138
No porosity—$FGM^{C(A1/\beta_0=0)}_{M(a=1/b=0/c=1/d=1/p=3)}$											
1	451.781	451.501	451.534	448.906	462.894	452.107	451.642	452.372	452.166	452.201	452.039
2	519.538	517.717	517.715	509.846	525.372	518.660	518.305	518.763	518.598	518.638	518.523
3	535.374	533.864	533.739	523.202	536.867	534.696	534.405	534.635	534.497	534.583	534.483
4	538.391	535.279	535.246	524.545	540.035	536.767	536.411	536.657	536.482	536.548	536.447
5	662.572	662.475	662.427	659.170	687.048	663.790	662.631	664.497	664.063	664.073	663.753
6	685.450	683.984	683.980	677.519	707.023	686.083	684.674	686.633	686.142	686.209	685.806
7	688.034	684.154	684.149	681.943	713.398	686.625	685.394	686.703	686.157	686.244	685.929
8	834.811	832.808	832.852	822.027	850.681	834.755	833.798	835.171	834.764	834.839	834.552
9	956.361	956.203	956.086	952.050	978.732	957.737	955.765	959.038	958.392	958.373	957.892
10	975.541	973.946	973.810	967.568	998.015	975.310	974.865	975.238	975.040	975.089	974.959
Linear I porosity—$FGM^{C(A1-L-I/\beta_0=0.8)}_{M(a=1/b=0/c=1/d=1/p=3)}$											
1	469.470	477.699	475.686	476.367	486.489	479.177	479.177	477.182	476.579	476.187	475.817
2	523.189	521.742	520.926	517.204	526.687	524.090	524.090	521.870	521.613	521.408	521.250
3	532.641	529.751	529.472	524.632	534.430	533.526	533.526	530.635	530.465	530.253	530.190
4	533.988	533.774	533.015	528.978	539.312	536.198	536.198	533.905	533.669	533.504	533.376
5	695.715	710.296	704.905	707.876	728.703	713.215	713.215	708.486	707.137	706.204	705.297
6	708.629	710.876	707.806	708.251	728.728	716.552	716.552	711.358	710.376	709.387	708.846
7	722.360	734.717	729.321	733.743	755.159	739.123	739.123	733.412	731.880	730.899	729.785
8	851.123	857.151	853.871	851.341	872.261	861.364	861.364	856.795	855.840	855.127	854.469
9	965.340	981.396	980.466	978.247	985.608	984.187	984.187	981.636	981.376	981.183	981.022
10	997.843	1022.899	1012.488	1020.393	1052.175	1026.901	1026.901	1018.437	1016.076	1014.640	1012.828

(Continued)

Table 5.5. (*Continued*)

Revolution hyperbolic hyperboloid (FCFC)											
Mode f(Hz)	3D FEM	FSDT	TSDT	ED1	EDZ1	ED2	EDZ2	ED3	EDZ3	ED4	EDZ4
DOFs	1804 563	5046	10 092	5046	7569	7569	10 092	10 092	12 615	12 615	15 138
Linear II porosity—$FGM_{M(a=1/b=0/c=1/d=1/p=3)}^{C(A1-L-II/\beta_0=0.8)}$											
1	426.271	433.181	433.790	432.014	437.709	436.179	434.764	436.121	435.409	435.333	434.867
2	494.291	498.368	498.595	494.885	502.066	501.446	500.300	501.349	499.959	500.781	499.669
3	513.355	516.455	516.465	512.056	520.170	520.349	518.934	520.237	518.059	519.612	517.879
4	514.445	517.621	517.403	513.054	520.248	520.527	519.290	520.263	518.379	519.720	518.162
5	611.787	619.004	620.373	617.441	629.520	625.822	622.408	625.639	624.395	623.856	623.124
6	642.694	647.988	648.482	645.113	657.361	655.387	651.520	654.714	652.236	652.863	651.222
7	646.351	649.943	652.093	649.064	659.613	656.871	653.092	657.394	655.869	655.476	654.634
8	789.466	797.296	798.421	792.499	805.450	803.933	800.979	804.131	801.626	802.684	800.819
9	882.296	890.863	893.540	888.759	905.579	899.862	895.143	900.384	899.236	897.849	897.254
10	933.411	960.519	960.554	957.865	965.484	965.358	962.963	964.995	962.891	963.913	962.391
Linear III porosity—$FGM_{M(a=1/b=0/c=1/d=1/p=3)}^{C(A1-L-III/\beta_0=0.8)}$											
1	447.169	451.883	452.144	450.915	457.201	453.550	452.862	453.635	453.563	453.563	452.915
2	502.425	502.999	502.988	500.126	506.670	504.503	503.846	504 346	504.155	504.155	503.715
3	513.693	515.389	515.314	511.308	519.280	517.427	516.426	516.690	516.413	516.413	516.221
4	514.533	516.547	516.220	512.765	519.462	517.745	517.046	517.360	517.123	517.123	516.848
5	647.011	650.240	650.801	649.062	663.092	654.672	652.950	654.537	654.419	654.419	652.820
6	666.717	671.285	671.143	668.248	684.560	675.697	673.557	674.387	674.122	674.122	673.178
7	690.903	693.136	694.119	692.110	705.735	696.847	694.846	697.393	697.254	697.254	696.014
8	815.187	817.510	817.997	813.546	826.457	820.964	819.567	821.072	820.787	820.787	819.689
9	925.944	935.138	936.224	934.380	953.843	941.247	938.530	941.915	941.801	941.801	939.185
10	940.653	962.599	962.330	959.602	967.106	964.415	963.338	963.687	963.465	963.465	963.259

Geometric inputs: $\alpha_1^0 = -1$, $\alpha_1^1 = 1$, $\alpha_2^0 = 0$, $\alpha_2^1 = \pi/2$, $h_1 = h_3 = 0.005m$, $h_2 = 0.05$m

Lamination scheme: $\left(70/FGM_{M(a=1/b=0/c=1/d=1/p=3)}^{C(A1/\beta_0=0.8)}/30\right)$, first layer: triclinic material, second layer: FGM material, third layer: graphite epoxy

Computational grid: CGL distribution with $I_N \times I_M = 31 \times 31$

Table 5.6. A free vibration analysis of a revolution hyperbolic hyperboloid with a porous FGM layer within the lamination scheme; the analysis accounts for a symmetric trigonometric porosity distribution.

Revolution hyperbolic hyperboloid (FCFC)										
f(Hz)	Symmetric cosine II porosity									
β_0	0	0.1	0.2	0.3	0.4	0.5	0.6	0.7	0.8	0.9
$p = 0$										
1	462.062	465.014	468.198	471.631	475.329	479.313	483.605	488.229	493.208	498.561
2	538.377	539.602	540.955	542.436	544.044	545.779	547.634	549.595	551.634	553.694
3	559.146	559.141	559.235	559.428	559.723	560.120	560.628	561.254	562.016	562.894
4	562.003	561.749	561.603	561.560	561.613	561.753	561.969	562.247	562.565	562.947
5	671.429	676.122	681.235	686.796	692.839	699.400	706.520	714.239	722.594	731.608
6	696.262	701.206	706.634	712.572	717.420	721.458	725.886	730.732	736.025	741.788
7	704.699	707.403	710.414	713.746	719.054	726.116	733.799	742.144	751.188	760.953
8	860.111	863.125	866.431	870.043	873.978	878.255	882.891	887.904	893.304	899.088
9	976.662	982.828	989.645	997.148	1005.379	1014.383	1024.205	1034.887	1046.459	1055.092
10	1019.667	1023.876	1028.103	1032.326	1036.514	1040.631	1044.624	1048.430	1051.959	1058.913
$p = 0.2$										
1	457.084	460.014	463.182	466.604	470.297	474.282	478.580	483.210	488.187	493.507
2	533.903	534.998	536.206	537.524	538.946	540.461	542.048	543.670	545.259	546.703
3	554.874	554.691	554.591	554.574	554.641	554.793	555.037	555.384	555.860	555.722
4	557.727	557.275	556.914	556.635	556.426	556.273	556.158	556.055	555.928	556.501
5	664.000	668.775	673.997	679.697	685.913	692.685	700.054	708.058	716.717	726.011
6	687.459	692.432	697.911	703.926	709.502	713.466	717.834	722.636	727.898	733.628
7	697.183	699.776	702.681	705.916	710.514	717.713	725.564	734.101	743.341	753.257
8	851.711	854.581	857.744	861.213	865.002	869.124	873.587	878.392	883.521	888.915
9	964.858	971.115	978.050	985.702	994.115	1003.334	1013.400	1024.338	1036.135	1040.988
10	1012.344	1016.155	1019.938	1023.660	1027.277	1030.733	1033.954	1036.837	1039.246	1048.684

(Continued)

Table 5.6. (*Continued*)

	Revolution hyperbolic hyperboloid (FCFC)									
f(Hz)	Symmetric cosine II porosity									
β_0	0	0.1	0.2	0.3	0.4	0.5	0.6	0.7	0.8	0.9
	$p = 0.5$									
1	452.673	455.507	458.565	461.859	465.398	469.191	473.238	477.522	481.988	486.492
2	529.209	530.077	531.024	532.036	533.090	534.153	535.167	536.039	536.620	536.685
3	549.980	549.544	549.164	548.835	548.555	548.320	548.132	547.993	547.275	546.087
4	552.791	552.066	551.399	550.774	550.172	549.565	548.919	548.180	547.908	547.849
5	657.932	662.688	667.890	673.565	679.742	686.442	693.676	701.419	709.565	717.819
6	680.201	685.098	690.493	696.410	701.791	705.468	709.500	713.886	718.587	723.477
7	690.359	692.756	695.450	698.457	702.873	709.899	717.491	725.614	734.145	742.759
8	843.725	846.308	849.155	852.265	855.636	859.255	863.088	867.067	871.047	874.726
9	954.740	960.919	967.756	975.279	983.509	992.453	1002.085	1012.304	1021.832	1021.006
10	1003.820	1007.039	1010.147	1013.090	1015.800	1018.182	1020.110	1021.410	1022.844	1033.062
	$p = 1$									
1	449.997	452.713	455.615	458.699	461.944	465.313	468.717	471.970	474.635	475.503
2	524.657	525.189	525.729	526.237	526.652	526.880	526.773	526.103	524.508	521.324
3	544.414	543.632	542.862	542.092	541.308	540.491	539.613	537.841	535.469	532.232
4	547.109	546.002	544.897	543.763	542.560	541.231	539.698	538.614	537.320	535.228
5	655.514	660.252	665.397	670.950	676.891	683.157	689.595	695.837	700.968	702.273
6	677.023	681.860	687.144	692.870	695.985	699.130	702.411	705.648	708.380	709.203
7	685.750	687.927	690.362	693.052	699.006	705.477	712.108	718.500	723.680	724.840
8	837.600	839.782	842.145	844.656	847.258	849.852	852.256	854.113	854.660	851.906
9	949.398	955.477	962.126	969.319	976.984	984.964	992.922	1000.113	996.808	990.567
10	993.910	996.264	998.359	1000.103	1001.369	1001.986	1001.707	1000.173	1004.762	1001.827

Geometric inputs: $\alpha_1^0 = -1$, $\alpha_1^1 = 1$, $\alpha_2^0 = 0$, $\alpha_2^1 = \pi/2$, $h_1 = h_3 = 0.005m$, $h_2 = 0.05$m

Lamination scheme: $\left(70/\mathrm{FGM}_{M(a=1/b=0/c=1/d=1/p)}^{C(\mathrm{A1-SC-II}/\beta_0)}/30\right)$, first layer: triclinic material, second layer: FGM material, third layer: graphite epoxy

Computational grid: CGL distribution with $I_N \times I_M = 31\times31$, EDZ4 displacement field assumption

Table 5.7. A free vibration analysis of a revolution hyperbolic hyperboloid with a porous FGM layer within the lamination scheme; the analysis accounts for an asymmetric trigonometric porosity distribution.

Revolution hyperbolic hyperboloid (FCFC)										
f(Hz)	Asymmetric cosine II porosity									
β_0	0	0.1	0.2	0.3	0.4	0.5	0.6	0.7	0.8	0.9
					$p = 0$					
1	462.062	463.612	465.183	466.751	468.282	469.731	471.037	472.113	472.837	473.032
2	538.377	538.961	539.574	540.202	540.824	541.414	541.937	542.343	542.562	542.492
3	559.146	558.997	558.893	558.830	558.800	558.794	558.799	558.518	558.196	557.850
4	562.003	561.293	560.665	560.114	559.635	559.217	558.850	558.795	558.752	558.626
5	671.429	673.520	675.691	677.901	680.092	682.191	684.093	685.653	686.665	686.824
6	696.262	697.695	699.286	700.979	702.704	704.232	704.204	704.038	703.615	702.757
7	704.699	704.410	704.255	704.201	704.210	704.368	705.851	706.985	707.534	707.158
8	860.111	860.625	861.245	861.939	862.665	863.363	863.955	864.325	864.312	863.674
9	976.662	980.073	983.627	987.248	990.834	994.254	997.334	999.833	1001.410	1001.575
10	1019.667	1022.864	1026.008	1029.076	1032.037	1034.852	1037.470	1039.823	1041.815	1043.313
					$p = 0.2$					
1	457.084	458.012	458.833	459.486	459.888	459.923	459.431	458.177	455.808	451.762
2	533.903	534.060	534.169	534.193	534.084	533.775	533.174	532.147	530.493	527.888
3	554.874	554.459	554.049	553.628	553.175	552.661	552.043	551.259	550.171	548.713
4	557.727	556.770	555.861	554.985	554.125	553.256	552.344	551.341	550.211	548.741
5	664.000	665.065	665.992	666.673	666.964	666.666	665.494	663.033	658.646	651.300
6	687.459	687.784	688.029	688.067	687.733	686.803	684.963	681.756	676.499	668.098
7	697.183	696.153	695.116	694.000	692.710	691.121	689.055	686.255	682.321	676.589
8	851.711	851.337	850.890	850.288	849.416	848.115	846.158	843.208	838.743	831.897
9	964.858	966.726	968.406	969.724	970.450	970.271	968.754	965.278	958.924	948.271
10	1012.344	1014.846	1017.186	1019.315	1021.165	1022.646	1023.630	1023.925	1017.065	1004.492

(*Continued*)

Table 5.7. *(Continued)*

f(Hz)	Revolution hyperbolic hyperboloid (FCFC)									
	Asymmetric cosine II porosity									
β_0	0	0.1	0.2	0.3	0.4	0.5	0.6	0.7	0.8	0.9
	$p = 0.5$									
1	452.673	452.824	452.670	452.082	450.878	448.796	445.440	440.167	431.724	416.797
2	529.209	528.815	528.251	527.442	526.282	524.620	522.228	518.733	513.413	501.507
3	549.980	549.206	548.372	547.443	546.369	545.080	543.464	541.342	538.324	532.884
4	552.791	551.494	550.188	548.841	547.414	545.848	544.053	541.882	539.050	534.436
5	657.932	657.711	657.006	655.588	653.133	649.167	642.959	633.294	617.707	586.008
6	680.201	679.125	677.596	675.358	672.055	667.171	659.927	649.046	632.019	597.968
7	690.359	688.367	686.142	683.533	680.332	676.232	670.757	663.087	651.456	626.352
8	843.725	842.225	840.370	837.977	834.781	830.393	824.202	815.148	800.960	768.684
9	954.740	954.662	953.886	952.064	948.715	943.155	934.375	920.797	899.577	858.489
10	1003.820	1005.453	1006.760	1007.640	1007.944	1002.491	991.960	975.534	948.848	891.892
	$p = 1$									
1	449.997	449.322	448.064	445.976	442.690	437.629	429.784	416.461	374.784	327.082
2	524.657	523.612	522.222	520.342	517.762	514.147	508.903	500.494	425.878	381.020
3	544.414	543.172	541.773	540.143	538.180	535.722	532.492	527.664	468.466	477.609
4	547.109	545.359	543.513	541.511	539.268	536.650	533.422	529.026	504.102	509.757
5	655.514	653.970	651.458	647.534	641.525	632.343	617.995	592.765	520.480	518.216
6	677.023	674.482	670.966	665.989	658.827	648.327	632.419	605.457	526.827	531.293
7	685.750	682.697	679.090	674.636	668.897	661.152	650.016	631.212	572.310	562.415
8	837.600	834.891	831.428	826.851	820.606	811.780	798.601	775.566	585.045	687.999
9	949.398	947.230	943.641	937.977	929.277	916.061	895.817	862.401	616.043	742.142
10	993.910	994.501	994.544	993.845	990.075	974.458	949.880	905.713	630.841	782.737

Geometric inputs: $\alpha_1^0 = -1$, $\alpha_1^1 = 1$, $\alpha_2^0 = 0$, $\alpha_2^1 = \pi/2$, $h_1 = h_3 = 0.005m$, $h_2 = 0.05$m

Lamination scheme: $\left(70/\mathrm{FGM}_{M(a=1/b=0/c=1/d=1/p)}^{C(\mathrm{A1-UC-II}/\beta_0)}/30\right)$, first layer: triclinic material, second layer: FGM material, third layer: graphite epoxy

Computational grid: CGL distribution with $I_N \times I_M = 31\times31$, EDZ4 displacement field assumption

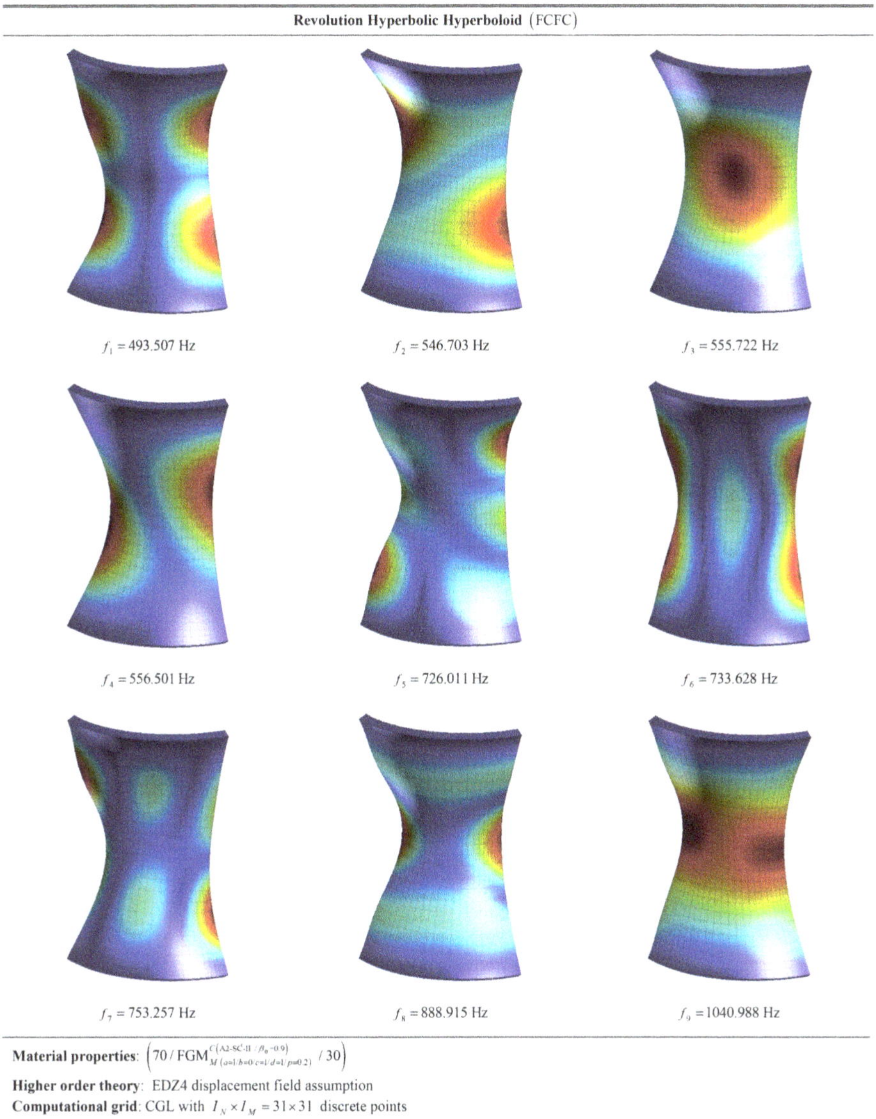

Figure 5.8. The first nine mode shapes of a revolution hyperbolic hyperboloid with a porous FGM layer within the lamination scheme, calculated using the EDZ4 higher-order theory. The through-the-thickness distribution of the constituent material is modeled by means of a symmetric trigonometric analytical expression.

Moreover, the introduction of orthotropic and anisotropic materials with general orientations influences the principal directions of vibration with respect to a corresponding isotropic structure.

Based on the analysis shown in table 5.6, four different configurations of central FGM core have been considered. It can be seen that a trigonometric distribution of porosity with a symmetric profile leads to an increase in the first natural frequencies as the scaling parameter β_0 increases. Furthermore, similar behavior is seen for different values of FGM power exponent $p^{(k)}$, since only a shift of the frequency values is seen. On the other hand, when an unsymmetric profile is selected for the porosity distribution (table 5.7), a reduction or increase in the natural frequencies can be observed with respect to the dense configuration, depending on the values assumed for the parameter $p^{(k)}$. The decrease of the frequency values becomes significant for the case $p^{(k)} = 1$, whereas in the isotropic configuration ($p^{(k)} = 0$), a slight increase in the natural frequencies can be noticed. In figure 5.9, the parametric

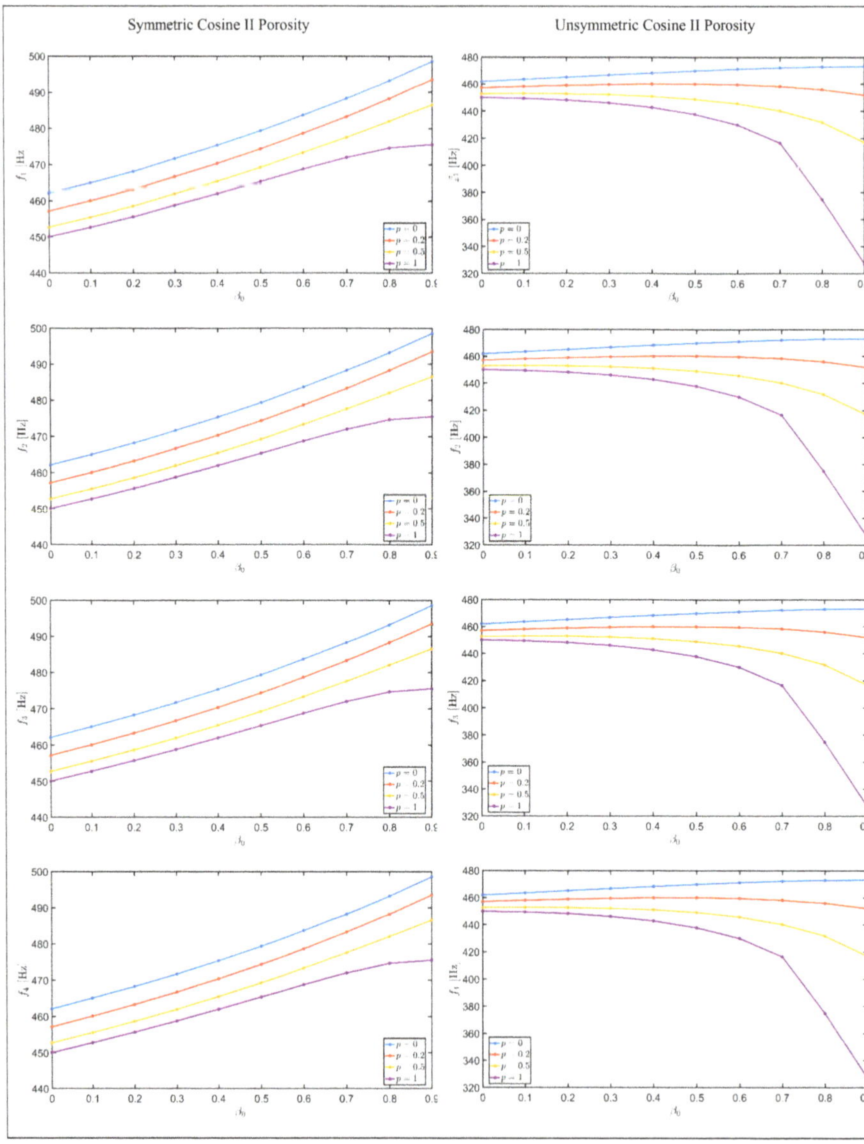

Figure 5.9. A parametric investigation of the natural frequencies of a revolution hyperbolic hyperboloid characterized by a central porous FGM core. The analysis focuses on two different through-the-thickness porosity distributions; it takes account of the A1 homogenization approach.

investigation is represented by useful graphs for the first four modes, highlighting the influence of the porosity parameter β_0 in a laminated doubly curved shell with an FGM central porous core.

5.8 Conclusions

In the present chapter, an innovative method has been presented for the free vibration analysis of doubly curved shells laminated with anisotropic materials and porous FGM layers. An efficient homogenization procedure based on the rule of mixture has been presented, accounting for a generalized five-parameters power law to assess the volume fraction of the constituent materials along the shell thickness. Furthermore, several profiles have been provided for the dispersion of porosity along the shell thickness of the FGM layers. Following the ESL approach, a generalized higher-order two-dimensional model has been provided for the assessment of the displacement field variable, accounting also for the interlaminar zigzag effects. The fundamental governing equations have been derived from the Hamiltonian principle directly in the strong form, and they have been numerically solved by means of the GDQ method. The accuracy of the formulation has been checked for different validating examples, involving structures with different curvatures and lamination schemes. The mode frequencies and shapes have been derived with the present two-dimensional model, compared to predictions from refined three-dimensional finite element simulations. The presented continuum-based approach allows one to model easily curved and layered structures made of FGM materials, accounting for a smooth variation of the material properties along the thickness. Unlike other previous works on the topic, this ESL model also provides an efficient methodology to model very complicated porosity distributions with polynomial and trigonometric profiles. Furthermore, the present two-dimensional model seems to predict accurately and easily a series of effects that are usually carried out only by refined three-dimensional simulations.

References

[1] Reddy J N 1997 *Mechanics of Laminated Composite Plates.* (Boca Raton, FL: CRC Press)

[2] Pilato L A and Michno M J 1994 *Advanced Composite Materials* (Berlin: Springer Science & Business Media)

[3] Cheng L, Thomas A, Glancey J L and Karlsson A M 2011 Mechanical behavior of bio-inspired laminated composites *Composites* A **42** 211–20

[4] Catapano A, Desmorat B and Vannucci P 2015 Stiffness and strength optimization of the anisotropy distribution for laminated structures *J. Optim. Theory Appl.* **167** 118–46

[5] Vannucci P and Verchery G 2001 Stiffness design of laminates using the polar method *Int. J. Solids Struct.* **38** 9281–94

[6] Pedersen P 1989 On optimal orientation of orthotropic materials *Struct. Optim.* **1** 101–6

[7] Setoodeh S, Abdalla M M and Gürdal Z 2006 Design of variable-stiffness laminates using lamination parameters *Composites* B **37** 301–9

[8] Tornabene F, Viscoti M, Dimitri R and Aiello M A 2021 Higher-order modeling of anisogrid composite lattice structures with complex geometries *Eng. Struct.* **244** 112686

[9] Kapania R K, Soliman H E, Vasudeva S, Hughes O and Makhecha D P 2008 Static analysis of sandwich panels with square honeycomb core *AIAA J.* **46** 627–34

[10] Bitzer T N 1997 *Honeycomb Technology: Materials, Design, Manufacturing, Applications and Testing* (Berlin: Springer Science & Business Media)

[11] Garg A, Belarbi M O, Chalak H D and Chakrabarti A 2021 A review of the analysis of sandwich FGM structures *Compos. Struct.* **258** 113427

[12] Murin J, Aminbaghai M and Kutiš V 2010 Exact solution of the bending vibration problem of FGM beams with variation of material properties *Eng. Struct.* **32** 1631–40

[13] Natarajan S and Manickam G 2012 Bending and vibration of functionally graded material sandwich plates using an accurate theory *Finite Elem. Anal. Des.* **57** 32–42

[14] Tornabene F 2009 Free vibration analysis of functionally graded conical, cylindrical shell and annular plate structures with a four-parameter power-law distribution *Comput. Methods Appl. Mech. Eng.* **198** 2911–35

[15] Tornabene F, Viola E and Inman D J 2009 2-D differential quadrature solution for vibration analysis of functionally graded conical, cylindrical shell and annular plate structures *J. Sound Vib.* **328** 259–90

[16] Fazzolari F A 2018 Generalized exponential, polynomial and trigonometric theories for vibration and stability analysis of porous FG sandwich beams resting on elastic foundations *Composites* B **136** 254–71

[17] Tornabene F and Reddy J N 2013 FGM and laminated doubly-curved and degenerate shells resting on nonlinear elastic foundations: a GDQ solution for static analysis with *a posteriori* stress and strain recovery *J. Indian Inst. Sci.* **93** 635–88

[18] Liu D Y, Wang C Y and Chen W Q 2010 Free vibration of FGM plates with in-plane material inhomogeneity *Compos. Struct.* **92** 1047–51

[19] Amirpour M, Das R and Flores E I S 2017 Bending analysis of thin functionally graded plate under in-plane stiffness variations *Appl. Math. Model.* **44** 481–96

[20] Tornabene F and Bacciocchi M 2018 *Anisotropic Doubly-Curved Shells: Higher-Order Strong and Weak Formulations for Arbitrarily Shaped Shell Structures* (Bologna: Esculapio)

[21] Hughes T J and Liu W K 1981 Nonlinear finite element analysis of shells: part I. Three-dimensional shells *Comput. Methods Appl. Mech. Eng.* **26** 331–62

[22] Reddy J N 1993 An evaluation of equivalent-single-layer and layerwise theories of composite laminates *Compos. Struct.* **25** 21–35

[23] Reddy J N and Robbins D H Jr. 1994 Theories and computational models for composite laminates *Appl. Mech. Rev.* **47** 147–69

[24] Liu D and Li X 1996 An overall view of laminate theories based on displacement hypothesis *J. Compos. Mater.* **30** 1539–61

[25] Soldatos K P and Hadjigeorgiou V P 1990 Three-dimensional solution of the free vibration problem of homogeneous isotropic cylindrical shells and panels *J. Sound Vib.* **137** 369–84

[26] Love A E H 2013 *A Treatise on the Mathematical Theory of Elasticity* 4th edn (Cambridge: Cambridge University Press) https://www.cambridge.org/gb/academic/subjects/mathematics/historical-mathematical-texts/treatise-mathematical-theory-elasticity-4th-edition

[27] Wang C M, Lim G T, Reddy J N and Lee K H 2001 Relationships between bending solutions of Reissner and Mindlin plate theories *Eng. Struct.* **23** 838–49

[28] Phan N D and Reddy J N 1985 Analysis of laminated composite plates using a higher-order shear deformation theory *Int. J. Numer. Methods Eng.* **21** 2201–19

[29] Reddy J N Analysis of functionally graded plates *Int. J. Numer. Methods Eng.* **47** 663–84

[30] Reddy J N 1987 A generalization of two-dimensional theories of laminated composite plates *Commun. Appl. Numer. Methods.* **3** 173–80
[31] Washizu K 1975 *Variational Methods in Elasticity and Plasticity* 2nd edn (Oxford: Pergamon)
[32] Tornabene F, Viola E and Fantuzzi N 2013 General higher-order equivalent single layer theory for free vibrations of doubly-curved laminated composite shells and panels *Compos. Struct.* **104** 94–117
[33] Tornabene F, Viscoti M, Dimitri R and Reddy J N 2021 Higher order theories for the vibration study of doubly-curved anisotropic shells with a variable thickness and isogeometric mapped geometry *Compos. Struct.* **267** 113829
[34] Tornabene F and Brischetto S 2018 3D capability of refined GDQ models for the bending analysis of composite and sandwich plates, spherical and doubly-curved shells *Thin-Walled Struct.* **129** 94–124
[35] Li X and Liu D 1995 Zigzag theory for composite laminates *AIAA J.* **33** 1163–5
[36] Murakami H 1986 Laminated composite plate theory with improved in-plane responses *ASME J. Appl. Mech.* **53** 661–6
[37] Groh R M J and Weaver P M 2015 On displacement-based and mixed-variational equivalent single layer theories for modelling highly heterogeneous laminated beams *Int. J. Solids Struct.* **59** 147–70
[38] Tornabene F, Viscoti M and Dimitri R 2022 Equivalent single layer higher order theory based on a weak formulation for the dynamic analysis of anisotropic doubly-curved shells with arbitrary geometry and variable thickness *Thin-Walled Struct.* **174** 109119
[39] Tessler A, Di Sciuva M and Gherlone M 2009 A refined zigzag beam theory for composite and sandwich beams *J. Compos. Mater.* **43** 1051–81
[40] Tessler A, Di Sciuva M and Gherlone M 2010 A consistent refinement of first-order shear deformation theory for laminated composite and sandwich plates using improved zigzag kinematics *J. Mech. Mater. Struct.* **5** 341–67
[41] Tornabene F, Viscoti M and Dimitri R 2021 Higher order formulations for doubly-curved shell structures with a honeycomb core *Thin-Walled Struct.* **164** 107789
[42] Liew K M, Pan Z Z and Zhang L W 2019 An overview of layerwise theories for composite laminates and structures: development, numerical implementation and application *Compos. Struct.* **216** 240–59
[43] Tornabene F, Viscoti M and Dimitri R 2022 Generalized higher order layerwise theory for the dynamic study of anisotropic doubly-curved shells with a mapped geometry *Eng. Anal. Bound. Elem.* **134** 147–83
[44] Tornabene F, Fantuzzi N, Bacciocchi M and Reddy J N 2016 An equivalent layer-wise approach for the free vibration analysis of thick and thin laminated and sandwich shells *Appl. Sci.* **7** 17
[45] Quarteroni A and Valli A 1999 *Domain Decomposition Methods for Partial Differential Equations* (Oxford: Oxford University Press) https://global.oup.com/academic/product/domain-decomposition-methods-for-partial-differential-equations-9780198501787
[46] Chan T F and Mathew T P 1994 Domain decomposition algorithms *Acta Numer.* **3** 61–143
[47] Hussaini M Y, Kopriva D A and Patera A T 1989 Spectral collocation methods *Appl. Numer. Math.* **5** 177–208
[48] Subich C J, Lamb K G and Stastna M 2013 Simulation of the Navier–Stokes equations in three dimensions with a spectral collocation method *Int. J. Numer. Methods Fluids* **73** 103–29

[49] Tornabene F, Fantuzzi N, Ubertini F and Viola E 2015 Strong formulation finite element method based on differential quadrature: a survey *ASME Appl. Mech. Rev.* **67** 020801
[50] Shu C and Richards B E 1992 Application of generalized differential quadrature to solve two-dimensional incompressible Navier–Stokes equations *Int. J. Numer. Methods Fluids* **15** 791–8
[51] Shu C and Du H 1997 A generalized approach for implementing general boundary conditions in the GDQ free vibration analysis of plates *Int. J. Solids Struct.* **34** 837–46
[52] Shu C 2012 *Differential Quadrature and Its Application in Engineering* (Berlin: Springer Science & Business Media)
[53] Shu C, Chen W, Xue H and Du H 2001 Numerical study of grid distribution effect on accuracy of DQ analysis of beams and plates by error estimation of derivative approximation *Int. J. Numer. Methods Eng.* **51** 159–79
[54] Shu C and Chen W 1999 On optimal selection of interior points for applying discretized boundary conditions in DQ vibration analysis of beams and plates *J. Sound Vib.* **222** 239–57
[55] Tornabene F, Fantuzzi N and Bacciocchi M 2018 *DiQuMASPAB: Differential Quadrature for Mechanics of Anisotropic Shells, Plates, Arches and Beams* (Bologna: Esculapio)

IOP Publishing

Functionally Graded Structures
Modelling and computation of static and dynamical problems
Snehashish Chakraverty, Subrat Kumar Jena and Ömer Civalek

Chapter 6

Assessing the effects of porosity on the buckling of functionally graded beams

Youcef Tlidji, Rabia Benferhat, Kada Draiche and Hassaine Daouadji Tahar

List of symbols

E	Young's modulus
E_c	Young's modulus at the top
E_m	Young's modulus at the bottom
ν	Poisson's ratio
δ	Variational
L	Beam length
h	Beam thickness
Π	Strain energy
V	External work
k	The material parameters
$V(z)$	The volume fraction
α	The porosity coefficient
N_0	The buckling load
$\bar{N}$	The dimensionless critical buckling load
U	Axial displacement components
W	Transverse displacement components
u_0	Axial displacement components of a point on the beam's neutral axis
w	Transverse displacement components of a point on the beam's neutral axis
φ	The rotational angle of the cross section about the y-axis
$'$	Partial differentiation of the quantities to x
λ_i	Eigenvalues associated with the matrix T
$[E]$	Eigenvectors associated with the matrix T
T	Matrix
K	Constant column vector

doi:10.1088/978-0-7503-5301-4ch6

6.1 Introduction

Functionally graded (FG) beams form a group of composite materials whose properties, such as the modulus of elasticity, density, Poisson's ratio, etc. vary in the thickness direction, the longitudinal direction, or in both directions. These types of material have been used in many fields such as the aerospace, mechanical, and medical sectors in order to reduce stress concentration and residual stress and to improve connection strengths. In the literature, various numerical and analytical methods have been proposed to analyze the buckling behavior of FG beams. Farhatnia *et al* [1] used two numerical methods to analyze the buckling of thick FG beams. Vo *et al* [2], developed a finite element method solution to obtain the critical buckling load of sandwich FG beams. In 2016, Trinh *et al* [3] used a state-space approach to determine the critical buckling loads of FG beams with various boundary conditions under mechanical and thermal loads. In addition, Şimşek [4] investigated the buckling of two-dimensional functionally graded beams under different boundary conditions using the Ritz method within the framework of Timoshenko beam theory. Shvartsman and Majak [5] analyzed the buckling of FG beams resting on an elastic foundation using Euler–Bernoulli beam theory and the initial parameter method. Ait Atmane *et al* [6] used a Navier-type solution method to obtain the critical buckling load of simply supported porous FG beams resting on elastic foundations using an efficient hyperbolic beam theory. Using a modified first-order shear deformation theory and the finite element method, Nam *et al* [7] studied the buckling behavior of variable-thickness FG Beams. Utilizing the Timoshenko beam and the generalized differential quadrature method (GDQM), Yas and Rahimi [8] calculated the critical thermal gradient of FG porous nanocomposite beams. Bellifa *et al* [9] studied the effect of porosity on the nonlinear thermal stability response of power-law functionally graded Euler–Bernoulli beams under general boundary conditions. Harsha *et al* [10] investigated the effects of porosity and axial profile loading on the elastic buckling and free vibration of a FG porous beam using the Ritz method in conjunction with the shear and normal deformable beam theory (SNDBT). Harsha *et al* [11] used the finite element method to study the buckling of FG Timoshenko porous beams resting on elastic foundations. They assumed that the material properties varied in the thickness direction according to the power law. Two different types of porosity pattern were considered.

6.2 Theoretical formulation

We consider a straight FG beam made from a mixture of aluminum (Al) and alumina (Al_2O_3). The beam's length is L and its thickness is h, as shown in figure 6.1. The material properties of the FG beam are expressed in the following general form:

$$E(z) = (E_c - E_m)V(z) + E_m, \tag{6.1}$$

where the subscripts c and m represent the ceramic and the metal, respectively. It should be noted that Poisson's ratio μ is assumed to be constant. For a power-law FG beam, the volume fraction $V(z)$ can be expressed as

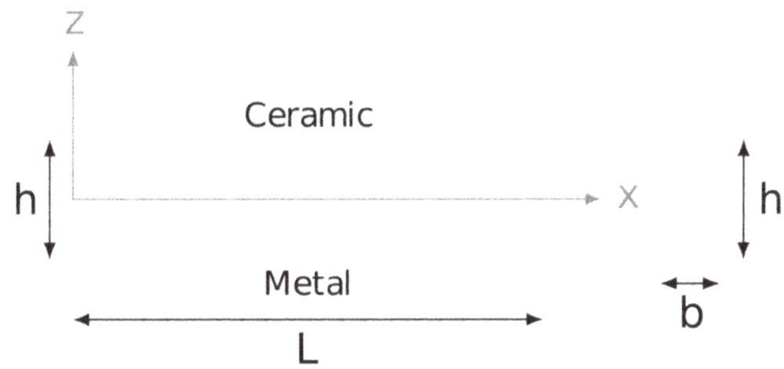

Figure 6.1. The geometry and coordinates of the FG beam.

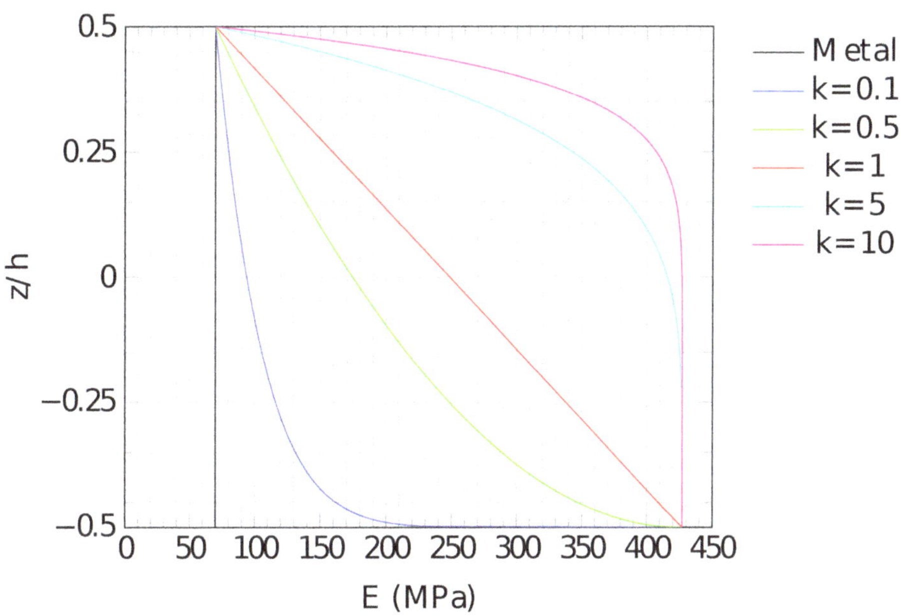

Figure 6.2. The variations of Young's modulus through the thickness of the FG beam.

$$V(z) = \left(\frac{1}{2} + \frac{z}{h}\right)^k. \tag{6.2}$$

Figure 6.2 present sample results for the variations of Young's modulus through the thickness of the FG beam. In the case of a porous FG beam, the rule of mixtures must be modified. For even and uneven distributions of porosity, the modified rule of mixtures takes the following forms (see Tlidji *et al* [12]):

Even distribution:

$$P(z) = (P_c - P_m)V(z) + P_m - \frac{\alpha}{2}(P_c + P_m) \tag{6.3}$$

Uneven distributions:

- X-shape

$$P(z) = (P_c - P_m)V(z) + P_m - \frac{\alpha}{2}(P_c + P_m)\left(\frac{2z}{h}\right) \tag{6.4}$$

- O-shape

$$P(z) = (P_c - P_m)V(z) + P_m - \frac{\alpha}{2}(P_c + P_m)\left(1 - \frac{2|z|}{h}\right) \tag{6.5}$$

- V-shape

$$P(z) = (P_c - P_m)V(z) + P_m - \frac{\alpha}{2}(P_c + P_m)\left(\frac{1}{2} - \frac{z}{h}\right) \tag{6.6}$$

where α is the porosity coefficient.

6.3 Kinematics

We assume that the deformation of FG beam only takes place in the (x, z) plane and let $U(x, z)$ and $W(x, z)$ be the axial and transverse displacement components of an arbitrary point. These components can be expressed as shown in equation (6.7):

$$U(x, z) = u_0(x) - zw'(x) + f(z)\varphi(x) \tag{6.7}$$

$$W(x, z) = w(x),$$

where u_0 and w denote the displacement components of a point on the beam's neutral axis in the x and z directions, while φ is the rotational angle of the cross section about the y-axis compared to the undeformed position. The shape function considered in this paper is that proposed by Daikh *et al* [13]. For brevity, prime (′) represents the partial differentiation of the quantities to x.

The strains related to the displacement field in equation (6.7) are:

$$\epsilon_x = \frac{\partial U(x, z)}{\partial x} = u_0'(x) - zw''(x) + f(z)\varphi'(x) \tag{6.8}$$

$$\gamma_{xz} = \frac{\partial U(x, z)}{\partial z} + \frac{\partial W(x, z)}{\partial x} = \frac{df(z)}{dz}\varphi(x) = g(z)\varphi(x).$$

6.4 Equilibrium equations

The principle of virtual work is used in order to derive the equilibrium equations and boundary conditions:

$$\int \delta(\Pi - V)dv = 0, \tag{6.9}$$

where $\delta\Pi$ and δV denote the virtual variation of the strain energy and external work. The virtual variation of the strain energy is given by:

$$\delta\Pi = \int_{-\frac{L}{2}}^{\frac{L}{2}} \int_{0}^{b} \int_{-\frac{h}{2}}^{\frac{h}{2}} [\sigma_x \delta x + \sigma_{xz} \delta z] \mathrm{d}z \mathrm{d}y \mathrm{d}x \tag{6.10}$$

$$= \int_{-\frac{L}{2}}^{\frac{L}{2}} \left[N_x \delta u_0 - M_x \delta w'' + P_x \delta\varphi' + Q_{xz} g(z) \delta\varphi \right] \mathrm{d}x,$$

in which:

$$N_x = \int_{-\frac{h}{2}}^{\frac{h}{2}} \sigma_x b \mathrm{d}z \tag{6.11}$$

$$M_x = \int_{-\frac{h}{2}}^{\frac{h}{2}} \sigma_x z b \mathrm{d}z$$

$$P_x = \int_{-\frac{h}{2}}^{\frac{h}{2}} \sigma_x f(z) b \mathrm{d}z$$

$$Q_{xz} = \int_{-\frac{h}{2}}^{\frac{h}{2}} \sigma_{xz} b \mathrm{d}z.$$

The virtual variation of the external work in linear buckling analysis can be determined as follows:

$$\delta V = - \int_{-\frac{L}{2}}^{\frac{L}{2}} [N_0 w' \delta w'] \mathrm{d}x. \tag{6.12}$$

If we assume that the FG beam obeys Hooke's law, the stress–strain relation can be written as

$$\sigma_x = E(z) \epsilon_x \tag{6.13}$$

$$\sigma_{xz} = \frac{E(z)}{2[1 + \nu]} \gamma_{xz} = Q_{55} \gamma_{xz}.$$

If we substitute equation (6.13) into equation (6.11), the stress resultants can be extended as shown below

$$\begin{Bmatrix} N_x \\ M_x \\ P_x \\ Q_{xz} \end{Bmatrix} = \begin{bmatrix} A & B & C & 0 \\ B & D & F & 0 \\ C & F & H & 0 \\ 0 & 0 & 0 & A_s \end{bmatrix} \begin{Bmatrix} u_0' \\ w'' \\ \varphi' \\ g(z)\varphi \end{Bmatrix}, \tag{6.14}$$

where

$$(A, B, D, C, F, H) = \int_{-\frac{h}{2}}^{\frac{h}{2}} E(z)(1, z, z^2, f(z), zf(z), f(z)^2)\mathrm{d}z \tag{6.15}$$

$$A_s = \int_{-\frac{h}{2}}^{\frac{h}{2}} Q_{55}\mathrm{d}z.$$

We now substitute equations (6.4) and (6.12) into equation (6.9), integrate the equation by parts, and collect the coefficients of δu_0, δw, and $\delta\varphi$. The equilibrium equations of the FG beam can be written as follows:

$$\delta u_0\colon N_x{}' = 0 \tag{6.16}$$

$$\delta w\colon M_x^{''} - N_0 w^{''} = 0$$

$$\delta\varphi\colon P_x{}' - Q_{xz}g(z) = 0.$$

Using the state-space approach [3, 14], the displacement components can be expressed as:

$$\begin{Bmatrix} u_0(x) \\ w(x) \\ \varphi(x) \end{Bmatrix} = \begin{bmatrix} U(x) \\ W(x) \\ \Phi(x) \end{bmatrix}. \tag{6.17}$$

We substitute equation (6.17) into equation (6.16) to obtain the following system of ordinary differential equations:

$$U^{''} = b_1\Phi + b_2 W'' \tag{6.18}$$

$$\Phi^{''} = b_3\Phi + b_4 W''$$

$$W^{iv} = b_5\Phi' + b_6 W'',$$

where b_n are the constant coefficients: $b_1 = \frac{-CA_s}{AH - C^2}$, $b_2 = \frac{BH - CF}{AH - C^2}$, $b_3 = \frac{AA_s}{AH - C^2}$, $b_4 = \frac{AF - BC}{AH - C^2}$, $b_5 = \frac{Bb_1 - Fb_3}{Bb_2 + Fb_4 - D}$, and $b_6 = \frac{N_0}{Bb_2 + Fb_4 - D}$. The system of equations given in (6.18) can be converted into matrix form as follows:

$$Z'(x) = TZ(x), \tag{6.19}$$

where

$$Z(x) = \{\Phi, \Phi', W, W', W', W'''\}T \text{ a matrix.}$$

$$\{T\} = \begin{bmatrix} 0 & 1 & 0 & 0 & 0 & 0 \\ b_3 & 1 & 0 & 0 & 0 & b_4 \\ 0 & 0 & 0 & 1 & 0 & 0 \\ 0 & 0 & 0 & 0 & 1 & 0 \\ 0 & 0 & 0 & 0 & 0 & 1 \\ 0 & b_5 & 0 & 0 & b_6 & 0 \end{bmatrix}$$

A formal solution of equation (6.19) is given by:

$$Z(x) = e^{Tx}\ K. \tag{6.20}$$

Here, K is a constant column vector determined from the boundary conditions at $x = \pm\frac{L}{2}$, and e^{Tx} is the general matrix solution:

$$e^{Tx} = [E]\begin{bmatrix} 1 & x & 0 & 0 & 0 & 0 \\ x & 1 & 0 & 0 & 0 & 0 \\ 0 & 0 & e^{\lambda_3 x} & 0 & 0 & 0 \\ 0 & 0 & 0 & e^{\lambda_4 x} & 0 & 0 \\ 0 & 0 & 0 & 0 & e^{\lambda_5 x} & 0 \\ 0 & 0 & 0 & 0 & 0 & e^{\lambda_6 x} \end{bmatrix}[E]^{-1}; \tag{6.21}$$

λ_i, $i = 3, \dots, 6$ are the eigenvalues and $[E]$ the eigenvectors associated with the matrix T.

6.4.1 Boundary conditions

In this chapter, classical boundary conditions are investigated. Boundary conditions can be expressed in terms of unknown functions as follows:

$$\text{Clamped}: U = W = W' = 0 \tag{6.22}$$

$$\text{Hinged}: \Phi = \Phi' = W'' = 0$$

$$\text{Free}: \Phi' = W'' = (Bb_1 + Fb_3)\Phi - N_0 W' + (Bb_2 + Fb_6 - D)W''' = 0$$

If we substitute equation (6.19) into equations (6.22), the following homogeneous system of equations is obtained:

$$e^{Tx}K = 0. \tag{6.23}$$

The nontrivial solutions for equation (6.23) require the determinant of $|e^{Tx}|$ to be zero; they allow the buckling load to be determined. It should be noted that a trial-and-error procedure must be used to obtain the buckling load values due to the presence of the unknown N_0 in matrix T.

6.5 Numerical results and discussion

This section presents tables and figures showing the dimensionless critical buckling load, $\bar{N} = N_0\frac{12L^2}{E_m h^3}$, of an FG beam made from aluminum (Al) and alumina (Al_2O_3) under different boundary conditions: hinged–hinged (HH), clamped–clamped (CC), and clamped–free (CF). The following material properties are employed here: $E_m = 70\ GPa$, $E_c = 380\ GPa$, and $\nu = 0.23$. Tables 6.1 and 6.2 compare the dimensionless critical buckling loads of perfect functionally graded material (PFGM) beams. The material parameters are $k = 0$, 1, 5, and 10; these are used under CC, CF, and HH boundary conditions for two values of the length-to-

Table 6.1. The dimensionless critical buckling loads of perfect FG beams for $L/h = 5$.

BC	Theory	k				
		0	0.5	1	5	10
SS	Present	48.8427	32.0100	24.6904	15.7303	14.1425
	Trinh (2016)	48.648	31.987	24.648	15.664	14.098
	Vo (2014)	48.8401	32.0094	24.6911	15.7400	14.1468
CC	This chapter	154.5938	103.7657	80.6090	47.6763	41.7754
	Trinh (2016)	154.534	103.732	80.566	47.468	41.759
	Vo (2014)	154.5500	103.7490	80.608	47.7562	41.8042
CF	This chapter	13.0772	8.5020	6.5427	4.2769	3.8817
	Trinh (2016)	13.043	8.293	6.500	4.270	3.848
	Vo (2014)	13.0771	8.5020	6.5428	4.2776	3.8821

Table 6.2. The dimensionless critical buckling loads of perfect FG beams for $L/h = 10$.

BC	Theory	k				
		0	0.5	1	5	10
SS	This chapter	52.3089	34.0080	26.1710	17.1075	15.5270
	Trinh (2016)	52.204	33.478	26.102	17.023	15.321
	Vo (2014)	52.3082	34.0087	26.1727	17.1118	15.5291
CC	This chapter	195.3706	128.0400	98.7615	62.9211	56.5701
	Trinh (2016)	195.179	127.876	98.625	62.902	56.431
	Vo (2014)	195.3610	128.0500	98.7868	62.9786	56.5971
CF	This chapter	13.3138	8.6369	6.6425	4.3728	3.9794
	Trinh (2016)	13.173	8.523	6.586	4.262	3.874
	Vo (2014)	13.3742	8.6714	6.6680	4.3976	4.0046

thickness ratio, namely $\frac{L}{h} = 5, \quad 10$. All the results obtained agree well with those of Vo *et al* [2] and Trinh *et al* [3]. It can be seen that the dimensionless critical buckling load reduces significantly as the gradient index k increases.

Tables 6.3, 6.4, 6.5, 6.6, 6.7, and 6.8 depict the effect of the volume fraction of porosity on the nondimensional critical buckling load of imperfect FG beams under various boundary conditions for two slenderness ratios. The values of the volume fraction of porosity are set to 5%, 10%, 15%, and 20%. For all boundary conditions and slenderness ratios, the dimensionless buckling load decreases as the porosity volume fraction increases for the H, O, and V shapes of porosity distribution, however the nondimensional critical buckling load increases as the porosity volume fraction increases for the X-shaped porosity distribution. These tables also reveal

Table 6.3. The dimensionless critical buckling loads of imperfect HH FG beams for $L/h = 5$.

α	Form	k				
		0	0.5	1	5	10
5%	U	47.3967	30.4675	23.0149	13.8910	12.4325
	O	48.4293	31.5623	24.1812	15.0983	13.5524
	V	48.1164	31.4410	24.1694	15.2091	13.5933
	X	48.8297	32.3947	25.2893	16.4516	14.6824
10%	U	45.9507	28.9162	21.3127	11.9643	10.6485
	O	48.0145	31.1117	23.6655	14.4404	12.9303
	V	47.3833	30.8690	23.6471	14.6849	13.0368
	X	48.7906	32.7422	25.8410	17.0890	15.1255
15%	U	44.5047	27.3548	19.5787	9.9158	8.7538
	O	47.5984	30.6581	23.1428	13.7523	12.2675
	V	46.6430	30.2939	23.1236	14.1570	12.4710
	X	48.7256	33.0527	26.3459	17.6452	15.4743
20%	U	43.0587	25.7819	17.8064	7.6879	6.6824
	O	47.1808	30.2012	22.6126	13.0282	11.5510
	V	45.8954	29.7154	22.5985	13.6248	11.8938
	X	48.6344	33.3264	26.8047	18.1221	15.7306

Table 6.4. The dimensionless critical buckling loads of imperfect CC FG beams for $L/h = 5$.

α	Form	k				
		0	0.5	1	5	10
5%	U	150.0170	102.2400	75.3995	42.1483	36.4921
	O	152.9346	102.0834	78.7752	45.3248	39.4030
	V	152.2972	101.8781	78.8501	45.8125	39.7956
	X	154.5612	104.7799	82.1978	49.2981	42.9454
10%	U	145.4402	94.0493	70.1086	36.4191	31.0316
	O	151.2647	100.3880	76.9195	42.8802	36.8900
	V	149.9835	99.9824	77.0876	43.9285	37.7779
	X	154.4634	105.6924	83.6490	50.7033	43.8950
15%	U	140.8634	89.1480	64.7186	30.4056	25.3162
	O	149.5837	98.6785	75.0405	40.3279	34.2020
	V	147.6519	98.0784	75.3210	42.0209	35.7147
	X	154.3004	106.5050	84.9669	51.9091	44.6398
20%	U	136.2867	84.2117	59.2065	23.9562	19.2109
	O	147.8907	96.954 00	73.1365	37.6491	31.2904
	V	145.3017	96.1653	73.5501	40.0853	33.5959
	X	154.0719	107.2191	86.1558	52.9296	45.1915

Table 6.5. The dimensionless critical buckling loads of imperfect CF FG beams for $L/h = 5$.

α	Form	k				
		0	0.5	1	5	10
5%	U	12.6901	8.0881	6.0920	3.7754	3.4207
	O	12.9767	8.3894	6.4124	4.1195	3.7423
	V	12.8827	8.3520	6.4062	4.1445	3.7437
	X	13.0735	8.6105	6.7112	4.4916	4.0449
10%	U	12.3029	7.6719	5.6340	3.2481	2.9379
	O	12.8761	8.2761	6.2804	3.9560	3.5970
	V	12.6862	8.2012	6.2694	4.0119	3.6045
	X	13.0623	8.7086	6.8668	4.6827	4.1795
15%	U	11.9157	7.2529	5.1675	2.6850	2.4224
	O	12.7754	8.1622	6.1466	3.7854	3.4445
	V	12.4877	8.0495	6.1321	3.8789	3.4639
	X	13.0436	8.7963	7.0096	4.8506	4.2857
20%	U	11.5286	6.8308	4.6909	2.0700	1.8537
	O	12.6745	8.0476	6.0108	3.6065	3.2826
	V	12.2870	7.8969	5.9945	3.7455	3.3217
	X	13.0175	8.8738	7.1397	4.9952	4.3637

Table 6.6. The dimensionless critical buckling loads of imperfect HH FG beams for $L/h = 10$.

α	Form	k				
		0	0.5	1	5	10
5%	U	50.7603	32.3525	24.3675	15.1015	13.6827
	O	51.9069	33.5575	25.6497	16.4781	14.9693
	V	51.5307	33.4082	25.6250	16.5780	14.9746
	X	52.2939	34.4419	26.8449	17.9662	16.1796
10%	U	49.2117	30.6874	22.5355	12.9922	11.7518
	O	51.5043	33.1046	25.1215	15.8241	14.3882
	V	50.7449	32.8049	25.0774	16.0474	14.4178
	X	52.2492	34.834 39	27.4673	18.7310	16.7179
15%	U	47.6630	29.0115	20.6697	10.7398	9.6898
	O	51.1015	32.6489	24.5863	15.1418	13.7781
	V	49.9509	32.1982	24.5285	15.5155	13.8557
	X	52.1746	35.1854	28.0385	19.4023	17.1427
20%	U	46.1145	27.3231	18.7634	8.2802	7.4146
	O	50.6981	32.1905	24.0430	14.4261	13.1306
	V	49.1482	31.5876	23.9781	14.9818	13.2867
	X	52.0701	35.4953	28.5586	19.980 96	17.4547

Table 6.7. The dimensionless critical buckling loads of imperfect CC FG beams for $L/h = 10$.

α	Form	k				
		0	0.5	1	5	10
5%	U	189.5866	121.8699	92.0597	55.5641	49.7302
	O	193.7170	126.2493	96.7248	60.3933	54.2095
	V	192.4654	125.7641	96.6775	60.8364	54.3734
	X	195.3187	129.5787	101.1571	65.8063	58.7295
10%	U	183.8026	115.6649	85.2509	47.8574	42.5940
	O	192.0581	124.4470	94.6620	57.7618	51.7214
	V	189.5331	123.4762	94.5886	58.7397	52.1470
	X	195.1625	130.9687	103.3639	68.3561	60.5021
15%	U	178.0186	109.4194	78.3149	39.6633	35.0153
	O	190.3937	122.6325	92.5713	55.0092	49.0702
	V	186.572 23	121.175 66	92.494 29	56.628 17	49.8842
	X	194.9023	132.2107	105.3838	70.5807	61.8972
20%	U	172.2346	103.1277	71.2258	30.7515	26.7297
	O	188.723 14	120.804 69	90.450 34	52.112 74	46.2039
	V	183.5815	118.8615	90.3941	54.4990	47.5754
	X	194.5378	133.3054	107.2186	72.4884	62.9225

Table 6.8. The dimensionless critical buckling loads of imperfect CF FG beams for $L/h = 10$.

α	Form	k				
		0	0.5	1	5	10
5%	U	12.9195	8.2154	6.1829	3.8597	3.5091
	O	13.2142	8.5243	6.5114	3.8431	3.8430
	V	13.1157	8.4850	6.5043	4.2401	3.8415
	X	13.3099	8.7489	6.8162	4.5977	4.1510
10%	U	12.5254	7.7914	5.7161	3.3195	3.0163
	O	13.1148	8.4110	6.3787	4.0534	3.7017
	V	12.9156	8.3321	6.3658	4.1074	3.7029
	X	13.2984	8.8502	6.9768	4.7983	4.2927
15%	U	12.1313	7.3646	5.2406	2.7422	2.4891
	O	13.0152	8.2971	6.2441	3.8839	3.5542
	V	12.7135	8.1783	6.2270	3.9744	3.5631
	X	13.2790	8.9407	7.1242	4.9747	4.4047
20%	U	11.7372	6.9346	4.7549	2.1108	1.9061
	O	12.9156	8.1826	6.1075	3.7061	3.3991
	V	12.5090	8.0235	6.0876	3.8414	3.4221
	X	13.2519	9.0207	7.2586	5.1269	4.4868

that the X-shaped porosity distribution yields significantly higher values of $\bar{N}$ than the other porosity distribution shapes.

The effects of the porosity distribution shape on the dimensionless buckling loads of imperfect FGM beams under various boundary conditions are shown in figures 6.3, 6.4, and 6.5.

From these figures, it can be seen that the effect of the porosity distribution shape on the dimensionless buckling load becomes more important as the volume fraction of porosity increases under different boundary conditions.

Figures 6.6 and 6.7 illustrate the dimensionless buckling loads of imperfect FG beams with several types of porosity distribution pattern versus varying span-to-depth ratios (L/h) for two boundary conditions (CC, CF).

The material parameter k is set to two and the porosity volume fraction is set to 0.1 and 0.20. Increasing the thickness ratio leads to an increase in the dimensionless buckling load.

6.6 Conclusions

The buckling behaviors of FG porous beams were analyzed in this chapter using a state-space approach. The number of primary variables used in this theory is even less than the number of variables used in higher-order shear deformation beam

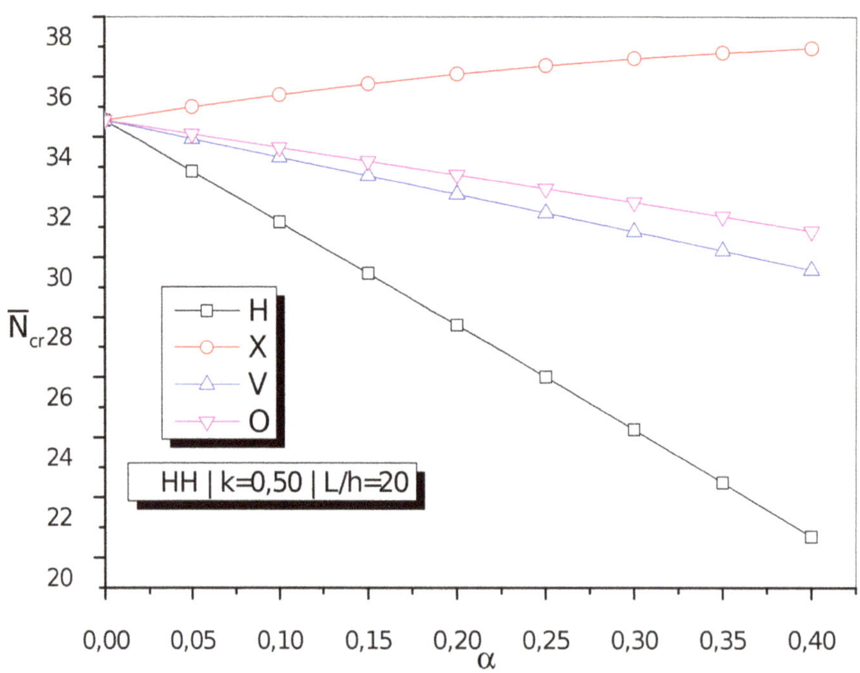

Figure 6.3. The variation of the dimensionless buckling load versus the porosity volume fraction of an HH porous FG beam.

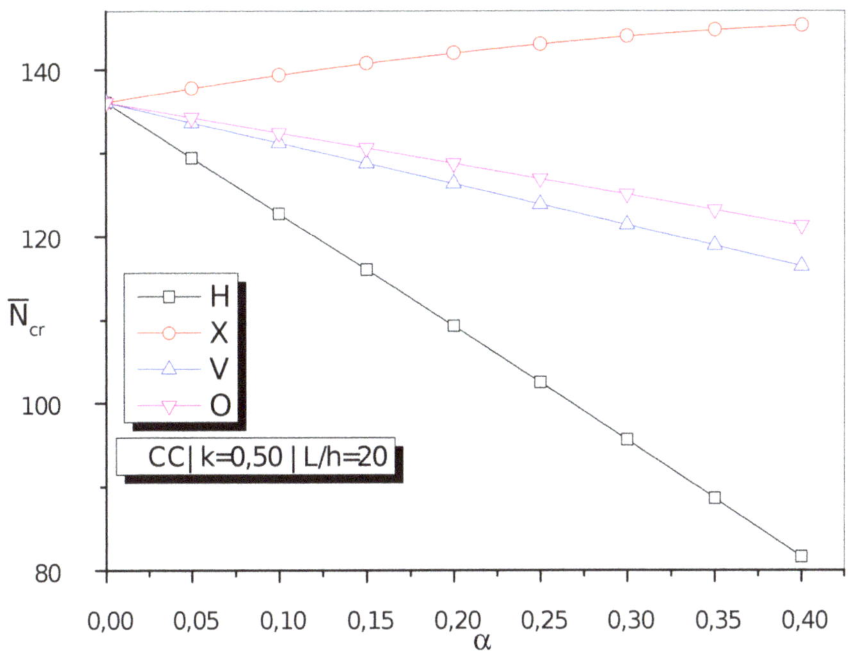

Figure 6.4. The variation of the dimensionless buckling load versus the porosity volume fraction of a CC porous FG beam.

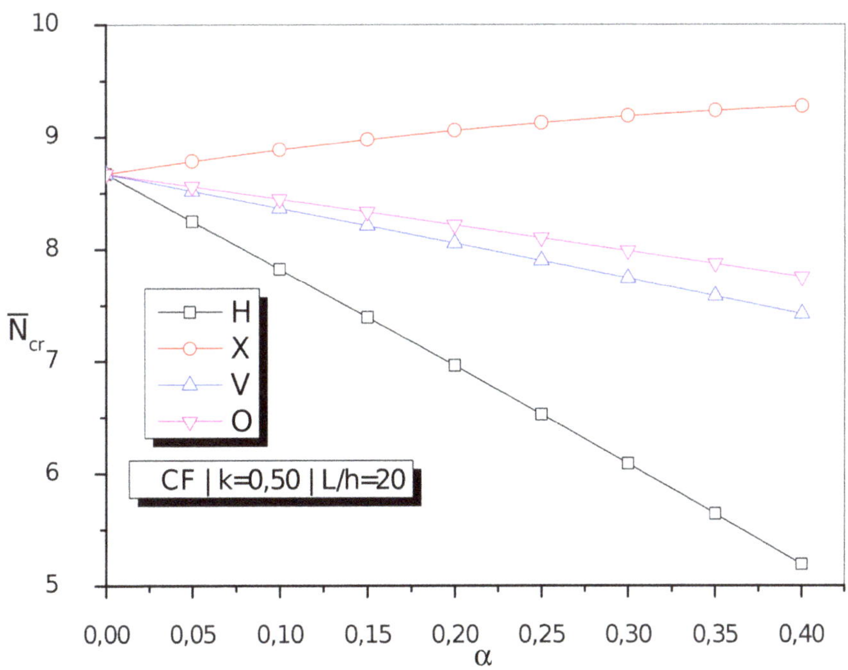

Figure 6.5. The variation of the dimensionless buckling load versus the porosity volume fraction of a CF porous FG beam.

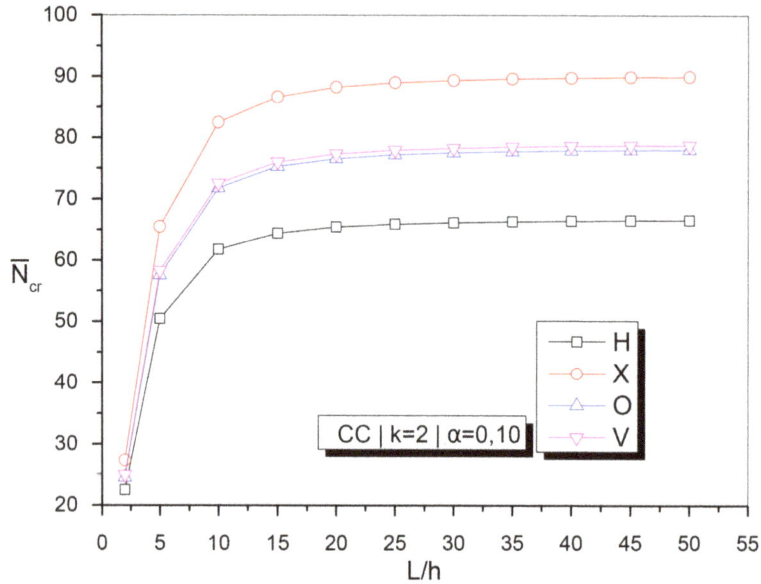

(a)

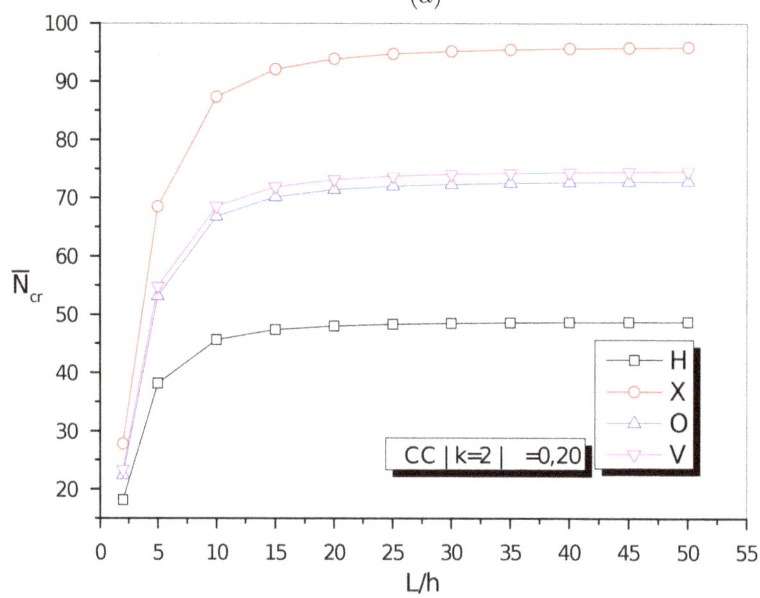

(b)

Figure 6.6. The variation of the dimensionless buckling load of a CC porous FG beam versus the span-to-depth ratio.

theories. This theory gives the parabolic distribution of the transverse shear strains and satisfies the zero traction boundary conditions on the surfaces of the beam without using shear correction factors.

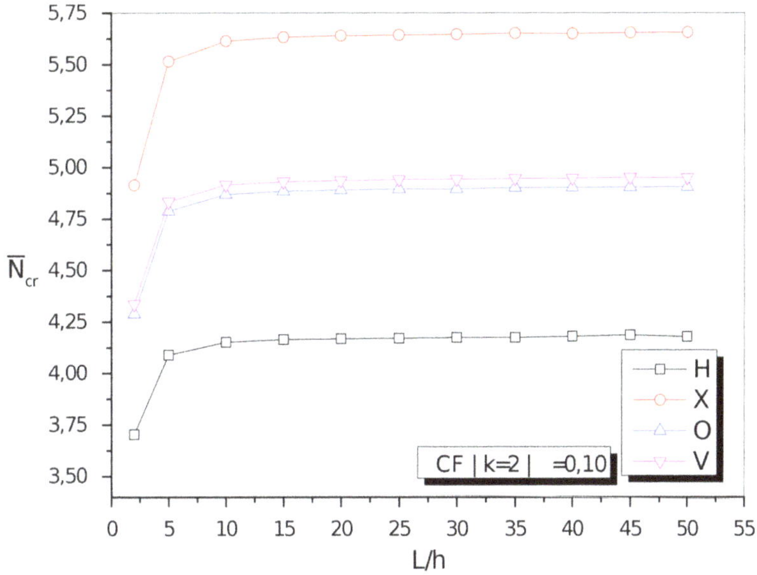

(a)

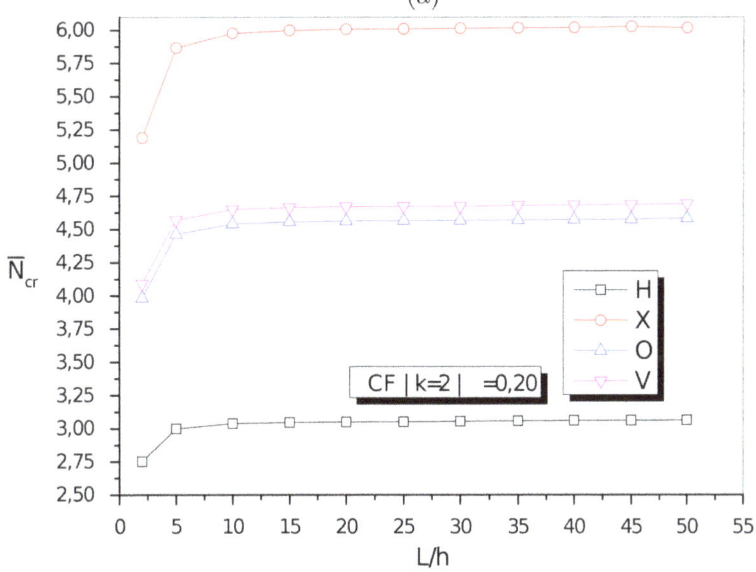

(b)

Figure 6.7. The variation of the dimensionless buckling load of a CF porous FG beam versus the span-to-depth ratio.

The material properties of FGM layers, taking porosity into account, are assumed to vary continuously through the thickness according to a simple power-law distribution in terms of the volume fractions of the constituents. From the numerical results, it can be concluded that the shape of the porosity distribution has a significant effect on the buckling of FGM beams.

References

[1] Farhatnia F, Bagheri M A and Ghobadi A 2012 Buckling analysis of FGM thick beam under different boundary conditions using GDQM *Adv. Mater. Res.* **433–40** 4920–4

[2] Vo T P, Thai H-T, Nguyen T-K, Maheri A and Lee J 2014 Finite element model for vibration and buckling of functionally graded sandwich beams based on a refined shear deformation theory *Eng. Struct.* **64** 12–22

[3] Trinh L C, Vo T P, Thai H-T and Nguyen T-K 2016 An analytical method for the vibration and buckling of functionally graded beams under mechanical and thermal loads *Composites* B **100** 152–63

[4] Şimşek M 2016 Buckling of Timoshenko beams composed of two-dimensional functionally graded material (2D-FGM) having different boundary conditions *Compos. Struct.* **149** 304–14

[5] Shvartsman B and Majak J 2016 Numerical method for stability analysis of functionally graded beams on elastic foundation *Appl. Math. Model.* **40** 3713–19

[6] Ait Atmane H, Tounsi A and Bernard F 2017 Effect of thickness stretching and porosity on mechanical response of a functionally graded beams resting on elastic foundations *Int. J. Mech. Mater. Des.* **13** 71–84

[7] Nam V H, Vinh P V, Chinh N V, Thom D V and Hong T T 2019 A new beam model for simulation of the mechanical behaviour of variable thickness functionally graded material beams based on modified first order shear deformation theory *Materials* **12** 404

[8] Yas M-H and Rahimi S 2020 Thermal buckling analysis of porous functionally graded nanocomposite beams reinforced by graphene platelets using generalized differential quadrature method *Aerosp. Sci. Technol.* **107** 106261

[9] Bellifa H, Selim M M, Chikh A, Bousahla A A, Bourada F, Tounsi A, Benrahou K H, Al-Zahrani M M and Tounsi A 2021 Influence of porosity on thermal buckling behavior of functionally graded beams *Smart Struct. Syst.* **27** 719–28

[10] Harsha B P, Jeyaraj P and Lenin Babou M C 2021 Effect of porosity and profile axial loading on elastic buckling and free vibration of functionally graded porous beam *IOP Conf. Ser.: Mater. Sci. Eng.* **1128** 012025 International Conference on Applications in Computational Engineering and Sciences

[11] Harsha B P, Jeyaraj P and Lenin B M C 2021 Buckling and free vibration of porous functionally graded metal ceramic beams under thermal and mechanical loading: a comparative study *J. Inst. Eng. (India):* C **102** 1107–17

[12] Tlidji Y, Benferhat R and Tahar H D 2021 Study and analysis of the free vibration for FGM microbeam containing various distribution shape of porosity *Struct. Eng. Mech.* **77** 217–29

[13] Daikh A A and Zenkour A M 2019 Free vibration and buckling of porous power-law and sigmoid functionally graded sandwich plates using a simple higher-order shear deformation theory *Mater. Res. Express.* **6** 115707

[14] Tlidji Y, Benferhat R, Trinh L C, Tahar H D and Tounsi A 2021 New state state-space approach to dynamic analysis of porous FG beam under different boundary conditions *Adv. Nano Res.* **11** 347–59

Chapter 7

Vibration and buckling problems of functionally graded structures

Chenxi Wei and Yin Zhang

Functionally graded materials (FGMs) are an advanced class of materials whose properties vary continuously in a certain direction. The application of FGM structures can greatly improve the performance of structural components in engineering. Functionally graded structures are more often used in moving or rotating devices; therefore their dynamic and stability problems have been widely studied. The central topics of this chapter are the vibration and buckling problems of FGM beams and FGM plates. Different mathematic models and solutions (both analytical and numerical) are discussed in the chapter. The effects of material property variations, the structure's size, the boundary conditions, and different loads are comprehensively discussed. FGMs show great prospects for use in engineering applications and are attracting worldwide attention.

List of abbreviations

ASDBT	Aydogdu shear deformation theory
CBT	Classical beam theory
CPT	Classical plate theory
DQM	Differential quadrature method
ESDBT	Exponential shear deformation beam theory
FEM	Finite element method
FGM	Functionally graded material
FSDBT	First-order shear deformation beam theory
FSDPT	First-order shear deformation plate theory
HSDBT	Hyperbolic shear deformation beam theory
HSDPT	Higher-order shear deformation plate theory
LRVE	Local representative volume elements
PSDBT	Parabolic shear deformation beam theory
PSDPT	Parabolic shear deformation plate theory

doi:10.1088/978-0-7503-5301-4ch7

RVE	Representative volume element
SIF	Stress intensity factor
TBT	Timoshenko beam theory
TSDPT	Trigonometric shear deformation plate theory

7.1 Functionally graded materials and homogenization

FGMs are revolutionary materials. Compared with pure or composite materials, FGMs have a continuously varying mechanical property in a certain direction. FGMs have been discovered in many components of species in nature, such as many plant roots, animal bones, and cells [1, 2]; in such cases, the gradient meets the mechanical demand. The concept of FGMs was proposed by Japanese scientists in 1984 and used to develop a compound of a ceramic and a metal for thermal barriers [3]. Designing an FGM provides greater freedom for engineers to improve the performance of FGM structures, for example, by achieving higher strength and reduced weight [4]. FGM structures are widely used in aerospace, bioengineering, nuclear energy, coatings, electronics, and smart structures [5].

Based on the form of the gradient used, FGMs are categorized into four types: the fraction gradient type, the shape gradient type, the orientation gradient type and the size gradient type, as shown in figure 7.1 [6]. In general, it is assumed that a typically sized representative volume element (RVE) can be used to simulate the bulk properties of FGMs [7]. Under this assumption, the FGM properties vary gradually in the spatial dimension, and the stress and strain of the structure are continuously distributed under load. Under this local homogenization assumption, the micro-mechanical modeling methods used for FGMs are as the Voigt model, the Mori–Tanaka method, self-consistent models, etc [8, 9].

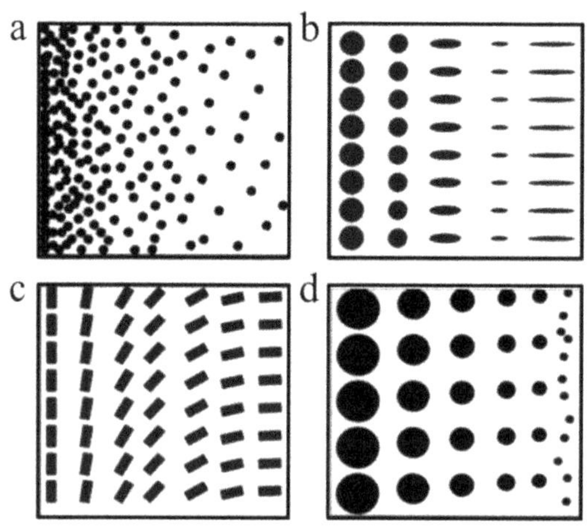

Figure 7.1. Different types of FGM: (a) the fraction gradient type; (b) the shape gradient type; (c) the orientation gradient type; and (d) the size gradient type [6] (2015), reprinted by permission of the publisher (Taylor & Francis Ltd, http://www.tandfonline.com).

The Voigt model is a simple mixture model used to characterize the properties of FGM structures. In the Voigt model, the effective Young's modulus, mass density, and Poisson's ratio are computed as a summation of the different material properties weighted by the volume fraction, which are expressed as:

$$P_{\mathrm{eff}} = \sum_{j=1}^{N} P_j V_{\mathrm{f}j}, \tag{7.1}$$

where P denotes the material properties of different material components and the mixture; $V_{\mathrm{f}j}$ is the volume fraction of each material, with $\sum_{j=1}^{k} V_{\mathrm{f}j} = 1$; the subscript j is the component number and N is the number of materials in the mixture. For a metal–ceramic material, the effective property can be calculated using the following equation [10]:

$$P_{\mathrm{eff}} = P_{\mathrm{m}} V_{\mathrm{m}} + P_{\mathrm{c}}(1 - V_{\mathrm{m}}), \tag{7.2}$$

where the subscripts m and c denote the metal and the ceramic, respectively; V_{m} is the volume fraction of the metal.

The Mori–Tanaka scheme is based on calculating the average internal stress in the matrix of a material with inclusions [11] and has inspired many studies in the field of composites. The uniform average stress within the matrix follows from a consideration of the interaction between the inclusions and the free boundary. Benveniste reformulated the Mori–Tanaka theory and applied it to compute the effective properties of composites, expressed as follows [12]:

$$\begin{cases} \dfrac{K - K_1}{K_2 - K_1} = \dfrac{V_{\mathrm{f}}}{1 + \dfrac{3V_{\mathrm{f}}(K_2 - K_1)}{3K_1 + 4G_1}}, \\ \dfrac{G - G_1}{G_2 - G_1} = \dfrac{V_{\mathrm{f}}}{1 + \dfrac{V_{\mathrm{f}}(G_2 - G_1)}{G_1 + f_1}}, \\ f_1 = \dfrac{G_1(9K_1 - 8G_1)}{6(K_1 + 2G_1)}, \end{cases} \tag{7.3}$$

where K and G are the local bulk and shear moduli, respectively. The subscripts 1 and 2 represent the base and reinforcement phases.

The self-consistent model is based on the solution to an auxiliary elastic problem, which is a uniform loading problem with an ellipsoidal inhomogeneity embedded in an infinite space [13]. It is assumed that the average stress and strain of the inclusion are equal to the overall stress and strain. The continuum of the self-consistent model was proposed by Hill. The scalar formula for the effective shear modulus of a spherical inclusion is expressed as [13]:

$$\frac{G - G_2}{G_1 - G_2} = \left[1 + \beta\left(\frac{G_1}{G} - 1\right)\right]^{-1} V_{\mathrm{f}1}, \tag{7.4}$$

where the phase value β is obtained using the following equation:

$$(1 - \beta)G^2 + [\beta(G_1 + G_2) - (V_{f1}G_1 + V_{f2}G_2)]G - \beta G_1 G_2 = 0. \tag{7.5}$$

The distributions of the phase fractions in a given direction are most frequently described by the power law model, exponential law model, the sigmoid law model, etc [8]. The power law model is often used in research into the dynamics of FGM structures. The volume fraction V_f of the phase follows a power law function in a given direction, which is given as follows [14]:

$$V_f(z) = \left(\frac{1}{2} + \frac{z}{h}\right)^n, \tag{7.6}$$

where n is the index of this power law which indicates property variation in the given direction. The fraction V_f is monotonic in the z-direction. The reader may note that $V_f = 0$ with $z/h = -1/2$ and $V_f = 1$ with $z/h = 1/2$.

The exponential law model is commonly used in fracture studies of functionally graded structures and coatings [15]. The FGM property is expressed as follows [16]:

$$P(z) = P_t \exp\left[-\delta\left(1 - \frac{2z}{h}\right)\right], \quad \delta = \frac{1}{2}\log\left(\frac{P_t}{P_b}\right), \tag{7.7}$$

where the subscripts t and b denote the top and the bottom of the FGM layers, respectively. By applying the exponential law in the axial direction of an FGM beam, the analytical solution can be derived. The stability of vibration is discussed in reference [17].

In addition, dozens of the FGM homogenization schemes have been studied [8, 18]. In a theoretical analysis of a functionally graded composite material with widely varying reinforcement volume fractions, the Mori–Tanaka, self-consistent, and Tamura models were used separately and their predictions for the effective properties were compared [19]. The study found that because the self-consistent model uses the fuzzy logic technique to produce a smooth transition with different phase fractions, it produces a better first-order estimation of the effective properties over the entire fraction range. Comparative studies of the Mori–Tanaka, self-consistent, and finite element models were performed for the static load and thermoelastic responses of a functionally graded structure [20, 21]. The elastic constants predicted by the Mori–Tanaka model are more accurate for the 'well-defined' continuous matrix and discontinuous inclusion phase, while the self-consistent model is more suitable for predicting the properties of skeletal microstructures in a wide transition zone between the matrix phases. The effective elastic properties were derived from a micromechanics-based elastic model for comparison with the two models above [22]. By simulating the RVE and capturing the gradient effect of the phase volume fraction distribution, the model can be reduced to the Mori–Tanaka model if the particle interaction terms are neglected. Although the interactions between the inclusions are included in the self-consistent model, it may fail to accurately predict the overall elastic properties of the composite materials by neglecting the phase to be treated as the matrix [22]. Therefore, the self-consistent model overestimated the shear modulus in a glass–epoxy composite with a high inclusion phase fraction and further violated the upper bounds at a low inclusion fraction [23, 24]. As a result, a combined self-consistent and Mori–Tanaka method has been proposed for the prediction of the effective elastic properties of FGM composites [25]. In this model, the parameter n is emphasized to separate the inclusions into two groups, whose effective properties are evaluated by the self-consistent model and the Mori–Tanaka model, respectively [25]. The size effect of the particle inclusions is considered in this model through the indirect relationship with the introduced parameter n [25].

The Voigt model is often adopted for the vibration analysis of FGM structures due to its convenient formula for computing the effective properties [26]. The influences of the micromechanical model of FGMs on the vibration performance need to be investigated. The vibrations of an FGM plate with horizontal side cracks were analyzed using the Mori–Tanaka and Voigt models [27]. The largest difference between the natural frequencies generated by the two models was 6%, which was for the case of the very thick plate ($h/b = 0.2$) with $n = 1$ in the power law distribution in equation (7.6) [27], where h is the thickness and b is the width. The small- and large-amplitude vibrations of thinner plates ($h/b = 0.125$ and 0.0125) resting on elastic foundations in thermal environments have been analyzed [28]. The frequency results obtained using the Mori–Tanaka and Voigt models were compared in the analysis. The results showed that the differences between the nonlinear and linear vibration frequencies predicted by the Mori–Tanaka and Voigt models were all very small [28]. This implies that the micromechanical model adopted may have a relatively small influence on the vibration of a thin FGM structure when the gradation is in the thickness direction. However, a comprehensive analysis of FGM beam vibration behavior was carried out using the different homogenization models of the Voigt, Reuss, Mori–Tanaka, Hashin–Shtrikman bounds, Tamura, and cubic local representative volume elements (LRVE) models [29]. This study showed that the Voigt and Mori–Tanaka schemes may provide inappropriate results for the effective elastic modulus estimation [29], which was analyzed in the previous analysis. The predicted effective properties were compared with different schemes as shown in figure 7.2. According to this analysis, the Voigt and Mori–Tanaka models both overestimated the frequencies, but the error of Mori–Tanaka model was smaller [29]. The Reuss model underestimated the frequencies of the FGM structure. This study recommended the LRVE and Tamura models because their results were very close and within the lower and upper frequencies of the Hashin–Shtrikman model, especially for relatively high stiffness gradients [29]. The thermal postbuckling of FGM circular plates was analyzed using the Voigt, Mori–Tanaka, and self-consistent models [30]; the self-consistent model underestimated the critical buckling temperature, and the maximum difference between these estimations produced by different models was about 3% at a high temperature level [30]. The mechanical buckling of thin conical panels on an elastic foundation was comprehensively investigated separately by eight different micromechanical models: the Voigt, Reuss, Mori–Tanaka, Hashin–Shtrikman, modified Wakashima–Tsukamoto, Halpin–Tsai, Tamura, and LRVE methods [31]. It was found that the FGM distribution law in the material gradient direction and the non-homogeneity index

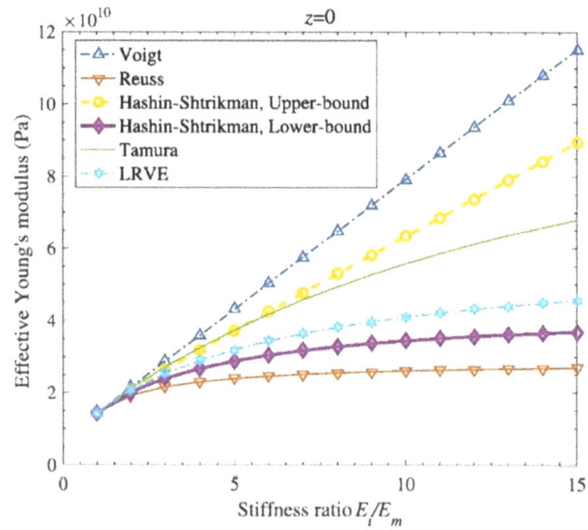

Figure 7.2. A comparison of the effective Young's moduli predicted by different homogenization schemes for an FGM beam. E_i and E_m are the Young's moduli of the two phases. Reprinted from [29], Copyright (2019), with permission from Elsevier.

n play important roles in the estimation accuracy of the critical loads predicted by different models [31].

In general, the homogenization models of FGM can be used to simulate functionally graded structures. However, the models need to be developed by the addition of the parameters related to the micromechanical interaction in order to fit more complex situations. In addition, experimental verification is important to improve the accuracy of FGM models based on the homogenization mechanism.

7.2 The vibration and buckling problems of functionally graded beams

Functionally graded material technology offers great potential to improve the performance of engineering structures [32]. The reason for this is that their mechanical properties, such as their modulus, Poisson's ratio, and density, can be modified by varying the volume fraction of inclusions in a certain direction [32]. FGM beams have many advantages in reducing the stress in the thickness direction, mitigating the residual stress, improving the thermal properties, increasing the fracture resistance and improving the dynamic characteristics [7, 33–35]. In this section, we review beam models and solutions, together with the FGM parameters and different types of load and boundary.

7.2.1 The development of a model for functionally graded beams

Starting from the Euler–Bernoulli beam assumptions, classical beam theory (CBT) has been employed to solve the FGM beam problem, in which the beam properties vary in the thickness direction [36]. In this model, the shear strain is neglected, causing the cross section to be kept in a plane before and after bending. This model results in high errors when used to predict the natural frequencies of FGM beams [37]. To improve on the Euler–Bernoulli beam theory, the Timoshenko beam theory was proposed, which considers the effects of shear deformation and inertia [38]; it is thus called the first-order shear deformation beam theory (FSDBT) or Timoshenko beam theory (TBT) [39]. However, in TBT the slope of the shear deformation is constant in the direction of the beam thickness, which requires the shear factor to modify the theory [40]. The reason for this is the improper shear strain distribution assumption in TBT. To solve this, the parabolic shear deformation beam theory (PSDBT) was developed, in which the shear deformation is modified by a parabolic term based on TBT [41, 42]. Furthermore, the higher-order shear deformation theories known as the exponential shear deformation beam theory (ESDBT), the hyperbolic shear deformation beam theory (HSDBT), the trigonometric shear deformation beam theory, and the Aydogdu shear deformation theory (ASDBT) have been proposed [43–46].

Figure 7.3 shows a rectangular beam in which the graded property varying in the direction of the thickness. To illustrate the deformation, the axial displacement u_x^b and the transverse displacement of the beam u_z^b are expressed as follows [37, 47]:

$$\begin{cases} u_x^b(x,\ z,\ t) = u_0^b(x,\ t) - zw_{0,\,x}^b(x,\ t) + \Phi^b(z)s^b(x,\ t), \\ u_z^b(x,\ z,\ t) = w_0^b(x,\ t), \end{cases} \tag{7.8}$$

where u_0^b and w_0^b are the axial displacement and the transverse displacement of the middle axis of the beam, respectively; s^b is the shear strain on the middle axis, which is given by

$$s^b(x,\ t) = w_{0,\,x}^b(x, t) - \phi_0^b(x,\ t). \tag{7.9}$$

Here, ϕ_0^b denotes the rotation angle of a cross section of the middle axis; t is time; $()_{,x}$ is the derivative with respect to x; and $\Phi^b(z)$ is the shape function characterizing the shear distribution in the direction of the thickness of the beam, given below [37, 47]:

$$\text{CBT:} \quad \Phi^b(z) = 0, \tag{7.10}$$

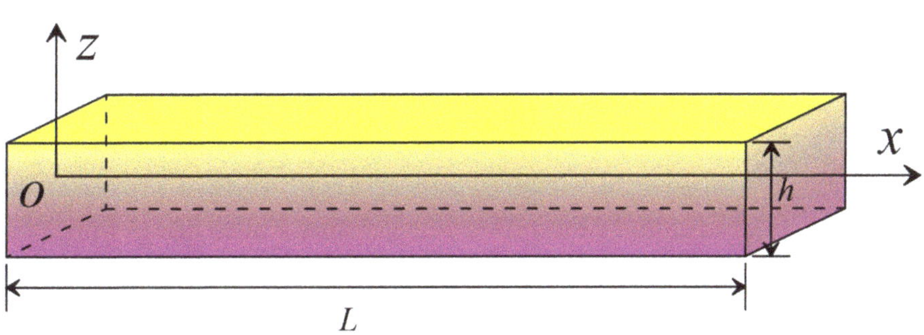

Figure 7.3. A diagram of a rectangular FGM beam.

$$\text{TBT:} \quad \Phi^b(z) = z, \tag{7.11}$$

$$\text{PSDBT:} \quad \Phi^b(z) = z\left(1 - \frac{4z^2}{3h^2}\right), \tag{7.12}$$

$$\text{TSDBT:} \quad \Phi^b(z) = \frac{h}{\pi}\sin\left(\frac{\pi z}{h}\right), \tag{7.13}$$

$$\text{HSDBT:} \quad \Phi^b(z) = h\sinh\left(\frac{z}{h}\right) - z\cosh\left(\frac{1}{2}\right), \tag{7.14}$$

$$\text{ESDBT:} \quad \Phi^b(z) = z\, \mathrm{e}^{-2(z/h)^2}, \tag{7.15}$$

$$\text{ASDBT:} \quad \Phi^b(z) = z\, \alpha^{-2(z/h)^2/\ln\alpha} \text{ with } \alpha = 3. \tag{7.16}$$

Comparisons between the natural frequency results computed by the different shear deformation theories are presented in references [37, 47]. The frequency difference between the CBT and higher-order shear deformation theories is significant when the slenderness ratio $L/h < 20$ [47], as shown in figure 7.4. For slender beams ($L/h > 20$), the effect of the slenderness ratio on the frequencies is negligible. Further models based on the higher-order shear deformation theory have been proposed to characterize FGM beam vibration [48–52]. Those various shear deformation models are reviewed in reference [53].

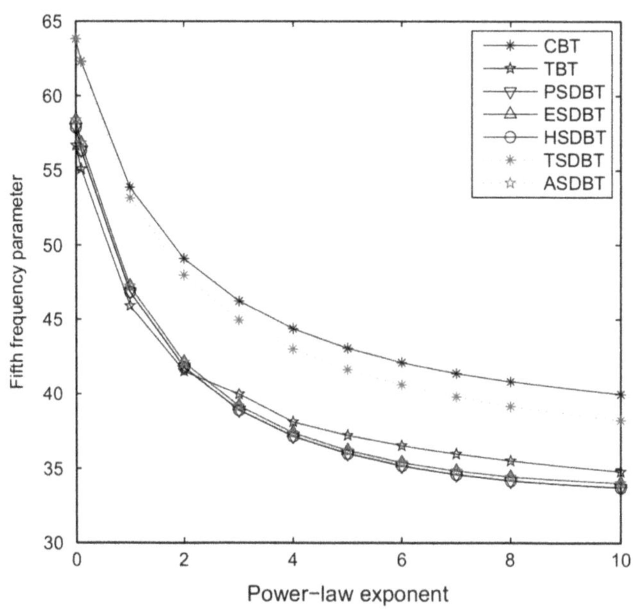

Figure 7.4. A comparison of the dimensionless frequencies for short beams (L/h=5) calculated using different shear deformation theories. Reprinted from [37], Copyright (2014), with permission from Elsevier.

7.2.2 Solutions of FGM beams

Based on the displacement kinematic assumption of the beam theory of equation (7.8), the governing equations of the FGM beam free vibration and buckling can be obtained by the variational method [54, 55]. The vibration and buckling problems can be solved by analytical and numerical methods, which are reviewed in this section.

7.2.2.1 Analytical solutions

The exact or closed-form solutions are important for the vibration and buckling analysis of FGM beams. The exact solutions can serve as the basic solutions for the numerical method for some complicated problems. Lu and Chen [56] obtained the exact solutions of the free vibration response of orthotropic functionally graded beams, based on two-dimensional theory. The vibration problem of FGM beams with properties that vary exponentially in the direction of thickness was solved using a hybrid state space and differential quadrature method [56]. Based on this method, the exact bending and vibration solutions of FGM beams resting on Winkler–Pasternak foundations can be given [57]. The parametric influences of the gradient index, aspect ratios, and foundation moduli on the mechanical performance have also been discussed [57]. Navier-type solutions are simple solution forms for the structure problem and also effective for the vibration problems of FGM beams [58, 59]. Aydogdu and Taskin [60] employed Navier-type solutions to obtain the natural frequencies of a simply supported FGM beam with different shear deformation models, namely FSDBT, PSDBT, and ESDBT. Sina *et al* [55] proposed a new shear deformation model for FGM beams and obtained an analytical solution to the free vibration problem. The results show that the new model is a little different from the traditional one in predicting natural frequencies. Koochaki [61] used the PSDBT to analyze the free vibration of a simply supported functionally graded beam and obtained analytical results using the Navier solution. The results showed that this shear deformation theory is better for thick beams. Thai and Vo [62] analyzed the bending and free vibration of FGM beams with various higher-order shear deformation beam theories. Using the Navier solution, analytical solutions were obtained for a simply supported beam [62]. It was found that the incorporation of the shear deformation effects leads to an increase in deflections and a decrease in the natural frequencies [62], because shear deformation reduces the effective stiffness of a beam. Beg and Yasin [63] solved the static, free, and forced vibration responses of circular curved FGM beams using the Fourier series method. Hadji [64, 65] proposed a new higher-order shear deformation model for FGM beams and obtained the Navier solutions for flexure and free vibration analysis. Wang [17] studied the free vibration of FGM beams with variable material properties along the beam length and thickness. By assuming that the properties follow the exponential law distribution throughout the beam length, the characteristic equations were derived in a closed form, and analytical solutions for the clamped–free and hinged–hinged beams were also obtained [17]. This study found a critical frequency that depends on the gradient parameter, which induces an abrupt jump when the natural frequency crosses the critical frequency [17]. Wei *et al* [66] proposed an analytical method to solve the free vibration problem of a cracked FGM beam which included rotary inertia and the shear deformation effect. The solution method was a combination of the transfer matrix method with the rotational spring model [66]. Jiang and Ding [67] derived an analytical solution for FGM beams using the superposition principle and trial and error. The analytical method can serve as a benchmark for numerical methods such as the finite element method (FEM), the boundary element method, etc. Li [68] proposed a new unified approach for the static and dynamic analysis of an FGM beam that included rotary inertia and shear deformation. Two branches of flexural waves with different wave speeds propagate in FGM beams. The two wave frequencies may intersect with an increase in the radius-to-span ratio [68]. Wang and Li [69] obtained an approximate analytical solution of FGM beam vibration analysis based on the Levinson beam theory, which is an example of PSDBT. The solution modified the mode shapes of the Levinson beam theory, which is different from the TBT. Ke *et al* [70] analytically studied the nonlinear vibration of FGM beams based on the CBT and von Kármán geometric nonlinearity. They found that the FGM beams showed different vibrational behaviors at positive and negative amplitudes due to a nonlinear quadratic term [70]. Sarkar and Ganguli [71] investigated the free vibration of a nonuniform Euler–Bernoulli beam using an inverse problem approach. Atmane [72] proposed a new shear deformation beam theory for functionally graded porous beams and obtained analytical solutions for vibration.

Li and Batra [73] obtained the closed-form relations of the critical buckling loads between functionally graded TBT beams and the homogeneous Euler–Bernoulli beam. The boundaries were restricted to simply supported, clamped, and clamped–free boundaries [73]. The predictions obtained using these relations were in agreement with the results found by the numeric shooting method. Nguyen *et al* [74] analyzed the bending, free vibration, and buckling of axially loaded FGM beams using an approach based on FSDBT and obtained analytical solutions for displacement, natural frequencies, buckling loads, and load–frequency curves. Huang *et al* [75] proposed an analytical method to solve the buckling problem of an axially functionally graded Timoshenko beam with the help of an auxiliary power series function. Atmane *et al* [76] proposed a beam theory for bending, free vibration, and buckling analyses of FGM beams on a two-parameter elastic foundation and obtained analytical solutions using the Navier technique. Aydogdu [77] and Aydogdu *et al* [78] used a semi-inverse method based on CBT to solve the vibration and buckling problem of simply supported FGM beams. Rahimi *et al* [79] obtained the closed-form solution for the vibration and postbuckling problems of FGM Timoshenko beams under an axial load. Najafi *et al* [80] analyzed the non-central low-velocity impact of simply supported functionally graded beams resting on Winkler–Pasternak elastic foundations. An asymptotic approach based on a two-step perturbation technique was employed to obtain the solution [80].

7.2.2.2 Numerical solutions

The Ritz method is a convenient and effective approach for the dynamic analysis of FGM structures. Şimşek [81] used the Ritz method with the TBT and higher-order shear deformation beam theory to solve the static problem of a simply supported FGM beam loaded by a uniform transverse pressure. Şimşek [82] subsequently analyzed the FGM beam buckling problem using the Euler–Bernoulli and Timoshenko beam models. Their properties vary in both the axial and thickness directions according to the power law form [82]. The critical load was obtained using the Ritz method with trial functions assumed in polynomial forms [82]. Pradhan and Chakraverty [37, 39, 83] proposed a higher-order shear deformation beam theory and analyzed the free vibration of FGM beams with different shear deformation models using the Rayleigh–Ritz method. In the Ritz method, the number of polynomials is found to determine the convergence of the computed frequencies. Fazzolari [84] presented a free vibration analysis of three-dimensional metallic and FGM beams with arbitrary boundary conditions. The beam deformation was modeled by a quasi-3D theory expressed in a power series expansion form [84]. The weak form of the governing equations was obtained using the principle of virtual displacement and was solved using the Ritz method [84]. Chen *et al* [85] analyzed the free and forced vibrations of FGM beams with nonuniform porosity distributions using the TBT framework and solved them using the Ritz method. In addition, Chen *et al* [86] solved the elastic buckling and static bending problems of functionally graded Timoshenko beams using the Ritz method with polynomial trial functions. Zhang [87, 88] proposed a new model of FGM beams based on the physical neutral surface and high-order shear deformation theory. The model introduced a shift of the axis along the thickness to adapt the physical bending deformation [87, 88]. The nonlinear bending and buckling analyses were performed using the Ritz method. Wattanasakulpong [89] used the Ritz method to analyze the thermal buckling and vibration of functionally graded beams, based on the improved third-order shear deformation theory. Ghiasian *et al* [90] studied the nonlinear dynamic buckling and imperfection sensitivity of FGM Timoshenko beams exposed to a sudden uniform temperature rise. Jena *et al* [91] utilized Chebyshev polynomials as the shape function of the Rayleigh–Ritz method to study the vibration problem of functionally graded porous beams because the Chebyshev polynomials are orthogonal, which may reduce the computational effort and avoid ill-conditioning in the computations [91].

The differential quadrature method (DQM) is widely used in the dynamic analysis of FGM structure due to its better convergence characteristics [9]. Jin and Wang [92] proposed a novel quadrature weak-form functionally graded beam element based on the CBT and the differential quadrature rule. The Gauss–Lobatto–Legendre quadrature was used to obtain the element stiffness matrix and mass matrix in the free vibration analysis of FGM beams [92]. Khalili *et al* [93] utilized a mixed Ritz–DQM in the forced vibration analysis of FGM beams with moving loads, in which the Rayleigh–Ritz method and DQM were separately employed to discretize the spatial partial derivatives and temporal derivatives. Parashar and Sharma [94] used the generalized DQM [95] to solve the shear-induced flexural vibration of functionally graded piezoelectric material beams. Zahedinejad [96] analyzed the free vibration of FGM beams resting on elastic foundations in thermal environments, in which the DQM was used to solve the governing equation.

Patil and Kadoli [97] used the DQM to investigate the vibration response and control of bidirectional FGM beams. Esfahani *et al* [98] analyzed the thermal buckling and postbuckling of FGM Timoshenko beams resting on nonlinear elastic foundations; they used the generalized DQM to discretize the equilibrium equations. They found that the buckling responses of FGM beams are of the bifurcation type for doubly clamped boundaries but have a unique stable path for simply supported boundaries because the simply supported boundaries cannot provide the extra moment to the FGM beams [98].

The FEM is most commonly used in structural analysis. Alshorbagy [99] used the FEM to solve the free vibration problem of FGM beams based on the Euler–Bernoulli beam theory, which is also the CBT. Vo *et al* [100] developed an FEM for the static and vibrational analysis of FGM beams that included a refined shear deformation beam theory. Jing [101] utilized the finite volume method for the static and free vibration analyses of FGM beams using the TBT. The impacts of the different shear correction factors were investigated and compared [101]. Kapuria [102] analyzed the static and dynamic responses of layered FGM beams according to the third-order zigzag theory using the FEM. Murin *et al* [103, 104] proposed a beam finite element to solve the axial, flexural, and torsional vibration problems of FGM beams. Almeida [105] presented a bending analysis of FGM beams based on a geometric nonlinear effect using the total Lagrangian formulation. Using the third-order shear deformation theory and the FEM, Doroushi *et al* [106] analyzed the free vibration and transient response of functionally graded piezoelectric beams under thermo-electro-mechanical loads. Shahba *et al* [107] used the FEM to study the free vibration of axially functionally graded tapered beams based on the Euler–Bernoulli model. Mohanty *et al* [108] used the FEM to investigate the bending and buckling behaviors of FGM beams, in which the shape functions of the FEM were obtained from the exact solution of the bending scenarios. They found that FGM beams with the power law gradient are more stable than those with the exponential law gradient [108]. Kahya and Turan [109] proposed a five-noded beam element for the vibration and buckling analysis of FGM beams based on FSDBT. Anandrao *et al* [110, 111] presented a large-amplitude free vibration and thermal postbuckling analysis of FGM beams using the finite element model based on the TBT.

Many other methods have been used in the dynamic analysis of FGM structures, such as the meshless method, the shooting method, and Newmark's method. Yang *et al* [112] developed a mesh-free boundary-domain integral equation method to analyze the forced vibration of two-dimensional FGM beams. Qian and Ching [113] obtained the numerical solutions of the static deformation and the free and forced vibrations of an FGM cantilever beam using the meshless local Petrov–Galerkin method. In this method, an orthogonal transformation technique was used to enforce the nodal variables in the essential boundary areas [113], and the test function was chosen to equal the weight function of the moving least-squares approximation [114]. Li *et al* [115] analyzed the free vibration responses of the FGM beams according to the CBT and the TBT by using the shooting method to solve them. The equations of motion for the problems were derived by considering the shear deformation and the axial, transverse, rotational, and axial–rotational coupling inertia [115]. Ma and Lee [116, 117] studied the static and dynamic responses of FGM beams subjected to in-plane thermal loading using the shooting method. Sun *et al* [118] used the shooting method to investigate the thermomechanical buckling and postbuckling of FGM Timoshenko beams resting on elastic foundations. Şimşek [119] studied the nonlinear vibration of FGM Timoshenko beams under the action of a moving harmonic load. The coupled nonlinear equations of motion were solved using the Newmark-β method in conjunction with the direct iteration method [119].

7.2.3 FGM beam vibration

The free vibration of FGM beams has been studied using different shear deformation models and homogenization models in many studies. In these studies, the relationship between the natural frequency of the FGM Euler–Bernoulli beam and that of the homogenous Euler–Bernoulli beam is expressed as [69]:

$$\omega_{\mathrm{E}} = \sqrt{\frac{12}{c\phi_0}}\,\omega_{\mathrm{E}}^*, \tag{7.17}$$

where ω_{E} and ω_{E}^* are the natural frequencies of the FGM beam and the homogenous beam, respectively; the parameters c and ϕ_0 depend on the material gradient of the FGM beams. For thick beams whose slenderness ratios $L/h<20$, the frequency prediction according to CBT is an overestimate, while the effect of the slenderness ratio on the natural frequencies can be negligible if $L/h>20$ [47]. Among the boundary conditions of the FGM beams, the clamped–clamped beam is the most sensitive to an increase in frequency due to the increase in the L/h ratio [47]. If an elastic foundation is present under an FGM beam, this provides a larger stiffness for the system, which increases the natural frequencies [57]. If the Winkler–Pasternak elastic foundation effect is considered, the smallest effect of a change in n is found in cantilevered beams, while the largest effect is found in doubly clamped beams [120]. A change in the elastic foundation coefficients mainly affects the natural frequency of FGM beams, while the effect decreases with increasing mode numbers. The effect of a change in the elastic foundation stiffness coefficient on the dimensionless frequency is lowest for doubly clamped beams but highest for cantilevered beams. When FGM beams are loaded by a moving harmonic force, the dynamic response is related to the force velocity and the harmonic frequency [35, 119, 121]. Under a constant moving load, the maximum dynamic deflections increase with an increase in the index n of the power law but first increase and then decrease with an increase in the moving load velocity [121]. However, the maximum deflections become smaller with an increase in the index n of the power law when the forcing frequency exceeds the natural frequency of the FGM beam, as shown in figure 7.5 [121]. The dynamic phenomenon of an FGM beam with a geometric nonlinearity loaded by a moving harmonic force is more complicated [119]. The axial force in the beam contributes to the lateral stiffness resulting from the geometrical nonlinearity. The maximum deflection of the nonlinear model is smaller than that of the linear model when the load frequency is far from the natural frequency of the FGM beam. The

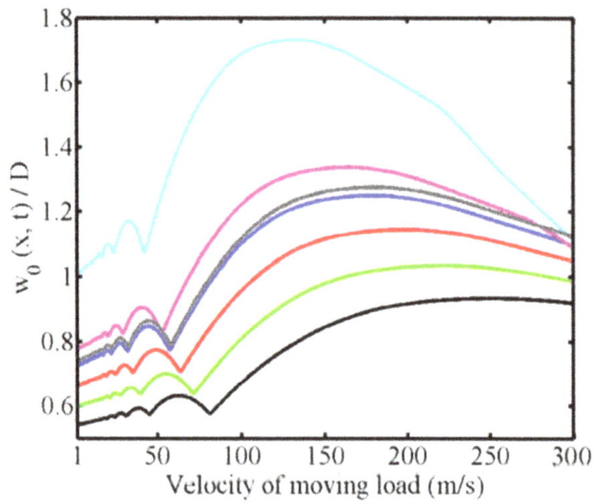

Figure 7.5. The variation of the maximum normalized dynamic deflection at the center of a beam with the velocity of the moving load for: (—) alumina only, (—) k=0.2, (—) k=0.5, (—) k=1, (—) k=2, (—) metal only, (—) exponential distribution. Reprinted from [121], Copyright (2009), with permission from Elsevier.

effect of geometrical nonlinearity is always stiffening, which causes a decrease in the magnitudes of the tensile and compressive stresses in the central section [119].

The nonlinear vibration behavior of the FGM beams was studied by the analytical method within the framework of the Euler–Bernoulli beam theory and the von Kármán geometric relationship [70]. The nonlinearity of the FGM beams exhibits the typical 'hardening' behavior at a large amplitude. The natural frequencies of nonlinear FGM beams are not affected by the sign of the vibration amplitude for doubly clamped graded beams, but are dependent on the signs of the vibration amplitudes for the hinged–hinged and clamped–hinged graded beams [70], as shown in figure 7.6. In the case of the forced vibration of nonlinear FGM beams, the amplitudes of the superharmonic and subharmonic resonances are affected by the parameters of the FGM model used for the perturbation solution [122]. The harvesting of piezoelectric energy from the nonlinear vibrations of FGM beams has been analyzed [123], and the excitation amplitude, material distribution, and load resistance value were shown to influence the harvested power. Furthermore, the nonlinear amplitude of bidirectional functionally graded microbeams has been studied [124]. Cyclic-fold bifurcations were discovered in a one-parameter bifurcation analysis as the frequency or amplitude of the external excitation was varied [124]. The loci of the cyclic-fold bifurcation occur in the two-parameter bifurcation analysis and the cusp bifurcation is found, as shown in figure 7.7 [124]. A parametric investigation showed that the size effect and variation of the material distribution do not change the type of nonlinearity but increase the quantitative difference between the results [124].

Bidirectional FGM beams are beams whose property gradation lies not only in the thickness direction but also along the axis of the beam. Şimşek [125] analyzed the free and forced vibrations of bidirectional functionally graded Timoshenko beams

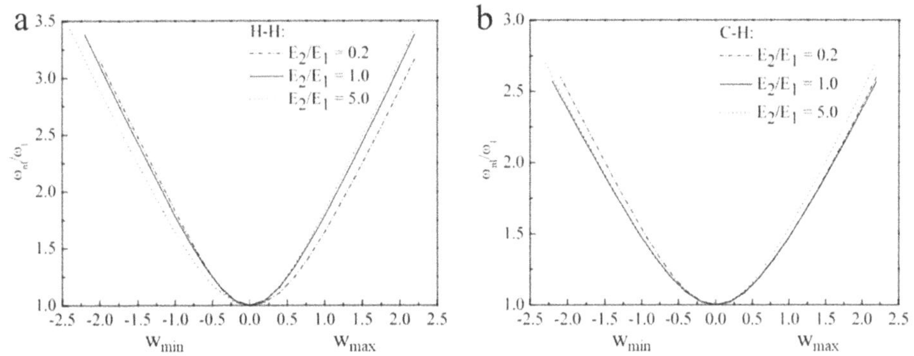

Figure 7.6. The variation of the nonlinear frequency ratio with the dimensionless amplitude for FGM beams with exponentially varying properties: (a) hinged–hinged and (b) clamped–hinged. Here, E_1 and E_2 are the Young's modulus of the beam top and bottom, respectively. Reprinted from [70] (2010) (Copyright © 2009, Springer Science Business Media B.V.) With permission of Springer.

under the action of a moving load. The index of the power law of the property gradient along the axis was found to be affected by the internal stress and the deflection of the FGM beam [125]. A semianalytical method has been proposed for bidirectional FGM beams [126]. Under a temperature load, the gradation of the material properties may reduce thermal stress and make it more uniform [126]. The analytical method has been used to study the free vibration of FGM beams with bidirectional property gradation according to the power law [17]. In this study, an abrupt jump was found for the natural frequencies of FGM beams, depending on the gradient parameter [17]. The nonlinear vibrations of bidirectional FGM beams have been analyzed [14, 127]. The asymmetric patterns induced by the axial functional gradation may affect the nonlinear vibration response of the beam, as shown in figure 7.8 [127].

FGMs can be used in the micro- and nanostructures of electromechanical systems or atomic force microscopes [128, 129]. The static and vibrational behaviors of FGM microbeams have been studied using a modified couple stress theory [130]. For FGM microbeams, the size-dependent theory is essential in order to reasonably predict the resonant frequency. The bending and vibrational responses of FGM microbeams have been analyzed using a modified couple stress theory and by considering the neutral axis [131]. This study showed that the incorporation of the couple stress effect makes the microbeam stiffer, leading to a decrease in the vertical displacement and an increase in the natural frequency. The impacts of the length scale on the dimensionless deflection and the first dimensionless natural frequency are shown in figure 7.9 [131]. The bending, buckling, and vibration of axially FGM beams have been studied using the nonlocal strain gradient theory [132]. The influences of the power law variation and size-dependent parameters on the bending, buckling, and vibration behaviors of axially graded beams were analyzed by this study [132]. The buckling and free vibration of nanocomposite beams reinforced by axially functionally graded graphene have been analyzed [133]. The reinforcing effect was found to be strongly dependent on both the graphene platelet density and the boundary conditions.

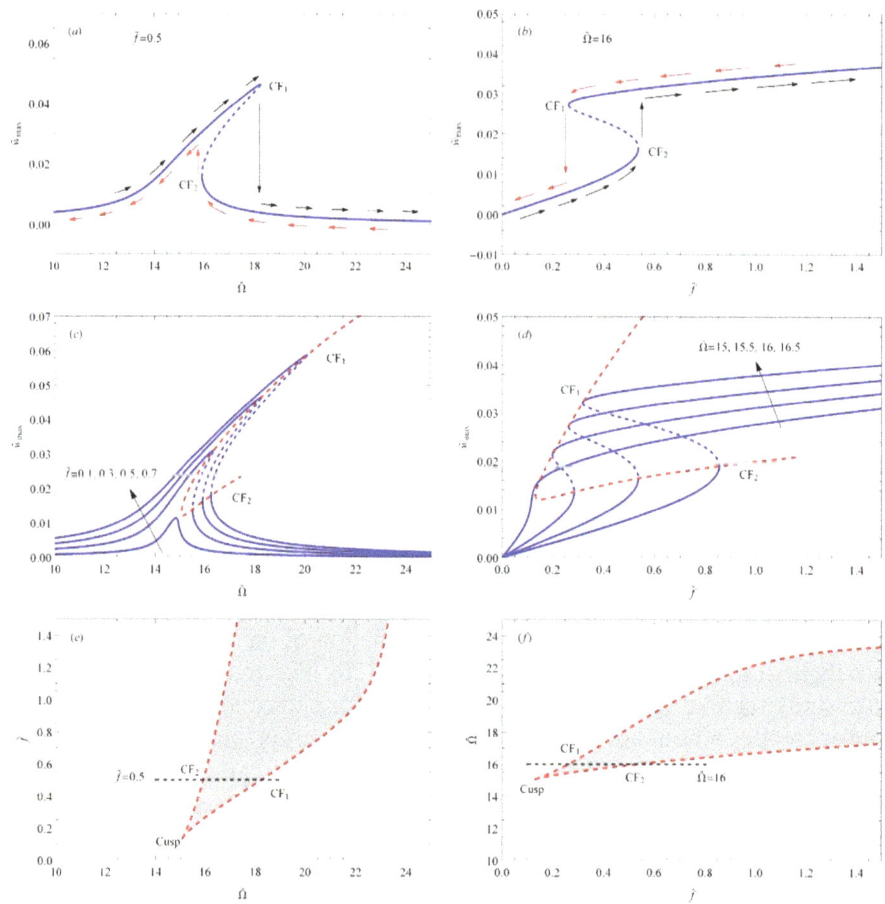

Figure 7.7. Graphs of the 'jump' phenomenon and the cyclic-fold bifurcation curves of bidirectional FGM beams. (a) and (b) The 'jump' phenomenon in the nonlinear resonance response. (c) and (d) The cyclic-fold bifurcation curves. (e) and (f) The two-parameter bifurcation diagrams. Reprinted from [124], Copyright (2019), with permission from Elsevier.

The elasticity problem for a nonhomogeneous plane with a crack was studied by Delale and Erdogan [134], in which the exponential index of the property was found to have a significant impact on the stress intensity factor (SIF). A cracked FGM beam was modeled as a rotational spring in the free vibration and elastic buckling analysis [135], in which the graded property exponentially varied along the thickness. The bending SIF was obtained by the investigation of Erdogan and Wu [136] for a flexural vibration analysis. The existence of a crack can be regarded as a local increase in the flexibility of the structure. Therefore, the frequency ratio and the buckling load ratio become much lower when the crack is located near the locus of the maximum bending moment. The vibration and buckling behaviors of FGMs with smaller modulus ratios E_b/E_t are more sensitive to the presence of cracks, where E_b and E_t are the Young's moduli of the bottom and top, respectively [136]. The nonlinear vibration of edge-cracked FGM Timoshenko beams has been analyzed

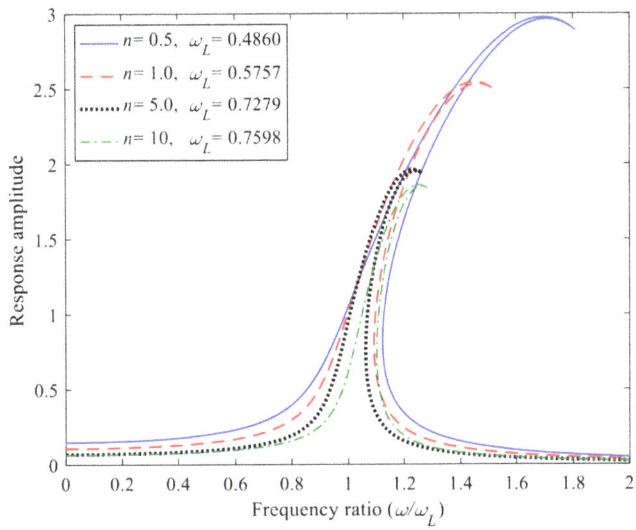

Figure 7.8. The influence of the axial functional gradient index on the responses of bidirectional FGM beams. Here, β is the exponential law index of the property and ω_L denotes the frequency of the linear model. Reprinted from [127] (2021) (Copyright © 2021, The Author(s), under exclusive licence to Springer Nature B.V.) With permission of Springer.

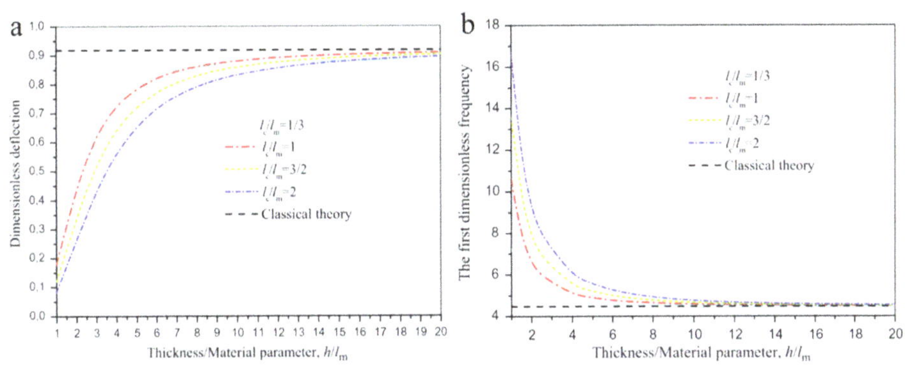

Figure 7.9. The dimensionless deflection and first dimensionless natural frequency for different length scale parameter ratios. Here, l_c and l_m are the length scale parameters. Reprinted from [131], Copyright (2015), with permission from Elsevier.

[137], and the influences of parameters such as the crack depth, crack location, and material property gradient were studied. It was found that the nonlinear frequency and mode shapes were much less affected by the change in crack depth than the linear ones [137]. The forced vibration responses of cracked FGM beams were studied using the modified couple stress theory and a damping effect [138]. The influences of the geometric parameters on the forced vibration responses of cracked FGM beams were discussed [138]. Most of the studies of cracked FGM beams have used a rotational spring model, but a continuous model of cracked FGM beams needs to be developed.

7.2.4 FGM beam buckling

For the simply supported, clamped–clamped and clamped–free beams, the closed-form relations between the critical buckling loads of the functionally graded Timoshenko and Euler–Bernoulli beams and those of a homogeneous Euler–Bernoulli beam were obtained and expressed as follows [73]:

$$\text{Euler–Bernoulli model:}\quad P_{\mathrm{Ecr}} = P^{*}_{\mathrm{Ecr}}/c_{\mathrm{b}}, \tag{7.18}$$

$$\text{Timoshenko model:}\quad P_{\mathrm{Tcr}} = \frac{P^{*}_{\mathrm{Ecr}}}{c_{\mathrm{b}} + c_{\mathrm{s}}P^{*}_{\mathrm{Ecr}}}, \tag{7.19}$$

where P_{Tcr}, P_{Ecr}, and P^{*}_{Ecr} are the critical buckling loads of the Timoshenko FGM beam, the Euler–Bernoulli FGM beam, and the homogeneous Euler–Bernoulli beam, respectively; c_{b} represents the inhomogeneity parameter, which is equal to one for a homogeneous beam; c_{s} reflects the effects of the shear deformations on the critical load. The vibration and buckling of FGM beams have been investigated under mechanical and thermal loads [139]. Exact solutions were obtained using the state space approach. The influences on the natural frequencies, critical buckling loads, critical temperatures, load–frequency curves, and temperature–frequencies curves were also investigated [139]. The buckling, free vibration, and dynamic stability of piezoelectric FGM beams were studied in a thermal environment [140]. The critical thermo-piezoelectric buckling loads of the beam were obtained, and it was shown that the material properties of the FGM beams were dependent on temperature [140]. The results showed it is necessary to consider the temperature-dependent material properties in high-temperature environments. The bending and buckling of FGM nanobeams were analyzed using the nonlocal Timoshenko and Euler–Bernoulli beam theories [141]. The nonlocal beam model predicted a larger deflection and a smaller buckling load than those of the local beam model [141]. The thermal buckling and elastic vibration due to the third-order shear deformation of FGM beams were analyzed using the Ritz method to solve the eigenvalue problem [89]. The results of a thermoelastic vibration analysis showed that the natural frequency decreases with an increase in temperature up to the critical point and increases in the postbuckling region [89]. The influences of the property distribution and boundary conditions were also investigated [89]. The critical buckling load of an FGM simply supported beam with a partially delaminated piezoelectric layer has been solved [142]. The effects induced by the delaminated length, asymmetry, piezoelectric thickness, voltage, and the volume fraction of FGM are discussed in reference [142].

7.3 The vibration and buckling problems of functionally graded plates

FGM plates have an improved structure compared to composite structures, as they avoid stress concentration at the interface. FGM plates have great prospects for use in engineering applications [143]. In this section, we discuss the model development, solutions, and vibration and buckling analyses of FGM plates.

7.3.1 The development of models for functionally graded plates

The classical plate theory (CPT) is based on the Kirchhoff–Love hypothesis that the straight lines of the cross section remain perpendicular to the middle plane before and after deformation [144], which ignores both shear and normal deformation effects. Reissner [145, 146] and Mindlin [147] developed the FSDPT by considering the effects of rotatory inertia and shear deformation; in this respect, the FSDPT is similar to the TBT. A shear correction factor is also needed in the FSDPT, but it is difficult to determine because it depends on the geometric parameters, loads, and boundaries. To more accurately simulate the shear deformation distribution of the plate structure, the higher-order shear deformation plate theories (HSDPTs) were developed [148, 149]. Among these, the parabolic shear deformation plate theory (PSDPT), also called the third-order shear deformation plate theory, is widely used in plate analysis [42, 150, 151]. Moreover, the trigonometric shear deformation plate theory (TSDPT) [43, 152, 153] and other HSDPTs have been developed and reviewed in the research literature [148, 154].

Figure 7.10 shows a rectangular plate in which the graded material property varies in the direction of the plate's thickness. Here a, b, and h are the plate lengths and thickness, respectively. To present the shear deformation theory, the following displacement field is assumed [154]:

$$\begin{cases} u_x^p(x, y, z, t) = u_0^p(x, y, t) - zw_{0,x}^p(x, y, t) + \Phi^p(z)s_1^p(x, y, t), \\ u_y^p(x, y, z, t) = v_0^p(x, y, t) - zw_{0,y}^p(x, y, t) + \Phi^p(z)s_2^p(x, y, t), \\ u_z^p(x, y, z, t) = w_0^p(x, y, t), \end{cases} \tag{7.20}$$

where u_0^p, v_0^p, and w_0^p are the displacements of the middle surface of the plate; s_1^p and s_2^p are the shear strain at the middle surface; t is time; $()_{,x}$ and $()_{,y}$ are the derivatives with respect to x and y; and $\Phi^p(z)$ is the shape function given in reference [154]

$$\text{CPT:} \quad \Phi^p(z) = 0, \tag{7.21}$$

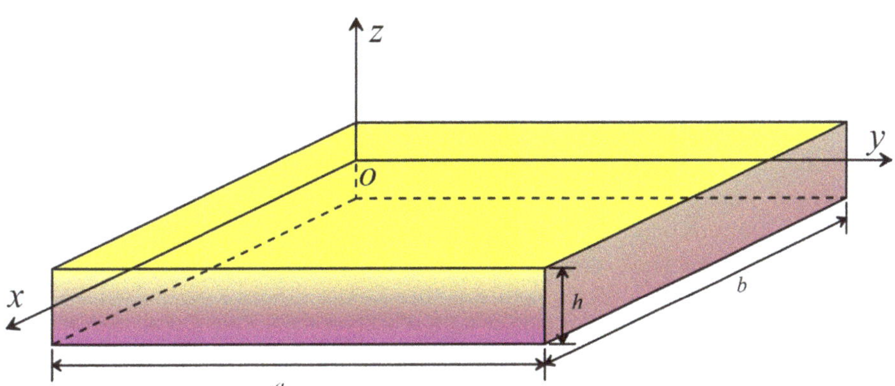

Figure 7.10. A diagram of a rectangular functionally graded plate.

$$\text{FSDPT:} \quad \Phi^p(z) = z, \tag{7.22}$$

$$\text{PSDPT:} \quad \Phi^p(z) = z\left(1 - \frac{4z^2}{3h^2}\right), \tag{7.23}$$

$$\text{TSDPT:} \quad \Phi^p(z) = \frac{h}{\pi}\sin\left(\frac{\pi z}{h}\right). \tag{7.24}$$

Yang and Shen [155] employed the CPT to study the transient response of initially stressed FGM plates resting on elastic foundations and loaded by lateral impacts. Abrate [144] used the CPT to investigate the free vibration behavior of an FGM plate with a continuous property varying in the thickness direction. The study showed that a suitable choice of the reference surface could eliminate the coupling between the in-plane and out-of-plane bending deformations. The equation of motion of FGM plates has the same formula as that of homogeneous plates [144]. Javaheri and Eslami [156] studied the buckling problem of FGM plates under in-plane compressive loading using the CPT. Nguyen *et al* [157] applied the first-order shear deformation plate theory to model FGM plates in which the power law property is distributed in the direction of the plate thickness. In reference [157], the identification of transverse shear factors was investigated using energy equivalence.

Hosseini-Hashemi *et al* [158] analyzed the free vibration of FGM rectangular plates supported by elastic foundations using the FSDPT. A new formula for the shear correction factors was obtained for FGM plates. Thai *et al* [159] proposed a new FSDPT for functionally graded sandwich plates composed of functionally graded face sheets and an isotropic homogeneous core; the shear correction factor is not necessary in this theory. Bellifa *et al* [160] developed a new FSDPT for the dynamic behavior and static bending of FGM plates. As a result of the neutral surface formulation, no stretching–bending coupling effect is present in the governing equations [160]. Nguyen *et al* [161] investigated shear correction factors for FGM plates, including a consideration of energy. The shear correction factor is a function of the material distribution and is noteworthy due to the high modulus ratio of the two phases [161]. Ferreira *et al* [162] studied the static response of simply supported FGM plates modeled by a third-order shear deformation theory; they used a meshless method to obtain the solution. Reddy and Kim [163] presented a general nonlinear third-order plate theory for FGM plates that considered the geometric nonlinearity and microstructure-dependent size effects. Bui *et al* [164] analyzed the high-temperature mechanical behaviors of heated FGM plates using the FEM and third-order shear deformation plate theory. Tounsi *et al* [165] utilized a refined trigonometric shear deformation theory to analyze the thermoelastic bending problem of FGM sandwich plates. They presented a comparison between the thermal stresses and deflections of this theory and those of other HSDPTs [165]. Belabed *et al* [166] presented a simple higher-order shear and normal deformation theory for the static bending and free vibration analyses of FGM plates. Their theory accounted for stretching and shear deformation effects and was more accurate for thick plates [166]. Zaoui *et al* [167] established two-dimensional and three-dimensional shear deformation theories for the free vibration of FGM plates resting on elastic foundations. Their new shear strain shape function included undetermined integral terms and unknowns which were related to both transverse shear and the thickness stretching effect [167].

7.3.2 Solutions for FGM plates

The solutions for FGM plates have to integrate deformations in the length and width directions of the structure, which are usually more complicated than those of FGM beams [168]. In this section, we discuss the analytical and numerical solutions for the vibration and buckling analyses of FGM plates.

7.3.2.1 Analytical solutions

Vel and Batra [169] presented a three-dimensional exact solution for the free and forced bending vibrations of simply supported functionally graded rectangular plates based on the power series method. The exact solution can be used for arbitrary variations of material properties in the thickness direction. The exact natural frequencies, displacements, and stresses were compared with the results of the CPT, the first-order shear deformation theory, and the third-order shear deformation theory for FGM plates [169]. Using the CPT, Baferani *et al* [170] analyzed the free vibration characteristics of thin functionally graded rectangular plates under different boundary conditions. To obtain the exact analytical solution, the Navier solution was used for simply supported rectangular plates, and the Levy-type solution was employed for various boundary conditions [170]. The effects of the aspect ratio, thickness–length ratio, and boundary conditions on the vibration responses were discussed [170]. Sayyad and Ghugal [171] analyzed the static bending and free vibration of doubly curved FGM shells, based on a generalized shell theory. The equations of motion and the solution of the problem were obtained using the Hamilton principle and the Navier method, respectively [171]. Wang and Zu [172] proposed an approximate analytical solution for use in the study of the nonlinear vibrational behavior of functionally graded rectangular plates with porosities moving in a thermal environment. The solution was obtained by using the Galerkin method to discretize the partial differential equations and the harmonic balanced method to solve the time-varying ordinary differential equations [172]. Woo *et al* [173] obtained an analytical solution for the nonlinear free vibration of FGM plates using mixed Fourier series. The effects of material properties, boundary conditions, and thermal loading on the dynamic behavior were investigated. Fallah *et al* [174] analyzed the free vibration of the moderately thick rectangular FGM plates resting on elastic foundations. A semianalytical solution was obtained by using the extended Kantorovich method together with an infinite power series.

Thai and Choi [175] proposed a simple refined model for the buckling analysis of FGM plates. Closed-form solutions of the critical loads for the rectangular plates were obtained using the Navier method. Yaghoobi and Torabi [176] investigated the buckling analysis of FGM plates resting on Pasternak foundations under thermal loads. The exact analytical solution was obtained by using the Levy-type solution to reduce the problem to ordinary differential equations, and the power series Frobenius method was used to solve these equations [176]. Javaheri and Eslami [177] derived the equilibrium and stability equations of functionally graded rectangular plates under thermal loads using the variation method and obtained approximate solutions for the critical buckling temperatures using Galerkin's formulation.

7.3.2.2 Numerical solutions

Nie and Zhong [178] solved the three-dimensional free and forced vibration problems of FGM circular plates under various boundary conditions using a semianalytical method. This analytical solution for the graded direction was obtained using the state space method and an approximate solution was obtained using the differential quadrature method in the radial direction. Huang *et al* [179] used the state space method to obtain the solution for FGM thick plates resting on Winkler–Pasternak elastic foundations. Pan and Han [180] presented an exact solution for multilayered rectangular plates made of functionally graded, anisotropic, and linear magneto-electro-elastic materials. For the multilayered plate structure, the propagation matrix method was employed [180]. The numerical results showed the influence of the property exponential factor, magneto-electro-elastic properties, and loading types on the results [180]. Jin *et al* [181] analyzed the free vibration of arbitrarily thick functionally graded rectangular plates under general boundary conditions, which was modeled using three-dimensional elasticity theory. An exact solution was obtained using the Rayleigh–Ritz procedure in the three-dimensional Fourier cosine series forms. Zhao *et al* [182] also used the Rayleigh–Ritz method to obtain an exact solution for the vibration of thick functionally graded porous rectangular plates under arbitrary boundary conditions. Zhao *et al* [183] employed the element-free *kp*-Ritz method to solve the mechanical and thermal buckling problem of FGM plates. The direct nodal integration method used in this study resulted in higher computational efficiency without shear locking for thin plates, compared to that of Gauss integration. Zhang *et al* [184] used a local Kriging meshless method to solve the thermal buckling problem of FGM plates, in which the shape functions were constructed by Kriging interpolation.

The DQM method is an effective method for plate analysis. Yang and Shen [185] utilized the one-dimensional differential quadrature technique, the Galerkin approach, and the modal superposition method to obtain the dynamic responses of FGM plates in a thermal environment. Wu *et al* [186] analyzed the large-amplitude vibration of functionally graded graphene nanocomposite annular plates in thermal environments. In their study, the DQM was used to solve the nonlinear partial differential governing equations of vibration. Mirtalaie and Hajabasi [187] used the DQM to analyze the free vibration of FGM plates and investigated the impact of boundary effects on the vibration response. Trabelsi *et al* [188] proposed the integration of a modified FSDPT-based four-node element into the FEM to analyze the thermal buckling behaviors of FGM plates and cylinder shells. In a different study, the phase-field method was used to model a crack in the FGM plate, and the FEM was employed to analyze the buckling behavior of cracked FGM plates [189].

7.3.3 FGM plate vibrations

Due to the property gradation in the thickness direction of FGM plates, shear deformation and the stretching effect are significant in vibration analyses. Compared to the exact solution for the vibration of FGM plates, the frequency results produced by CPT are obviously different from those computed by FSDPT and the third-order shear deformation theories for functionally graded thick plates [169]. According to research, FSDPT even performs better than the third-order shear deformation theory for frequency predictions. The relation between the natural frequencies of thin FGM plates and those of the corresponding homogeneous plates is given by [190]:

$$\omega_{\mathrm{p}} = \omega_{\mathrm{p}}^{*} \Big/ \sqrt{c_{\mathrm{p}}\overline{\Phi}_{0}}, \tag{7.25}$$

where ω_{p} and ω_{p}^{*} are the natural frequencies of the FGM plates and the corresponding homogeneous plates, respectively, and c_{p} is a dimensionless parameter that integrates the effects of the transverse inhomogeneity of the property. It should be noted that $c_{\mathrm{p}}=1$ for homogeneous plates. $\overline{\Phi}_{0}$ denotes the dimensionless average density in the thickness direction. A study that used the Levy solution for the vibration analysis of FGM rectangular plates concluded the following [170]: (i) the dimensionless frequency of the FGM plates decreases with an increase in the aspect ratio, except for the free edge boundaries; (ii) the frequencies of all modes decrease as the power law index of the property decreases, which is similar to the behavior of FGM beams; (iii) for higher values of the aspect ratio, the effect of the boundary conditions influences the response to a greater degree. The effect of a uniform elastic boundary on the free vibration of functionally graded porous rectangular plates has been studied [191]. In reference [191], the porosity distributions are separated into three types: type I (uneven symmetrical distribution), type II (uneven asymmetric distribution), and type III (even distribution). The results of this analysis showed that the type I rectangular plates are more sensitive to changes in parameters such as porosity coefficients, boundary conditions, and aspect ratios [191]. When the plates have the uneven porosity distributions of type I and type II, the second frequency parameter shows a more complicated variation with the porosity coefficient than that of type III plates [191].

The thermal effect is very important in the vibration analysis of FGM plates, because many FGM plates are designed to resist high temperature gradients [192]. The vibrations of two-dimensional FGM nanotubes have been studied in a thermal environment, based on the higher-order theory [193]. The results show that increasing the temperature leads to a reduction from the initial natural frequency until it reaches zero [193]. The zero natural frequency corresponds to buckling, which takes place at a zero effective stiffness. In the postbuckling region, the FGM shell regains stiffness and the natural frequency increases, as shown in figure 7.11 [193]. The thermoelastic vibrations of FGM microplates have been analyzed while

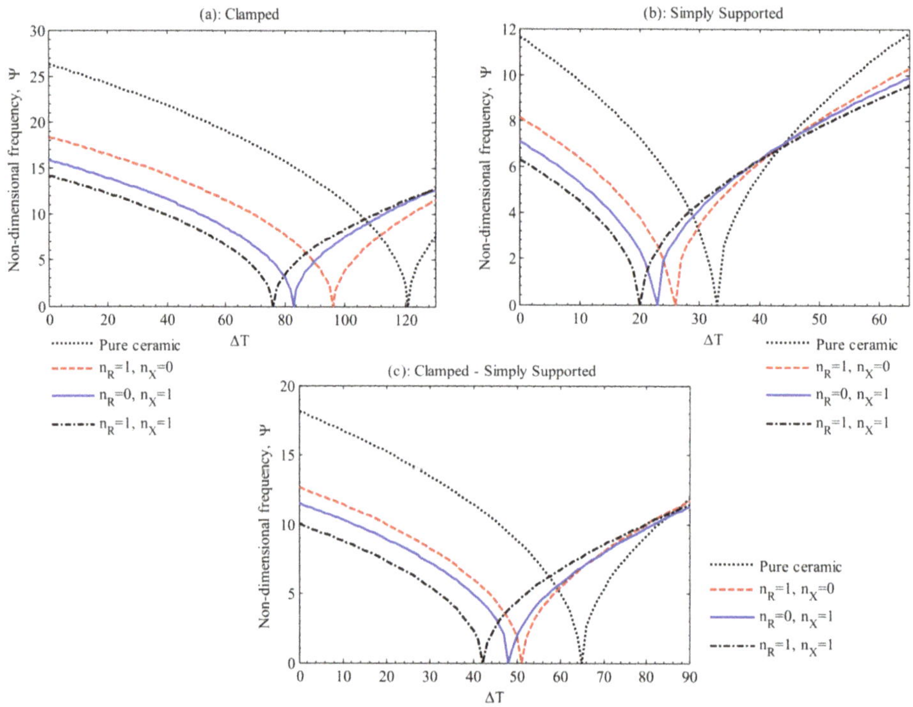

Figure 7.11. The effects of increasing temperatures on the dimensionless frequency of FGM tubes under different boundary conditions. Reprinted from [193], Copyright (2018), with permission from Elsevier.

taking account of the thermoelastic damping effect [194, 195]. The coupled problems of heat conduction and deformation have been numerically solved using Kirchhoff's plate theory and Levison's plate theory [195]. In this analysis, the material gradient had an obvious impact on the temperature distribution in the thickness direction and also on the thermoelastic damping [195]. The influences of the thickness and the power law index of the property on the natural frequency and damping are shown in figure 7.12 [195]. Temperature variation induces deformation of the structure. Accordingly, thermally induced vibrations of FGM plates and shell panels have been studied [196]. The effect of the panel geometry, volume fraction index, and boundary conditions on the vibration response have been investigated [196].

7.3.4 FGM plate buckling

Plate buckling is induced by compression. For FGM plates, the critical buckling load can be expressed as follows [190]:

$$P_{\text{pcr}} = P^{*}_{\text{pcr}}/c_{\text{p}}, \tag{7.26}$$

where P_{pcr} and P^{*}_{pcr} are the critical buckling loads of FGM plates and the corresponding homogeneous plates, respectively. Similarly, this expression has been used in the buckling analysis of FGM plates in the CPT [156] and a

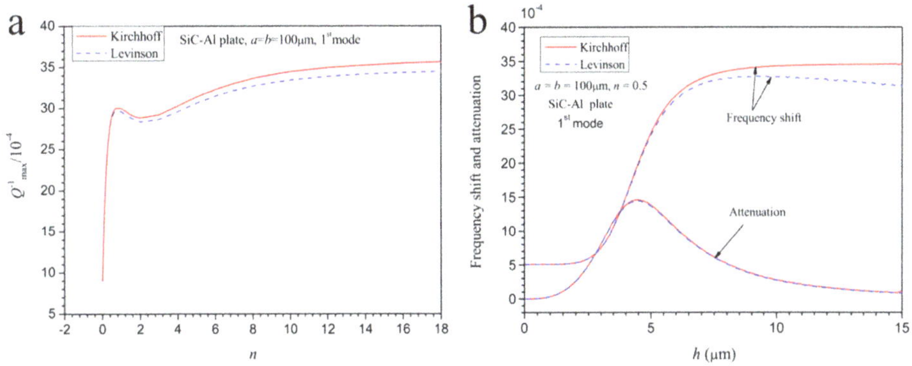

Figure 7.12. (a) The maximum thermoelastic damping of an FGM plate versus the power law index of the property gradation. (b) The frequency shift and attenuation versus the plate thickness of an FGM plate. Reprinted from [195], Copyright (2021), with permission from Elsevier.

proportional relationship has been validated [197]. The critical load of the FGM plate under combined compression and tension was greater than that under uniaxial and biaxial compression [156]. The buckling behaviors of FGM plates with stiffeners have been investigated [198]. The results showed that stiffeners are effective improvements that can enhance the critical buckling loads, as shown in figure 7.13, and thus reduce the weight of the structure [198].

Thermal buckling is important part of the buckling studies of FGM plates. One study obtained the closed-form solution, which gives the critical buckling temperature difference for simply supported FGM plates [199], using the higher-order shear deformation plate theory. In this analysis, the critical buckling temperature was decreased by increasing the length-to-thickness ratio or decreasing the aspect ratio [199]. This study also pointed out that the higher-order shear deformation theory underestimates the buckling load compared to the results obtained using CPT [199]. However, for thick FGM plates, the third-order shear deformation theory is recommended, because the classical theory overpredicts the critical buckling temperature loads [200]. The thermal buckling of FGM plates has been studied using on a four-variable refined plate theory [201]. This research investigated the uniform, linear, and nonlinear temperature rise distribution in the thickness direction, and found that their critical temperatures differed [201]. The thermal postbuckling and vibration behaviors of the FGM plates have been studied using the FEM [202]. The vibrational behaviors of a Si_3N_4/SUS304 FGM plate and an aluminum oxide/Ti–6Al–4V FGM plate are shown for increasing temperatures in figure 7.14 [202]. The natural frequency decreases with increasing temperatures in the prebuckling region, but it increases in the postbuckling region [202]. However, the fundamental frequencies of the FGM plates do not go to zero, because the postbuckling behavior of the FGM plate is not the bifurcation buckling behavior. Due to the material gradation in the thickness direction, rising temperatures create a bending moment on the FGM plate, which avoids the bifurcation phenomenon.

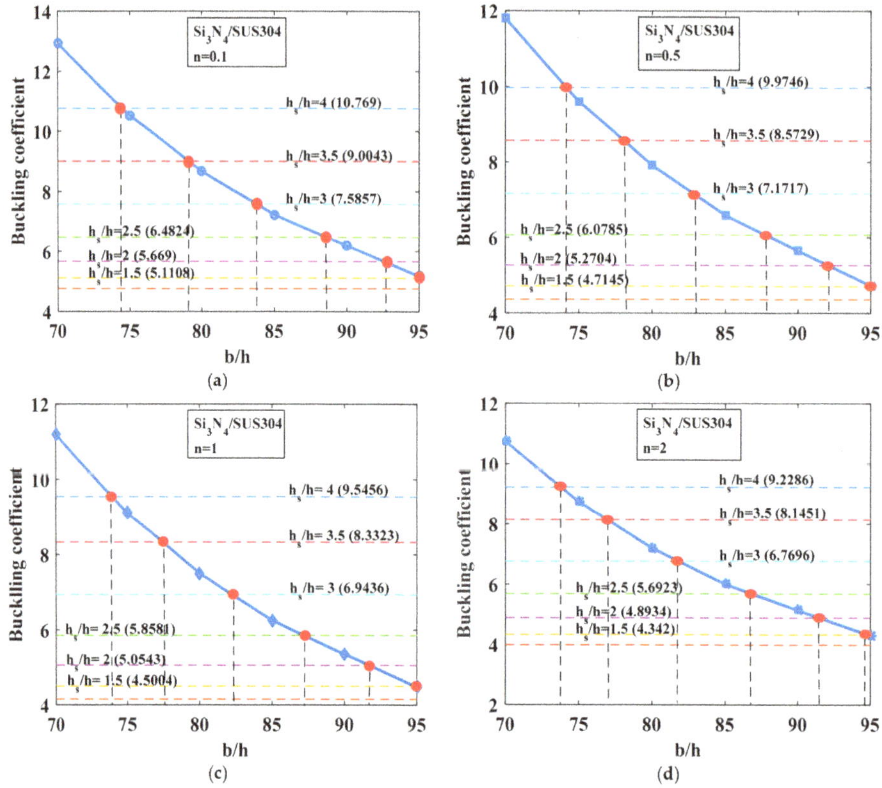

Figure 7.13. The buckling coefficients of FGM plates for different geometries of the plate and the stiffener. Solid blue line: the buckling coefficient of the plate without a stiffener; horizontal dashed–dotted line: the buckling coefficient of the plate with a central stiffener. (a) The power law index n=0.1; (b) the power law index n = 0.5; (c) the power law index n = 1; and (d) the power law index n = 2. Reprinted from [198] CC BY 4.0 © 2019 by the authors. Licensee MDPI, Basel, Switzerland.

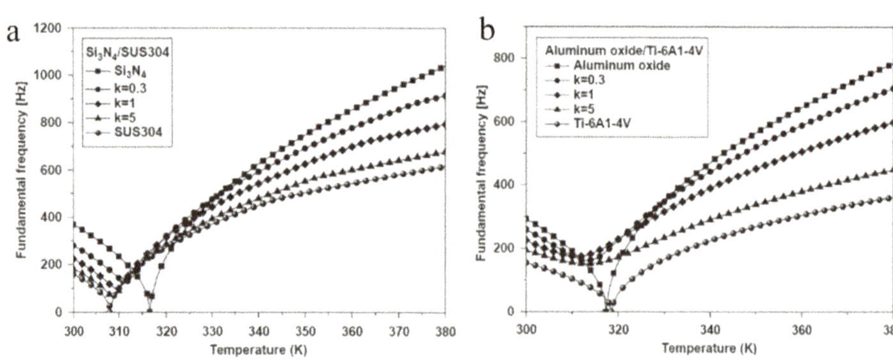

Figure 7.14. The vibrational behaviors of (a) a Si_3N_4/SUS304 FGM plate and (b) an aluminum oxide/Ti–6Al–4V FGM plate for increasing temperatures. Here, k is the gradient index of the FGM plate. Reprinted from [202], Copyright (2006), with permission from Elsevier.

7.4 Summary and prospects

FGM technology provides a new pathway for improving structural performance under different kinds of interaction. Studies of the modeling and loaded responses of FGM structures essentially provide the basis for FGM structural design and applications. This chapter reviewed different types of functionally graded material and discussed homogenization methods. To more accurately simulate the behavior of FGM, the interactions between the inclusions and the bulk materials in the constitutive relations should be emphasized and developed to further satisfy the requirements of structural design and manufacture. Due to the different types of FGM, it is important and urgent to develop different types of homogenization scheme for shape gradient, orientation gradient, and size gradient materials. The dynamic responses of FGM structures are sensitive to the gradation parameter. More experiments are needed to verify the modeling theories used for FGMs.

The vibration and buckling responses of FGM beams and plates were reviewed and discussed by comparing their model developments, analytical and numerical solutions, and dynamic characteristics. The higher-order shear deformation distribution theories are significant for FGM beams and plates because shear deformation and stretching effects can greatly influence their dynamic responses. As a result of the property gradation in the thickness direction, the neutral surface may shift, which is a noteworthy issue. Compared with traditional materials, FGMs have been found to exhibit higher performance in many fields, including thermal barriers, mechanical strength and nanomechanics. Because the composition and the design of the inclusions can be adjusted, FGMs provide engineers with the freedom to solve complex problems. The convolutions of FGMs with other effects, such as the thermal effect, nonlinearity, the size effect, and cracks, were discussed in detail. In some situations, the application of an FGM may complicate the response of the structure, which motivates researchers to study the multiple effects of FGM structures. In the future, FGMs are expected to have more and wider applications in the engineering field.

References

[1] Sola A, Bellucci D and Cannillo V 2016 Functionally graded materials for orthopedic applications—an update on design and manufacturing *Biotechnol. Adv.* **34** 504–31

[2] Sarathchandra D T, Kanmani Subbu S and Venkaiah N 2018 Functionally graded materials and processing techniques: an art of review *Mater. Today Proc.* **5** 21328–34

[3] Koizumi M 1997 FGM activities in Japan *Composites* B **28** 1–4

[4] Rajan T P D and Pai B C 2014 Developments in processing of functionally gradient metals and metal–ceramic composites: a review *Acta Metall. Sin.* **27** 825–38

[5] Saleh B, Jiang J, Fathi R, Al-hababi T, Xu Q, Wang L, Song D and Ma A 2020 30 Years of functionally graded materials: an overview of manufacturing methods, applications and future challenges *Composites* B **201** 108376

[6] Gupta K M 2015 *Engineering Materials: Research, Applications and Advances* (Boca Raton, FL: Taylor and Francis)

[7] Birman V and Byrd L W 2007 Modeling and analysis of functionally graded materials and structures *Appl. Mech. Rev.* **60** 195–216

[8] Boggarapu V, Gujjala R, Ojha S, Acharya S, Venkateswara Babu P, Chowdary S and Kumar Gara D 2021 State of the art in functionally graded materials *Compos. Struct.* **262** 113596

[9] Zahedinejad P, Zhang C, Zhang H and Ju S 2020 A comprehensive review on vibration analysis of functionally graded beams *Int. J. Struct. Stab. Dyn.* **20** 2030002

[10] Reddy J N and Chin C D 1998 Thermomechanical analysis of functionally graded cylinders and plates *J. Therm. Stresses* **21** 593–626

[11] Mori T and Tanaka K 1973 Average stress in matrix and average elastic energy of materials with misfitting inclusions *Acta Metall.* **21** 571–4

[12] Benveniste Y 1987 A new approach to the application of Mori–Tanaka's theory in composite materials *Mech. Mater.* **6** 147–57

[13] Hill R 1965 A self-consistent mechanics of composite materials *J. Mech. Phys. Solids* **13** 213–22

[14] Tang Y and Ding Q 2019 Nonlinear vibration analysis of a bi-directional functionally graded beam under hygro-thermal loads *Compos. Struct.* **225** 111076

[15] Zhang C, Savaidis A, Savaidis G and Zhu H 2003 Transient dynamic analysis of a cracked functionally graded material by a BIEM *Comput. Mater. Sci.* **26** 167–74

[16] Chakraborty A, Gopalakrishnan S and Reddy J N 2003 A new beam finite element for the analysis of functionally graded materials *Int. J. Mech. Sci.* **45** 519–39

[17] Wang Z, Wang X, Xu G, Cheng S and Zeng T 2016 Free vibration of two-directional functionally graded beams *Compos. Struct.* **135** 191–8

[18] Zaidi M, Joshi K K, Shukla A and Cherinet B 2021 A review of the various modelling schemes of unidirectional functionally graded material structures *AIP Conf. Proc* **2341** 020021

[19] Zuiker J R 1995 Functionally graded materials: choice of micromechanics model and limitations in property variation *Compos. Eng.* **5** 807–19

[20] Reiter T and Dvorak G J 1998 Micromechanical models for graded composite materials: II. Thermomechanical loading *J. Mech. Phys. Solids* **46** 1655–73

[21] Reiter T, Dvorak G J and Tvergaard V 1997 Micromechanical models for graded composite materials *J. Mech. Phys. Solids* **45** 1281–302

[22] Yin H M, Sun L Z and Paulino G H 2004 Micromechanics-based elastic model for functionally graded materials with particle interactions *Acta Mater.* **52** 3535–43

[23] Torquato S 1997 Effective stiffness tensor of composite media—I. Exact series expansions *J. Mech. Phys. Solids* **45** 1421–48

[24] Torquato S 1998 Effective stiffness tensor of composite media: II. Applications to isotropic dispersions *J. Mech. Phys. Solids* **46** 1411–40

[25] Peng X, Hu N, Zheng H and Fukunaga H 2009 Evaluation of mechanical properties of particulate composites with a combined self-consistent and Mori–Tanaka approach *Mech. Mater.* **41** 1288–97

[26] Shen H S 2009 *Functionally Graded Materials: Nonlinear Analysis of Plates and Shells* (Boca Raton, FL: CRC Press)

[27] Huang C S, McGee O G and Chang M J 2011 Vibrations of cracked rectangular FGM thick plates *Compos. Struct.* **93** 1747–64

[28] Shen H S and Wang Z X 2012 Assessment of Voigt and Mori–Tanaka models for vibration analysis of functionally graded plates *Compos. Struct.* **94** 2197–208

[29] Karami B, Shahsavari D, Janghorban M and Li L 2019 Influence of homogenization schemes on vibration of functionally graded curved microbeams *Compos. Struct.* **216** 67–79

[30] Kiani Y and Eslami M R 2015 Thermal postbuckling of imperfect circular functionally graded material plates: examination of Voigt, Mori–Tanaka, and self-consistent schemes *J. Press. Vessel Technol. Trans. ASME* **137** 021201

[31] Nemati A R and Mahmoodabadi M J 2020 Effect of micromechanical models on stability of functionally graded conical panels resting on Winkler–Pasternak foundation in various thermal environments *Arch. Appl. Mech.* **90** 883–915

[32] Gupta A and Talha M 2015 Recent development in modeling and analysis of functionally graded materials and structures *Prog. Aerosp. Sci.* **79** 1–14

[33] Huang Y and Li X-F 2010 A new approach for free vibration of axially functionally graded beams with non-uniform cross-section *J. Sound Vib.* **329** 2291–303

[34] Sankar B V and Tzeng J T 2002 Thermal stresses in functionally graded beams *AIAA J.* **40** 1228–32

[35] Şimşek M, Kocatürk T and Akbaş Ş D 2012 Dynamic behavior of an axially functionally graded beam under action of a moving harmonic load *Compos. Struct.* **94** 2358–64

[36] Sankar B V 2001 An elasticity solution for functionally graded beams *Compos. Sci. Technol.* **61** 689–96

[37] Pradhan K K and Chakraverty S 2014 Effects of different shear deformation theories on free vibration of functionally graded beams *Int. J. Mech. Sci.* **82** 149–60

[38] Weaver W, Timoshenko S P and Young D H 1990 *Vibration Problems in Engineering* 4th edn (New York: Wiley)

[39] Pradhan K K and Chakraverty S 2013 Free vibration of Euler and Timoshenko functionally graded beams by Rayleigh–Ritz method *Composites* B **51** 175–84

[40] Hutchinson J R 2001 Shear coefficients for timoshenko beam theory *J. Appl. Mech.* **68** 87–92

[41] Levinson M 1981 A new rectangular beam theory *J. Sound Vib.* **74** 81–7

[42] Reddy J N 1984 A simple higher-order theory for laminated composite plates *J. Appl. Mech.* **51** 745–52

[43] Touratier M 1991 An efficient standard plate theory *Int. J. Eng. Sci.* **29** 901–16

[44] Soldatos K P 1992 A transverse shear deformation theory for homogeneous monoclinic plates *Acta Mech.* **94** 195–220

[45] Karama M, Afaq K S and Mistou S 2003 Mechanical behaviour of laminated composite beam by the new multi-layered laminated composite structures model with transverse shear stress continuity *Int. J. Solids Struct.* **40** 1525–46

[46] Aydogdu M 2009 A new shear deformation theory for laminated composite plates *Compos. Struct.* **89** 94–101

[47] Şimşek M 2010 Fundamental frequency analysis of functionally graded beams by using different higher-order beam theories *Nucl. Eng. Des.* **240** 697–705

[48] Avcar M, Hadji L and Civalek Ö 2021 Natural frequency analysis of sigmoid functionally graded sandwich beams in the framework of high order shear deformation theory *Compos. Struct.* **276** 114564

[49] Nguyen T K, Truong-Phong Nguyen T, Vo T P and Thai H T 2015 Vibration and buckling analysis of functionally graded sandwich beams by a new higher-order shear deformation theory *Composites* B **76** 273–85

[50] Zohra Z, Lemya H, Abderahman Y, Mustapha M, Abdelouahed T and Djamel O 2017 Free vibration analysis of functionally graded beams using a higher-order shear deformation theory *Math. Model. Eng. Probl.* **4** 7–12
[51] Chen S, Geng R and Li W 2021 Vibration analysis of functionally graded beams using a higher-order shear deformable beam model with rational shear stress distribution *Compos. Struct.* **277** 114586
[52] Nguyen Q H, Nguyen L B, Nguyen H B and Nguyen-Xuan H 2020 A three-variable high order shear deformation theory for isogeometric free vibration, buckling and instability analysis of FG porous plates reinforced by graphene platelets *Compos. Struct.* **245** 112321
[53] Sayyad A S and Ghugal Y M 2019 Modeling and analysis of functionally graded sandwich beams: a review *Mech. Adv. Mater. Struct.* **26** 1776–95
[54] Akgöz B and Civalek Ö 2013 Buckling analysis of functionally graded microbeams based on the strain gradient theory *Acta Mech.* **224** 2185–201
[55] Sina S A, Navazi H M and Haddadpour H 2009 An analytical method for free vibration analysis of functionally graded beams *Mater. Des.* **30** 741–7
[56] Lü C F and Chen W Q 2005 Free vibration of orthotropic functionally graded beams with various end conditions *Struct. Eng. Mech.* **20** 465–76
[57] Ying J, Lü C F and Chen W Q 2008 Two-dimensional elasticity solutions for functionally graded beams resting on elastic foundations *Compos. Struct.* **84** 209–19
[58] Bert C W and Malik M 1994 Frequency equations and modes of free vibrations of rectangular plates with various edge conditions *Proc. Inst. Mech. Eng.* C **208** 307–19
[59] Eisenberger M and Godoy L A 2021 Navier type exact analytical solutions for vibrations of thin-walled shallow shells with rectangular planform *Thin-Walled Struct.* **160** 107356
[60] Aydogdu M and Taskin V 2007 Free vibration analysis of functionally graded beams with simply supported edges *Mater. Des.* **28** 1651–6
[61] Koochaki G R 2011 Free vibration analysis of functionally graded beams *Int. J. Mech. Mechatron* **74** 514–7
[62] Thai H T and Vo T P 2012 Bending and free vibration of functionally graded beams using various higher-order shear deformation beam theories *Int. J. Mech. Sci.* **62** 57–66
[63] Beg M S and Yasin M Y 2021 Bending, free and forced vibration of functionally graded deep curved beams in thermal environment using an efficient layerwise theory *Mech. Mater.* **159** 103919
[64] Hadji L, Khelifa Z and Abbes A B E 2016 A new higher order shear deformation model for functionally graded beams *KSCE J. Civ. Eng.* **20** 1835–41
[65] Hadji L, Daouadji T H, Meziane M A A, Tlidji Y and Bedia E A A 2016 Analysis of functionally graded beam using a new first-order shear deformation theory *Struct. Eng. Mech.* **57** 315–25
[66] Wei D, Liu Y and Xiang Z 2012 An analytical method for free vibration analysis of functionally graded beams with edge cracks *J. Sound Vib.* **331** 1686–700
[67] Jiang A and Ding H 2005 The analytical solutions for orthotropic cantilever beams (II): solutions for density functionally graded beams *J. Zhejiang Univ. Sci.* **6A** 155–8
[68] Li X F 2008 A unified approach for analyzing static and dynamic behaviors of functionally graded Timoshenko and Euler–Bernoulli beams *J. Sound Vib.* **318** 1210–29
[69] Wang X and Li S 2016 Free vibration analysis of functionally graded material beams based on Levinson beam theory *Appl. Math. Mech.* **37** 861–78

[70] Ke L-L, Yang J and Kitipornchai S 2010 An analytical study on the nonlinear vibration of functionally graded beams *Meccanica* **45** 743–52
[71] Sarkar K and Ganguli R 2013 Closed-form solutions for non-uniform Euler–Bernoulli free–free beams *J. Sound Vib.* **332** 6078–92
[72] Atmane H A, Tounsi A, Bernard F and Mahmoud S R 2015 A computational shear displacement model for vibrational analysis of functionally graded beams with porosities *Steel Compos. Struct.* **19** 369–84
[73] Li S R and Batra R C 2013 Relations between buckling loads of functionally graded timoshenko and homogeneous Euler–Bernoulli beams *Compos. Struct.* **95** 5–9
[74] Nguyen T K, Vo T P and Thai H T 2013 Static and free vibration of axially loaded functionally graded beams based on the first-order shear deformation theory *Composites* B **55** 147–57
[75] Huang Y, Zhang M and Rong H 2016 Buckling analysis of axially functionally graded and non-uniform beams based on Timoshenko theory *Acta Mech. Solida Sin.* **29** 200–7
[76] Ait Atmane H, Tounsi A and Bernard F 2017 Effect of thickness stretching and porosity on mechanical response of a functionally graded beams resting on elastic foundations *Int. J. Mech. Mater. Des.* **13** 71–84
[77] Aydogdu M 2008 Semi-inverse method for vibration and buckling of axially functionally graded beams *J. Reinf. Plast. Compos.* **27** 683–91
[78] Aydogdu M, Maróti G and Elishakoff I 2013 A note on semi-inverse method for buckling of axially functionally graded beams *J. Reinf. Plast. Compos.* **32** 511–2
[79] Rahimi G H, Gazor M S, Hemmatnezhad M and Toorani H 2013 On the postbuckling and free vibrations of FG Timoshenko beams *Compos. Struct.* **95** 247–53
[80] Najafi F, Shojaeefard M H and Saeidi Googarchin H 2017 Nonlinear dynamic response of FGM beams with Winkler–Pasternak foundation subject to noncentral low velocity impact in thermal field *Compos. Struct.* **167** 132–43
[81] Şimşek M 2009 Static analysis of a functionally graded beam under a uniformly distributed load by Ritz method *Int. J. Eng. Appl. Sci.* **1** 1–11 http://ijeas.akdeniz.edu.tr/_dinamik/160/14.pdf
[82] Şimşek M 2016 Buckling of Timoshenko beams composed of two-dimensional functionally graded material (2D-FGM) having different boundary conditions *Compos. Struct.* **149** 304–14
[83] Pradhan K K and Chakraverty S 2015 Generalized power-law exponent based shear deformation theory for free vibration of functionally graded beams *Appl. Math. Comput.* **268** 1240–58
[84] Fazzolari F A 2016 Quasi-3D beam models for the computation of eigenfrequencies of functionally graded beams with arbitrary boundary conditions *Compos. Struct.* **154** 239–55
[85] Chen D, Yang J and Kitipornchai S 2016 Free and forced vibrations of shear deformable functionally graded porous beams *Int. J. Mech. Sci.* **108–9** 14–22
[86] Chen D, Yang J and Kitipornchai S 2015 Elastic buckling and static bending of shear deformable functionally graded porous beam *Compos. Struct.* **133** 54–61
[87] Zhang D G 2013 Nonlinear bending analysis of FGM beams based on physical neutral surface and high order shear deformation theory *Compos. Struct.* **100** 121–6
[88] Zhang D G 2014 Thermal post-buckling and nonlinear vibration analysis of FGM beams based on physical neutral surface and high order shear deformation theory *Meccanica* **49** 283–93

[89] Wattanasakulpong N, Gangadhara Prusty B and Kelly D W 2011 Thermal buckling and elastic vibration of third-order shear deformable functionally graded beams *Int. J. Mech. Sci.* **53** 734–43

[90] Ghiasian S E, Kiani Y and Eslami M R 2015 Nonlinear thermal dynamic buckling of FGM beams *Eur. J. Mech.* A **54** 232–42

[91] Jena S K, Chakraverty S and Malikan M 2021 Application of shifted Chebyshev polynomial-based Rayleigh–Ritz method and Navier's technique for vibration analysis of a functionally graded porous beam embedded in Kerr foundation *Eng. Comput.* **37** 3569–89

[92] Jin C and Wang X 2015 Accurate free vibration analysis of Euler functionally graded beams by the weak form quadrature element method *Compos. Struct.* **125** 41–50

[93] Khalili S M R, Jafari A A and Eftekhari S A 2010 A mixed Ritz-DQ method for forced vibration of functionally graded beams carrying moving loads *Compos. Struct.* **92** 2497–511

[94] Parashar S K and Sharma P 2016 Modal analysis of shear-induced flexural vibration of FGPM beam using generalized differential quadrature method *Compos. Struct.* **139** 222–32

[95] Bert C W and Malik M 1996 Differential quadrature method in computational mechanics: a review *Appl. Mech. Rev.* **49** 1–28

[96] Zahedinejad P 2016 Free vibration analysis of functionally graded beams resting on elastic foundation in thermal environment *Int. J. Struct. Stab. Dyn.* **16** 1550029

[97] Patil M A and Kadoli R 2022 Effect of porosity and gradation of Galfenol-D on vibration suppression of bidirectional functionally graded beam *Mater. Today Proc.* **66** 1870–4

[98] Esfahani S E, Kiani Y and Eslami M R 2013 Non-linear thermal stability analysis of temperature dependent FGM beams supported on non-linear hardening elastic foundations *Int. J. Mech. Sci.* **69** 10–20

[99] Alshorbagy A E, Eltaher M A and Mahmoud F F 2011 Free vibration characteristics of a functionally graded beam by finite element method *Appl. Math. Model.* **35** 412–25

[100] Vo T P, Thai H T, Nguyen T K and Inam F 2014 Static and vibration analysis of functionally graded beams using refined shear deformation theory *Meccanica* **49** 155–68

[101] Jing L, Ming P, Zhang W, Fu L and Cao Y 2016 Static and free vibration analysis of functionally graded beams by combination Timoshenko theory and finite volume method *Compos. Struct.* **138** 192–213

[102] Kapuria S, Bhattacharyya M and Kumar A N 2008 Bending and free vibration response of layered functionally graded beams: a theoretical model and its experimental validation *Compos. Struct.* **82** 390–402

[103] Murin J, Aminbaghai M, Hrabovsky J, Gogola R and Kugler S 2016 Beam finite element for modal analysis of FGM structures *Eng. Struct.* **121** 1–18

[104] Murin J, Aminbaghai M, Hrabovský J, Kutiš V and Kugler S 2013 Modal analysis of the FGM beams with effect of the shear correction function *Composites* B **45** 1575–82

[105] Almeida C A, Albino J C R, Menezes I F M and Paulino G H 2011 Geometric nonlinear analyses of functionally graded beams using a tailored Lagrangian formulation *Mech. Res. Commun.* **38** 553–9

[106] Doroushi A, Eslami M R and Komeili A 2011 Vibration analysis and transient response of an FGPM beam under thermo-electro-mechanical loads using higher-order shear deformation theory *J. Intell. Mater. Sys. Struct.* **22** 231–43

[107] Shahba A, Attarnejad R and Hajilar S 2011 Free vibration and stability of axially functionally graded tapered Euler–Bernoulli beams *Shock Vib.* **18** 683–96

[108] Mohanty S C, Dash R R and Rout T 2012 Static and dynamic stability analysis of a functionally graded timoshenko beam *Int. J. Struct. Stab. Dyn.* **12** 1250025

[109] Kahya V and Turan M 2017 Finite element model for vibration and buckling of functionally graded beams based on the first-order shear deformation theory *Composites* B **109** 108–15

[110] Anandrao K S, Gupta R K, Ramchandran P and Rao G V 2012 Non-linear free vibrations and post-buckling analysis of shear flexible functionally graded beams *Struct. Eng. Mech.* **44** 339–61

[111] Anandrao K S, Gupta R K, Ramchandran P and Rao G V 2010 Thermal post-buckling analysis of uniform slender functionally graded material beams *Struct. Eng. Mech.* **36** 545–60

[112] Yang Y, Lam C C and Kou K P 2016 Forced vibration analysis of functionally graded beams by the meshfree boundary-domain integral equation method *Eng. Anal. Bound. Elem.* **72** 100–10

[113] Qian L and Ching H 2004 Static and dynamic analysis of 2-D functionally graded elasticity by using meshless local Petrov–Galerkin method *J. Chinese Inst. Eng.* **27** 491–503

[114] Lancaster P and Salkauskas K 1981 Surfaces generated by moving least squares methods *Math. Comput.* **37** 141–58

[115] Li S, Wan Z and Zhang J 2014 Free vibration of functionally graded beams based on both classical and first-order shear deformation beam theories *Appl. Math. Mech.* **35** 591–606

[116] Ma L S and Lee D W 2011 A further discussion of nonlinear mechanical behavior for FGM beams under in-plane thermal loading *Compos. Struct.* **93** 831–42

[117] Ma L S and Lee D W 2012 Exact solutions for nonlinear static responses of a shear deformable FGM beam under an in-plane thermal loading *Eur. J. Mech.* A **31** 13–20

[118] Sun Y, Li S R and Batra R C 2016 Thermal buckling and post-buckling of FGM Timoshenko beams on nonlinear elastic foundation *J. Therm. Stresses* **39** 11–26

[119] Şimşek M 2010 Non-linear vibration analysis of a functionally graded Timoshenko beam under action of a moving harmonic load *Compos. Struct.* **92** 2532–46

[120] Avcar M and Mohammed W K M 2018 Free vibration of functionally graded beams resting on Winkler–Pasternak foundation *Arab. J. Geosci.* **11** 232

[121] Şimşek M and Kocatürk T 2009 Free and forced vibration of a functionally graded beam subjected to a concentrated moving harmonic load *Compos. Struct.* **90** 465–73

[122] Shooshtari A and Rafiee M 2011 Nonlinear forced vibration analysis of clamped functionally graded beams *Acta Mech.* **221** 23–38

[123] Fatehi P and Farid M 2019 Piezoelectric energy harvesting from nonlinear vibrations of functionally graded beams: finite-element approach *J. Eng. Mech.* **145**

[124] Chen X, Lu Y, Zhu B, Zhang X and Li Y 2019 Nonlinear resonant behaviors of bi-directional functionally graded material microbeams: one-/two-parameter bifurcation analyses *Compos. Struct.* **223** 110896

[125] Şimşek M 2015 Bi-directional functionally graded materials (BDFGMs) for free and forced vibration of Timoshenko beams with various boundary conditions *Compos. Struct.* **133** 968–78

[126] Lü C F, Chen W Q, Xu R Q and Lim C W 2008 Semi-analytical elasticity solutions for bi-directional functionally graded beams *Int. J. Solids Struct.* **45** 258–75

[127] Tang Y, Wang T, Ma Z S and Yang T 2021 Magneto-electro-elastic modelling and nonlinear vibration analysis of bi-directional functionally graded beams *Nonlinear Dyn.* **105** 2195–227

[128] Witvrouw A and Mehta A 2005 The use of functionally graded poly-SiGe layers for MEMS applications *Mater. Sci. Forum.* *492–* **3** 255–60

[129] Rahaeifard M, Kahrobaiyan M H and Ahmadian M T 2009 Sensitivity analysis of atomic force microscope cantilever made of functionally graded materials *Proc. ASME Design Engineering Technical Conf., ASMEDC* pp 539–44

[130] Asghari M, Ahmadian M T, Kahrobaiyan M H and Rahaeifard M 2010 On the size-dependent behavior of functionally graded micro-beams *Mater. Des.* **31** 2324–9

[131] Al-Basyouni K S, Tounsi A and Mahmoud S R 2015 Size dependent bending and vibration analysis of functionally graded micro beams based on modified couple stress theory and neutral surface position *Compos. Struct.* **125** 621–30

[132] Li X, Li L, Hu Y, Ding Z and Deng W 2017 Bending, buckling and vibration of axially functionally graded beams based on nonlocal strain gradient theory *Compos. Struct.* **165** 250–65

[133] Liu D, Chen D, Yang J and Kitipornchai S 2021 Buckling and free vibration of axially functionally graded graphene reinforced nanocomposite beams *Eng. Struct.* **249** 113327

[134] Delale F and Erdogan F 1983 The crack problem for a nonhomogeneous plane *J. Appl. Mech. Trans. ASME.* **50** 609–14

[135] Yang J and Chen Y 2008 Free vibration and buckling analyses of functionally graded beams with edge cracks *Compos. Struct.* **83** 48–60

[136] Erdogan F and Wu B H 1997 The surface crack problem for a plate with functionally graded properties *J. Appl. Mech. Trans. ASME.* **64** 449–56

[137] Kitipornchai S, Ke L L, Yang J and Xiang Y 2009 Nonlinear vibration of edge cracked functionally graded Timoshenko beams *J. Sound Vib.* **324** 962–82

[138] Akbaş Ş D 2018 Forced vibration analysis of cracked nanobeams *J. Brazilian Soc. Mech. Sci. Eng.* **40** 39

[139] Trinh L C, Vo T P, Thai H-T and Nguyen T-K 2016 An analytical method for the vibration and buckling of functionally graded beams under mechanical and thermal loads *Composites* B **100** 152–63

[140] Fu Y, Wang J and Mao Y 2012 Nonlinear analysis of buckling, free vibration and dynamic stability for the piezoelectric functionally graded beams in thermal environment *Appl. Math. Model.* **36** 4324–40

[141] Şimşek M and Yurtcu H H 2013 Analytical solutions for bending and buckling of functionally graded nanobeams based on the nonlocal Timoshenko beam theory *Compos. Struct.* **97** 378–86

[142] Torki M E and Reddy J N 2016 Buckling of functionally-graded beams with partially delaminated piezoelectric layers *Int. J. Struct. Stab. Dyn.* **16** 1450104

[143] Jha D K, Kant T and Singh R K 2013 A critical review of recent research on functionally graded plates *Compos. Struct.* **96** 833–49

[144] Abrate S 2008 Functionally graded plates behave like homogeneous plates *Composites* B **39** 151–8

[145] Reissner E 1944 On the theory of bending of elastic plates *J. Math. Phys.* **23** 184–91

[146] Reissner E 1945 The effect of transverse shear deformation on the bending of elastic plates *J. Appl. Mech.* **12** A69–77

[147] Mindlin R D 1951 Influence of rotatory inertia and shear on flexural motions of isotropic, elastic plates *J. Appl. Mech.* **18** 31–8

[148] Ghugal Y M and Shimpi R P 2002 A review of refined shear deformation theories of isotropic and anisotropic laminated plates *J. Reinf. Plast. Compos.* **21** 775–813

[149] Khandan R, Noroozi S, Sewell P and Vinney J 2012 The development of laminated composite plate theories: a review *J. Mater. Sci.* **47** 5901–10

[150] Levinson M 1980 An accurate, simple theory of the statics and dynamics of elastic plates *Mech. Res. Commun.* **7** 343–50

[151] Murthy M V V 1981 *Improved transverse shear deformation theory for laminated anisotropic plates* 1903 NASA https://ntrs.nasa.gov/api/citations/19820003615/downloads/19820003615.pdf

[152] Levy M 1877 Mémoire sur la théorie des plaques élastiques planes *J. Math. Pures Appl.* **3** 219–306

[153] Stein M 1986 Nonlinear theory for plates and shells including the effects of transverse shearing *AIAA J.* **24** 1537–44

[154] Mantari J L, Oktem A S and Guedes Soares C 2012 A new trigonometric shear deformation theory for isotropic, laminated composite and sandwich plates *Int. J. Solids Struct.* **49** 43–53

[155] Yang J and Shen H-S 2001 Dynamic response of initially stressed functionally graded rectangular thin plates *Compos. Struct.* **54** 497–508

[156] Javaheri R and Eslami M R 2002 Buckling of functionally graded plates under in-plane compressive loading *Z. Angew. Math. Mech.* **82** 277–83

[157] Nguyen T K, Sab K and Bonnet G 2008 First-order shear deformation plate models for functionally graded materials *Compos. Struct.* **83** 25–36

[158] Hosseini-Hashemi S, Rokni Damavandi Taher H, Akhavan H and Omidi M 2010 Free vibration of functionally graded rectangular plates using first-order shear deformation plate theory *Appl. Math. Model.* **34** 1276–91

[159] Thai H T, Nguyen T K, Vo T P and Lee J 2014 Analysis of functionally graded sandwich plates using a new first-order shear deformation theory *Eur. J. Mech.* A **45** 211–25

[160] Bellifa H, Benrahou K H, Hadji L, Houari M S A and Tounsi A 2016 Bending and free vibration analysis of functionally graded plates using a simple shear deformation theory and the concept the neutral surface position *J. Braz. Soc. Mech. Sci. Eng.* **38** 265–75

[161] Nguyen T K, Sab K and Bonnet G 2007 Shear correction factors for functionally graded plates *Mech. Adv. Mater. Struct.* **14** 567–75

[162] Ferreira A J M, Batra R C, Roque C M C, Qian L F and Martins P A L S 2005 Static analysis of functionally graded plates using third-order shear deformation theory and a meshless method *Compos. Struct.* **69** 449–57

[163] Reddy J N and Kim J 2012 A nonlinear modified couple stress-based third-order theory of functionally graded plates *Compos. Struct.* **94** 1128–43

[164] Bui T Q, Van Do T, Ton L H T, Doan D H, Tanaka S, Pham D T, Nguyen-Van T A, Yu T and Hirose S 2016 On the high temperature mechanical behaviors analysis of heated functionally graded plates using FEM and a new third-order shear deformation plate theory *Composites* B **92** 218–41

[165] Tounsi A, Houari M S A, Benyoucef S and Bedia E A A 2013 A refined trigonometric shear deformation theory for thermoelastic bending of functionally graded sandwich plates *Aerosp. Sci. Technol.* **24** 209–20

[166] Belabed Z, Ahmed Houari M S, Tounsi A, Mahmoud S R and Anwar Bég O 2014 An efficient and simple higher order shear and normal deformation theory for functionally graded material (FGM) plates *Composites* B **60** 274–83

[167] Zaoui F Z, Ouinas D and Tounsi A 2019 New 2D and quasi-3D shear deformation theories for free vibration of functionally graded plates on elastic foundations *Composites* B **159** 231–47
[168] Chakraverty S and Pradhan K K 2016 *Vibration of Functionally Graded Beams and Plates* 1st edn (Waltham, MA: Elsevier)
[169] Vel S S and Batra R C 2004 Three-dimensional exact solution for the vibration of functionally graded rectangular plates *J. Sound Vib.* **272** 703–30
[170] Baferani A H, Saidi A R and Jomehzadeh E 2011 An exact solution for free vibration of thin functionally graded rectangular plates *Proc. Inst. Mech. Eng.* C **225** 526–36
[171] Sayyad A S and Ghugal Y M 2021 Static and free vibration analysis of doubly-curved functionally graded material shells *Compos. Struct.* **269** 114045
[172] Wang Y Q and Zu J W 2017 Vibration behaviors of functionally graded rectangular plates with porosities and moving in thermal environment *Aerosp. Sci. Technol.* **69** 550–62
[173] Woo J, Meguid S A and Ong L S 2006 Nonlinear free vibration behavior of functionally graded plates *J. Sound Vib.* **289** 595–611
[174] Fallah A, Aghdam M M and Kargarnovin M H 2013 Free vibration analysis of moderately thick functionally graded plates on elastic foundation using the extended Kantorovich method *Arch. Appl. Mech.* **83** 177–91
[175] Thai H T and Choi D H 2012 An efficient and simple refined theory for buckling analysis of functionally graded plates *Appl. Math. Model.* **36** 1008–22
[176] Yaghoobi H and Torabi M 2013 Exact solution for thermal buckling of functionally graded plates resting on elastic foundations with various boundary conditions *J. Therm. Stresses* **36** 869–94
[177] Javaheri R and Eslami M R 2002 Thermal buckling of functionally graded plates *AIAA J.* **40** 162–9
[178] Nie G J and Zhong Z 2007 Semi-analytical solution for three-dimensional vibration of functionally graded circular plates *Comput. Methods Appl. Mech. Eng.* **196** 4901–10
[179] Huang Z Y, Lü C F and Chen W Q 2008 Benchmark solutions for functionally graded thick plates resting on Winkler–Pasternak elastic foundations *Compos. Struct.* **85** 95–104
[180] Pan E and Han F 2005 Exact solution for functionally graded and layered magneto-electro-elastic plates *Int. J. Eng. Sci.* **43** 321–39
[181] Jin G, Su Z, Shi S, Ye T and Gao S 2014 Three-dimensional exact solution for the free vibration of arbitrarily thick functionally graded rectangular plates with general boundary conditions *Compos. Struct.* **108** 565–77
[182] Zhao J, Choe K, Xie F, Wang A, Shuai C and Wang Q 2018 Three-dimensional exact solution for vibration analysis of thick functionally graded porous (FGP) rectangular plates with arbitrary boundary conditions *Composites* B **155** 369–81
[183] Zhao X, Lee Y Y and Liew K M 2009 Mechanical and thermal buckling analysis of functionally graded plates *Compos. Struct.* **90** 161–71
[184] Zhang L W, Zhu P and Liew K M 2014 Thermal buckling of functionally graded plates using a local Kriging meshless method *Compos. Struct.* **108** 472–92
[185] Yang J and Shen H S 2002 Vibration characteristics and transient response of shear-deformable functionally graded plates in thermal environments *J. Sound Vib.* **255** 579–602
[186] Wu H, Zhu J, Kitipornchai S, Wang Q, Ke L L and Yang J 2020 Large amplitude vibration of functionally graded graphene nanocomposite annular plates in thermal environments *Compos. Struct.* **239** 112047

[187] Nie G J and Zhong Z 2008 Vibration analysis of functionally graded annular sectorial plates with simply supported radial edges *Compos. Struct.* **84** 167–76

[188] Trabelsi S, Frikha A, Zghal S and Dammak F 2019 A modified FSDT-based four nodes finite shell element for thermal buckling analysis of functionally graded plates and cylindrical shells *Eng. Struct.* **178** 444–59

[189] Vu Hoai N, Hong Doan D, Khoa N M, Van Do T and Thi Tran H 2019 Phase-field buckling analysis of cracked stiffened functionally graded plates *Compos. Struct.* **217** 50–9

[190] Li S R, Wang X and Batra R C 2015 Correspondence relations between deflection, buckling load, and frequencies of thin functionally graded material plates and those of corresponding homogeneous plates *J. Appl. Mech.* **82** 111006

[191] Zhao J, Wang Q, Deng X, Choe K, Zhong R and Shuai C 2019 Free vibrations of functionally graded porous rectangular plate with uniform elastic boundary conditions *Composites* B **168** 106–20

[192] Swaminathan K and Sangeetha D M 2017 Thermal analysis of FGM plates—a critical review of various modeling techniques and solution methods *Compos. Struct.* **160** 43–60

[193] Shafiei N and She G L 2018 On vibration of functionally graded nano-tubes in the thermal environment *Int. J. Eng. Sci.* **133** 84–98

[194] Li S R and Ma H K 2020 Analysis of free vibration of functionally graded material micro-plates with thermoelastic damping *Arch. Appl. Mech.* **90** 1285–304

[195] Li S-R, Xiang Y and Shen H S 2021 Modelling and evaluation of thermoelastic damping of FGM micro plates based on the Levinson plate theory *Compos. Struct.* **278** 114684

[196] Pandey S and Pradyumna S 2017 A finite element formulation for thermally induced vibrations of functionally graded material sandwich plates and shell panels *Compos. Struct.* **160** 877–86

[197] Abrate S 2006 Free vibration, buckling, and static deflections of functionally graded plates *Compos. Sci. Technol.* **66** 2383–94

[198] Nguyen H N, Tan T C, Luat D T, Phan V D, Van Thom D and Van Minh P 2019 Research on the buckling behavior of functionally graded plates with stiffeners based on the third-order shear deformation theory *Materials (Basel)* **12** 1262

[199] Javaheri R and Eslami M R 2002 Thermal buckling of functionally graded plates based on higher order theory *J. Therm. Stresses* **25** 603–25

[200] Shariat B A S and Eslami M R 2007 Buckling of thick functionally graded plates under mechanical and thermal loads *Compos. Struct.* **78** 433–9

[201] Bouiadjra M B, Ahmed Houari M S and Tounsi A 2012 Thermal buckling of functionally graded plates according to a four-variable refined plate theory *J. Therm. Stresses* **35** 677–94

[202] Park J S and Kim J H 2006 Thermal postbuckling and vibration analyses of functionally graded plates *J. Sound Vib.* **289** 77–93

IOP Publishing

Functionally Graded Structures
Modelling and computation of static and dynamical problems
Snehashish Chakraverty, Subrat Kumar Jena and Ömer Civalek

Chapter 8

Free vibration analysis of porous functionally graded sandwich beams

Mehmet Avcar, Lazreg Hadji and Ömer Civalek

8.1 Introduction

For millennia, composite materials (CMs) have been utilized to build superior constructions by mixing two or more components with different characteristics without dissolving or merging them. For example, ancient dwellings were created thousands of years ago using composite materials such as mud bricks reinforced with straw. Such materials are now used as specialist materials when traditional materials are unsuitable or their attributes need to be improved, since the use of CMs gives engineers numerous ways to determine the final composite's properties during the manufacturing process [1].

Sandwich composites are composite materials made of two strong upper and lower surface layers sandwiching a thick yet very light core material. Fairbairn [2] introduced the idea of sandwich structures in the middle of the 19th century. Due to the many sandwich constructions already in use, such as those of the Mosquito night bomber, sandwich structures assumed significant relevance during World War II [3–6].

Functionally graded materials (FGMs) are novel kinds of composite materials composed of two components, a ceramic and a metal. Such materials exhibit a progressive material property alteration in the thickness direction. In FGMs, the ceramic acts as a heat-resistant material, while the metal provides structural stiffness. FGMs are employed in various sectors, including aerospace, nuclear applications, and heat exchangers, due to their tolerance to high temperatures [7–15].

The use of FGMs in sandwich structures is becoming more common due to their ability to smoothly, gradually, and continuously change the material properties from one surface to another. As a result, sandwich structures can avoid delamination failure, matrix cracks, and increased stress levels [16–19].

Porous materials are a class of materials distinguished by their high porosity, low density, and large specific surface area, such as foams. The manufacture of FGM

doi:10.1088/978-0-7503-5301-4ch8

specimens through a multistep sequential infiltration technique combined with certain technical problems, such as a significant difference in the solidification temperatures between the material components during sintering, can lead to porosity in FGMs, noticeably altering their mechanical properties [20, 21]. Therefore, to guarantee the safe design of functionally graded (FG) structures, it is essential to investigate the impact of porosity on FGMs. In addition, functionally graded porous materials with continuous changes in microstructure and porosity in a given direction(s) can be produced by applying the FGM concept to porous materials. Therefore, it is possible to modify their material qualities to match the construction requirements for their use. Because of its significance, the study of the mechanical characteristics of sandwich beams constructed of porous (P)-FGMs has attracted several researchers' interest [22–27].

This chapter discusses the free vibration analysis of porous FG sandwich beams (FGSBs) consisting of two porous face sheets made of FGMs and a homogenous isotropic metal core. For the purposes of this discussion, the material characteristics of the face sheets of sandwich beams are presumed to vary continuously in the thickness direction, as indicated by the volume fractions of the constituents specified using the modified rule of mixtures, which include porosity volume fractions with different types of porosity distribution pattern across the cross section. A higher-order shear deformation theory, i.e. trigonometric shear deformation theory (TrSDT), which does not need shear correction factors, is used. The equations of motion are derived by employing Hamilton's principle. For simply supported boundary conditions at both ends, Navier's method is used to analytically obtain results for the free vibration of porous/nonporous (P/NP) FG sandwich beams. To validate this formulation, comparisons are made with results published in the research literature. Numerical examples are presented to illustrate the influences of the skin–core–skin thickness ratio, power-law index, span-to-depth ratio, porosity distribution pattern, and porosity volume fraction on the natural frequencies of FG sandwich beams.

8.2 The fundamental relations of FGSBs

Figure 8.1 shows the geometry of a P-FGSB that has porous FG face sheets and a homogeneous isotropic metal core, whose axes are defined to run along the length, width, and height of the FGSB, in that order. The FGSB is made up of three elastic layers from bottom to top, namely 'Layer 1,' 'Layer 2,' and 'Layer 3.' The two interfaces between the layers and the bottom and top vertical positions are respectively denoted by $h_1 = -h/2$, h_2, h_3, $h_4 = +h/2$. The skin–core–skin thickness ratio (SCSTR), which describes the thickness of each layer from bottom to top, is denoted by a combination of three numbers, such as 1-8-1, 2-1-2, and so on, due to the necessity for simplicity.

The layer schemes of FGSBs are listed by SCSTR in table 8.1.

In this study, four different porosity distribution patterns (PDPs), i.e. H-PDP, O-PDP, V-PDP, and X-PDP, are considered, as shown in figure 8.2.

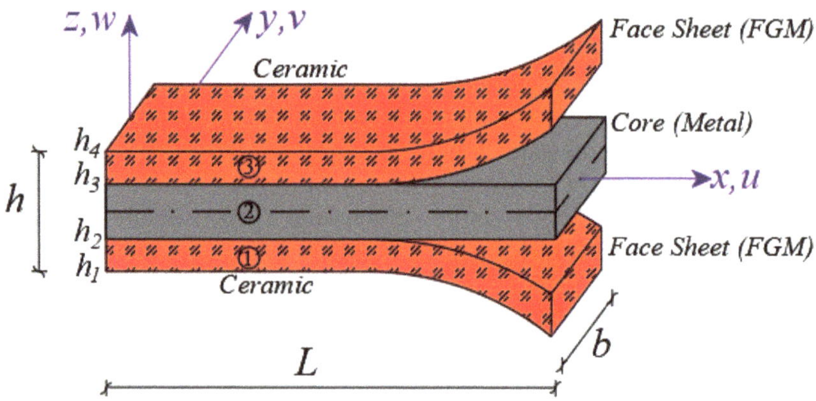

Figure 8.1. The geometry of a P-FGSB.

Table 8.1. The layer schemes of FGSBs.

Layer scheme	h_1	h_2	h_3	h_4
1-0-1	$-h/2$	0	0	$+h/2$
2-1-2	$-h/2$	$-h/10$	$+h/10$	$+h/2$
2-1-1	$-h/2$	0	$+h/4$	$+h/2$
1-1-1	$-h/2$	$-h/6$	$+h/6$	$+h/2$
2-2-1	$-h/2$	$-h/10$	$+3h/10$	$+h/2$
1-2-1	$-h/2$	$-h/4$	$+h/4$	$+h/2$
1-8-1	$-h/2$	$-4h/10$	$+4h/10$	$+h/2$

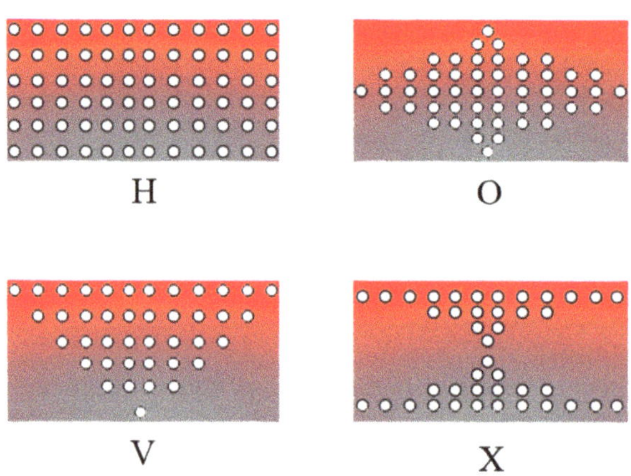

Figure 8.2. The PDPs of P-FGSBs.

It is presumed that the power-law function in the thickness direction governs the volume fractions of the FGMs, as follows:

$$V^{(1)}(z) = \left(\frac{z - h_1}{h_2 - h_1}\right)^p z \in [h_1, h_2] \tag{8.1a}$$

$$V^{(2)}(x, z) = 1, \quad z \in [h_2, h_3] \tag{8.1b}$$

$$V^{(3)}(x, z) = \left(\frac{z - h_4}{h_3 - h_4}\right)^p, \; z \in [h_3, h_4]. \tag{8.1c}$$

Here, $V^{(n)}(z)$ is the volume fraction function of the nth layer and can be defined in terms of the volume fraction index p $(0 \leqslant p \leqslant \infty)$ in the thickness direction of the P-FGSBs.

The following are the material's effective characteristics for each layer: for H-PDP:

$$P^{(1)}(z) = (P_m - P_c)V^{(1)}(z) + P_c - \frac{\alpha}{2}(P_m + P_c)$$

$$P^{(2)}(z) = (P_m - P_c)V^{(2)}(z) + P_c$$

$$P^{(3)}(z) = (P_m - P_c)V^{(3)}(z) + P_c - \frac{\alpha}{2}(P_m + P_c) \tag{8.2}$$

For O-PDP

$$P^{(1)}(z) = (P_m - P_c)V^{(1)}(z) + P_c - \frac{\alpha}{2}(P_m + P_c)\sqrt{\left(\frac{h_1 - h_2}{2}\right)^2 - \left(z - \left(\frac{h_1 + h_2}{2}\right)\right)^2}$$

$$P^{(2)}(z) = (P_m - P_c)V^{(2)}(z) + P_c$$

$$\begin{aligned} P^{(3)}(z) &= (P_m - P_c)V^{(3)}(z) + P_c \\ &- \frac{\alpha}{2}(P_m + P_c)\sqrt{\left(\frac{h_3 - h_4}{2}\right)^2 - \left(z - \left(\frac{h_3 + h_4}{2}\right)\right)^2} \end{aligned} \tag{8.3}$$

For V-PDP:

$$P^{(1)}(z) = (P_m - P_c)V^{(1)}(z) + P_c - \frac{\alpha}{2}(P_m + P_c)\left(\frac{z - h_2}{h_1 - h_2}\right)$$

$$P^{(2)}(z) = (P_m - P_c)V^{(2)}(z) + P_c$$

$$P^{(3)}(z) = (P_m - P_c)V^{(3)}(z) + P_c - \frac{\alpha}{2}(P_m + P_c)\left(\frac{z - h_4}{h_3 - h_4}\right) \tag{8.4}$$

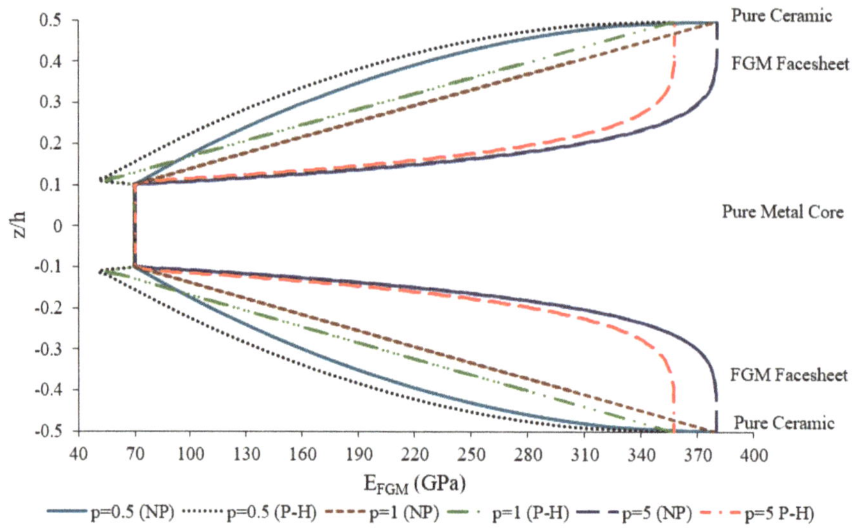

Figure 8.3. The variation of the Young's modulus of an FGSB that has an SCSTR of 2-1-2 versus the power-law index.

For X-PDP:

$$P^{(1)}(z) = (P_m - P_c)V^{(1)}(z) + P_c - \frac{\alpha}{2}(P_m + P_c)\left|\frac{2z - (h_1 + h_2)}{h_1 - h_2}\right|$$

$$P^{(2)}(z) = (P_m - P_c)V^{(2)}(z) + P_c$$

$$P^{(3)}(z) = (P_m - P_c)V^{(3)}(z) + P_c - \frac{\alpha}{2}(P_m + P_c)\left|\frac{2z - (h_3 + h_4)}{h_3 - h_4}\right|, \quad (8.5)$$

where α ($\alpha < 1$) expresses the porosity volume fraction and $P^{(i)}i = 1, 2, 3$ are the material features of the layers. In this chapter, Poisson's ratio (ν) is assumed to remain constant, since its influence has been determined to be negligible [28, 29]. Unless otherwise specified, it is assumed that the FGSBs are made of Al (metal) and Al_2O_3 (ceramic), whose material properties are: $E_m = 70$ GPa, $\nu_m = 0.3$, $\rho_m = 2702$ kg m^{-3}; $E_c = 380$ GPa, $\nu_c = 0.3$, $\rho_c = 3960$ kg m^{-3}. Figure 8.3 illustrates the variation of the Young's modulus of P-FGSBs with an SCSTR of 2-1-2 versus the power-law index for $\alpha = 0.1$ and H-PDP.

8.3 The governing equations of FGSBs

The following assumptions are made in this section:

- The Cartesian coordinate system begins at the middle surface of the FGSBs.
- Since the displacements relative to the height of the FGSB are insignificant, the stresses involved are also assumed to be negligible.
- The transverse displacements are composed of bending w_b and shear w_s components, which are only impacted by x and t:

$$w(x, z, t) = w_b(x, t) + w_s(x, t). \tag{8.6}$$

The transverse normal stress, σ_z, is negligible in contrast to in-plane stress, σ_x, and the axial displacement in the x-direction is composed of extension, bending, and shear components, as follows:

$$u = u_0 + u_b + u_s. \tag{8.7}$$

The classical beam theory is used to obtain the displacements of the bending component w_b and thus u_b is given by

$$u_b = -z\frac{\partial w_b}{\partial x}. \tag{8.8}$$

The shear stress (τ_{xz}) is zero at the top and bottom faces of the FGSB because the shear component (u_s) creates a hyperbolic variation in shear strain (γ_{xz}) when combined with (w_s). Consequently, u_s is given by

$$u_s = -f(z)\frac{\partial w_s}{\partial x}, \tag{8.9}$$

where $f(z)$ is specified as a trigonometric function [30]:

$$f(z) = z - \frac{h}{\pi}\sin\left(\frac{\pi z}{h}\right). \tag{8.10}$$

As a result of the assumptions outlined in the preceding section, equations (8.6)–(8.10) may be used to determine the displacement field of the TrSDT [31]:

$$u(x, z, t) = u_0(x, t) - z\frac{\partial w_b}{\partial x} - f(z)\frac{\partial w_s}{\partial x} \tag{8.11a}$$

$$w(x, z, t) = w_b(x, t) + w_s(x, t). \tag{8.11b}$$

The displacements in equations (8.11a) and (8.11b) are connected to the strains shown below:

$$\varepsilon_x = \varepsilon_x^0 + z\; k_x^b + f(z)k_x^s \tag{8.12a}$$

$$\gamma_{xz} = g(z)\gamma_{xz}^s. \tag{8.12b}$$

Here

$$\varepsilon_x^0 = \frac{\partial u_0}{\partial x},\quad k_x^b = -\frac{\partial^2 w_b}{\partial x^2},\; k_x^s = -\frac{\partial^2 w_s}{\partial x^2},\quad \gamma_{xz}^s = \frac{\partial w_s}{\partial x} \tag{8.12c}$$

$$g(z) = 1 - \frac{df(z)}{dz}. \tag{8.12d}$$

If we consider Hooke's law, the stresses become

$$\sigma_x = Q_{11}(z)\ \varepsilon_x \text{ and } \tau_{xz} = Q_{55}(z)\gamma_{xz}, \tag{8.13a}$$

where

$$Q_{11}(z) = E(z) \text{ and } Q_{55}(z) = \frac{E(z)}{2(1+\nu)}. \tag{8.13b}$$

The equations of motion are analytically derived through the use of Hamilton's principle [32]:

$$\int_{t_1}^{t_2} (\delta U - \delta K)dt = 0. \tag{8.14}$$

Here, δU corresponds to the virtual variation of the strain energy and δK denotes the virtual variation of the kinetic energy.

The strain energy change of the FGSB may be described as follows:

$$\begin{aligned}\delta\ U &= \int_0^L \int_{-\frac{h}{2}}^{\frac{h}{2}} (\sigma_x\delta\ \varepsilon_x + \tau_{xz}\delta\ \gamma_{xz})dzdx \\ &= \int_0^L \left(N_x\frac{d\delta\ u_0}{dx} - M_x^{\mathrm{b}}\frac{d^2\delta\ w_{\mathrm{b}}}{dx^2} - M_x^{\mathrm{s}}\frac{d^2\delta\ w_{\mathrm{s}}}{dx^2} + Q_{xz}\frac{d\delta\ w_{\mathrm{s}}}{dx}\right)dx,\end{aligned} \tag{8.15}$$

where N_x, M_x^{b}, M_x^{s}, and Q_{xz} are the stress resultants and are specified by

$$(N_x, M_x^{\mathrm{b}}, M_x^{\mathrm{s}}) = \int_{-\frac{h}{2}}^{\frac{h}{2}} (1, z, f(z))\sigma_x dz \text{ and } Q_{xz} = \int_{-\frac{h}{2}}^{\frac{h}{2}} g(z)\tau_{xz}dz. \tag{8.16}$$

The kinetic energy change of the FGSB may be described by

$$\begin{aligned}\delta\ K &= \int_0^L \int_{-\frac{h}{2}}^{\frac{h}{2}} \rho(z)[\dot{u}\delta\ \dot{u} + \dot{w}\delta\ \dot{w}]\ dz_{ns}dx \\ &= \int_0^L \Bigg\{ I_0[\dot{u}_0\delta\ \dot{u}_0 + (\dot{w}_{\mathrm{b}} + \dot{w}_{\mathrm{s}})(\delta\ \dot{w}_{\mathrm{b}} + \delta\ \dot{w}_{\mathrm{s}})] \\ &\quad - I_1\left(\dot{u}_0\frac{d\delta\ \dot{w}_{\mathrm{b}}}{dx} + \frac{d\ \dot{w}_{\mathrm{b}}}{dx}\delta\ \dot{u}_0\right) + I_2\left(\frac{d\dot{w}_{\mathrm{b}}}{dx}\frac{d\delta\ \dot{w}_{\mathrm{b}}}{dx}\right) \\ &\quad - J_1\left(\dot{u}_0\frac{d\delta\ \dot{w}_{\mathrm{s}}}{dx} + \frac{d\ \dot{w}_{\mathrm{s}}}{dx}\delta\ \dot{u}_0\right) \\ &\quad + K_2\left(\frac{d\dot{w}_{\mathrm{s}}}{dx}\frac{d\delta\ \dot{w}_{\mathrm{s}}}{dx}\right) + J_2\left(\frac{d\dot{w}_{\mathrm{b}}}{dx}\frac{d\delta\ \dot{w}_{\mathrm{s}}}{dx} + \frac{d\dot{w}_{\mathrm{s}}}{dx}\frac{d\delta\ \dot{w}_{\mathrm{b}}}{dx}\right)\Bigg\}dx.\end{aligned} \tag{8.17}$$

where the dot-superscript convention denotes differentiation with respect to the time variable t, $\rho(z)$ is the mass density, and $(I_0, I_1, J_1, I_2, J_2, K_2)$ are the mass inertias given by

$$(I_0, I_1, J_1, I_2, J_2, K_2) = \int_{-\frac{h}{2}}^{\frac{h}{2}} (1, z, f, z^2, zf, f^2)\rho(z)dz. \tag{8.18}$$

The equations of motion are obtained by substituting the expressions for δU and δT from equations (8.15), (8.17), and (8.18) into equation (8.14), integrating the displacements by parts, and setting the coefficients of δu_0, δw_b, and δw_s to zero separately:

$$\delta\ u_0\colon \frac{dN_x}{dx} = I_0\ddot{u}_0 - I_1\frac{d\ddot{w}_b}{dx} - J_1\frac{d\ddot{w}_s}{dx} \tag{8.19a}$$

$$\delta w_b\colon \frac{d^2M_b}{dx^2} = I_0(\ddot{w}_b + \ddot{w}_s) + I_1\frac{d\ddot{u}_0}{dx} - I_2\frac{d^2\ddot{w}_b}{dx^2} - J_2\frac{d^2\ddot{w}_s}{dx^2} \tag{8.19b}$$

$$\delta w_s\colon \frac{d^2M_s}{dx^2} + \frac{dQ_{xz}}{dx} = I_0(\ddot{w}_b + \ddot{w}_s) + J_1\frac{d\ddot{u}_0}{dx} - J_2\frac{d^2\ddot{w}_b}{dx^2} - K_2\frac{d^2\ddot{w}_s}{dx^2}. \tag{8.19c}$$

If we substitute equation (8.16) into equation (8.20), the equations of motion may be written as displacements (u_0,w_b,w_s), and the corresponding equations have the following structure:

$$A_{11}\frac{\partial^2 u_0}{\partial x^2} - B_{11}\frac{\partial^3 w_b}{\partial x^3} - B_{11}^s\frac{\partial^3 w_s}{\partial x^3} = I_0\ddot{u}_0 - I_1\frac{d\ddot{w}_b}{dx} - J_1\frac{d\ddot{w}_s}{dx} \tag{8.20a}$$

$$B_{11}\frac{\partial^3 u_0}{\partial x^3} - D_{11}\frac{\partial^4 w_b}{\partial x^4} - D_{11}^s\frac{\partial^4 w_s}{\partial x^4} = I_0(\ddot{w}_b + \ddot{w}_s) + I_1\frac{d\ddot{u}_0}{dx} - I_2\frac{d^2\ddot{w}_b}{dx^2} - J_2\frac{d^2\ddot{w}_s}{dx^2} \tag{8.20b}$$

$$\begin{aligned} &B_{11}^s\frac{\partial^3 u_0}{\partial x^3} - D_{11}^s\frac{\partial^4 w_b}{\partial x^4} - H_{11}^s\frac{\partial^4 w_s}{\partial x^4} + A_{55}^s\frac{\partial^2 w_s}{\partial x^2} \\ &= I_0(\ddot{w}_b + \ddot{w}_s) + J_1\frac{d\ddot{u}_0}{dx} - J_2\frac{d^2\ddot{w}_b}{dx^2} - K_2\frac{d^2\ddot{w}_s}{dx^2}, \end{aligned} \tag{8.20c}$$

where A_{11}, D_{11}, etc. are the stiffnesses of the FGSB and are defined by:

$$\begin{aligned} &\left(A_{ij}, A_{ij}^s, B_{ij}, D_{ij}, B_{ij}^s, D_{ij}^s, H_{ij}^s\right) \\ &= \int_{-\frac{h}{2}}^{\frac{h}{2}} Q_{ij}(1, g^2(z), z, z^2, f(z), z\ f(z), f^2(z))dz. \end{aligned} \tag{8.21}$$

8.4 An analytical solution

The solution of the equation of motion for FGSBs with simply supported ends can be found using Navier's method as follows:

$$w_b = w_s = 0 \ \text{at} \ x = 0, L. \tag{8.22}$$

The following definitions apply to the displacements that are appropriate for the simply supported ends:

$$\begin{Bmatrix} u_0 \\ w_\text{b} \\ w_\text{s} \end{Bmatrix} = \begin{Bmatrix} U_m X_m' e^{i\omega t} \\ W_{\text{b}m} X_m e^{i\omega t} \\ W_{\text{s}m} X_m e^{i\omega t} \end{Bmatrix}, \tag{8.23}$$

where ω is the eigenfrequency associated with the mth eigenmode, and U_m, $W_{\text{b}m}$, and $W_{\text{s}m}$ are random parameters to be found.

The admissible function that satisfies simply supported ends is as follows [1]:

$$X_m(x) = \sin(\beta_m x),\ \beta_m = m\frac{\pi}{L},\ m = 1, 2, \ldots. \tag{8.24}$$

The solution is found by substituting equation (8.23) into equations (8.20a)–(8.20d):

$$\left(\begin{bmatrix} a_{11}\ a_{12}\ a_{13} \\ a_{21}\ a_{22}\ a_{23} \\ a_{31}\ a_{32}\ a_{33} \end{bmatrix} - \omega^2 \begin{bmatrix} m_{11}\ m_{12}\ m_{13} \\ m_{21}\ m_{22}\ m_{23} \\ m_{31}\ m_{32}\ m_{33} \end{bmatrix}\right) \begin{Bmatrix} U_m \\ W_{\text{b}m} \\ W_{\text{s}m} \end{Bmatrix} = \begin{Bmatrix} 0 \\ 0 \\ 0 \end{Bmatrix}, \tag{8.25}$$

where

$$a_{11} = \int_0^L A_{11} X_m'' X_m' dx,\ a_{12} = \int_0^L -B_{11} X_m'' X_m' dx,\ a_{13} = \int_0^L -B_{11}^s X_m'' X_m' dx,$$

$$a_{21} = \int_0^L B_{11} X_m''' X_m dx,\quad a_{22} = \int_0^L (-D_{11} X_m''') X_m dx,$$

$$a_{23} = \int_0^L (-D_{11}^s X_m'') X_m dx,$$

$$a_{31} = \int_0^L -B_{11}^s X_m''' X_m dx,$$

$$a_{32} = \int_0^L (-D_{11}^s X_m''') X_m dx$$

$$a_{33} = \int_0^L (-H_{11}^s X_m''' + A_{55}^s X_m'') X_m dx \tag{8.26}$$

$$m_{11} = -I_0 \int_0^L X' X' dx,\quad m_{12} = I_1 \int_0^L X' X' dx,\ m_{13} = J_1 \int_0^L X' X' dx$$

$$m_{21} = -I_1 \int_0^L X'' X dx,\ m_{22} = -\int_0^L (I_0 X - I_2 X'') X dx,\ m_{23} = -\int_0^L (I_0 X + J_2 X'') X dx$$

$$m_{31} = -J_1\int_0^L X''Xdx,\ m_{32} = -\int_0^L (I_0X - J_2X'')Xdx,\ m_{33} = -\int_0^L (I_0X - K_2X'')Xdx. \tag{8.27}$$

8.5 Numerical results and discussion

This section utilizes numerical examples to illustrate the current topic. To begin, comparisons are made to establish the validity of the formulation given in this chapter. The effects of the porosity, number of modes, material property changes, and geometrical characteristics on the nondimensional natural frequencies (NDNFs) of P- and/or NP-FGSBs are next examined through the use of parametric research. Note that NDNF is defined as: $\varpi = \frac{\omega L^2}{h}\sqrt{\frac{\rho_m}{E_m}}$.

Table 8.2 compares the results for the variation of the nondimensional fundamental natural frequencies (NDFNFs) of NP-FGSBs versus the power-law index (p) with those of [33, 34]. As shown in table 8.2, the findings obtained using the method outlined in this chapter are consistent with the results of both investigations.

Table 8.3 compares the variation of the NDFNFs of P-FGBs versus the power-law index (p) with those of [35, 36]. The results of both the published research and the findings of this chapter agree, as indicated in table 8.3.

Table 8.4 shows the variation of the NDNFs of NP/P-FGSBs versus the first three mode numbers (MNs) for $L/h = 5,\ p = 1,\ \alpha = 0.1$, and H-PDP. It can be seen that the NDNFs rise in correlation with the MN. However, when porosity is considered, especially in the fundamental mode, the impact of mode number variation on the NDNFs of FGSBs becomes less noticeable.

Table 8.5 describes the variation of the NDFNFs of NP/P-FGSBs versus the porosity volume fraction (α) for $L/h = 5,\ p = 1$, and H-PDP. It shows that the effect of porosity increases with the growth of α. On the other hand, the NDFNFs of FGSBs change according to the SCSTR with an increase in α, i.e. except in the case of a FGSB without a soft core (1-0-1), the values of the NDFNFs decrease with increasing α. However, the effect of the variation of α on the NDFNFs strongly depends on the SCSTR.

The change of the NDFNFs of NP/P-FGSBs versus the power-law index (p) for $L/h = 5,\ \alpha = 0.1$ and H-PDP is reported in table 8.6. It shows that the DFNFs rise as p increases. The impact of the variation of the NDFNFs of FGSBs changes according to the SCSTR. In addition, considering the porosity increases the impact of the variation on the NDFNFs of FGSBs.

The change in the NDFNFs of NP/P-FGSBs versus the span-to-depth ratio for $p = 1,\ \alpha = 0.1$ and H-PDP is reported in table 8.7. It can be seen that the NDFNFs increase as L/h increases. The effect of the change in L/h on NDFNFs of FGSBs changes according to the SCSTR. In addition, when porosity is considered, the impact of the change in L/h on the NDFNFs of FGSBs is slightly diminished.

Table 8.2. The variation of the NDNFs of NP-FGSBs versus the power-law index.

L/h	p	Results	1-0-1	2-1-2	2-1-1	1-1-1	2-2-1	1-2-1
5	0	This chapter	2.6775	2.6775	2.6775	2.6775	2.6775	2.6775
		Nguyen and Nguyen [33]	2.6773	2.6773	2.6773	2.6773	2.6773	2.6773
		Vo *et al* [34]	2.6773	2.6773	2.6773	2.6773	2.6773	2.6773
	0.5	This chapter	4.4371	4.2973	4.1947	4.1773	4.0495	3.9890
		Nguyen and Nguyen [33]	4.4437	4.3052	4.1998	4.1844	4.0549	3.9926
		Vo *et al* [34]	4.4427	4.3046	4.1960	4.1839	4.0504	3.9921
	1	This chapter	4.8450	4.7055	4.5881	4.5728	4.4229	4.3576
		Nguyen and Nguyen [33]	4.8519	4.7168	4.5947	4.5848	4.4305	4.3656
		Vo *et al* [34]	4.8525	4.7178	4.5916	4.5858	4.4270	4.3663
	2	This chapter	5.0875	4.9821	4.8620	4.8554	4.6972	4.6302
		Nguyen and Nguyen [33]	5.0940	4.9958	4.8697	4.8725	4.7083	4.6447
		Vo *et al* [34]	5.0945	4.9970	4.8668	4.8740	4.7047	4.6459
	5	This chapter	5.1840	5.1469	5.0358	5.0497	4.8950	4.8338
		Nguyen and Nguyen [33]	5.1876	5.1592	5.0422	5.0687	4.9070	4.8547
		Vo *et al* [34]	5.1880	5.1603	5.0399	5.0703	4.9038	4.8564
	10	This chapter	5.1828	5.1854	5.0836	5.1106	4.9618	4.9079
		Nguyen and Nguyen [33]	5.1846	5.1957	5.0885	5.1286	4.9730	4.9307
		Vo *et al* [34]	5.1848	5.1966	5.0866	5.1301	4.9700	4.9326
20	0	This chapter	2.8371	2.8371	2.8371	2.8371	2.8371	2.8371
		Nguyen and Nguyen [33]	2.8371	2.8371	2.8371	2.8371	2.8371	2.8371
		Vo *et al* [34]	2.8371	2.8371	2.8371	2.8371	2.8371	2.8371
	0.5	This chapter	4.8574	4.7454	4.6049	4.6288	4.4610	4.4157
		Nguyen and Nguyen [33]	4.8594	4.7473	4.6065	4.6305	4.4630	4.4168
		Vo *et al* [34]	4.8579	4.7460	4.6050	4.6294	4.4611	4.4160
	1	This chapter	5.2984	5.2206	5.0538	5.1148	4.9117	4.8930
		Nguyen and Nguyen [33]	5.2990	5.2216	5.0544	5.1159	4.9124	4.8937
		Vo *et al* [34]	5.2990	5.2217	5.0541	5.1160	4.9121	4.8938
	2	This chapter	5.5233	5.5099	5.3385	5.4393	5.2235	5.2430
		Nguyen and Nguyen [33]	5.5239	5.5111	5.3392	5.4409	5.2245	5.2444
		Vo *et al* [34]	5.5239	5.5113	5.3390	5.4410	5.2242	5.2445
	5	This chapter	5.5642	5.6371	5.4831	5.6224	5.4158	5.4822
		Nguyen and Nguyen [33]	5.5645	5.6381	5.4836	5.6241	5.4169	5.4841
		Vo *et al* [34]	5.5645	5.6382	5.4834	5.6242	5.4166	5.4843
	10	This chapter	5.5300	5.6442	5.5070	5.6604	5.4660	5.5552
		Nguyen and Nguyen [33]	5.5302	5.6451	5.5074	5.6620	5.4670	5.5573
		Vo *et al* [34]	5.5302	5.6452	5.5073	5.6621	5.4667	5.5575

The change in the NDFNFs of NP/P-FGSBs versus several PDPs for $L/h = 5$, $p = 1$, $\alpha = 0.1$ is reported in table 8.8. It can be seen that the NDFNFs change considerably, depending on the PDP. Compared to the effects of changing the SCSTR, the effects of changes in PDP on the NDFNFs of FGSBs can be

Table 8.3. The variation of NDNFs of P-FGBs versus the power-law index.

			p				
L/h	α	Results	0	1	2	5	10
5	0	This chapter	5.1531	3.9907	3.6263	3.3998	3.2811
		Sayyad *et al* [35]	5.1527	3.9904	3.6264	3.4014	3.2816
		Zouatina *et al* [36]	5.1527	3.9900	3.4012	3.2816	2.6773
	0.1	This chapter	5.2227	3.9073	3.4417	3.1460	3.0281
		Sayyad *et al* [35]	5.2208	3.9062	3.4411	3.1466	3.0275
		Zouatina *et al* [36]	5.2223	3.9070	3.1478	3.2092	2.2354
	0.2	This chapter	5.3052	3.7867	3.1488	2.6941	2.5694
		Sayyad *et al* [35]	5.3032	3.7858	3.1483	2.6954	2.5704
		Zouatina *et al* [36]	5.3048	3.7865	2.6961	2.5718	1.8433
	0.3	This chapter	5.4045	3.6018	2.6143	1.5195	1.1732
		Sayyad *et al* [35]	5.4025	3.6011	2.6140	1.5198	1.1757
		Zouatina *et al* [36]	—	—	—	—	—
20	0	This chapter	5.4603	4.2051	3.8361	3.6484	3.5389
		Sayyad *et al* [35]	5.4603	4.2050	3.8361	3.6485	3.5390
		Zouatina *et al* [36]	5.4603	4.2050	3.6485	3.5389	2.8371
	0.1	This chapter	5.5341	4.1118	3.6334	3.3775	3.2809
		Sayyad *et al* [35]	5.5340	4.1117	3.6334	3.3776	3.2809
		Zouatina *et al* [36]	5.5340	4.1117	3.3767	3.2809	2.4960
	0.2	This chapter	5.6214	3.9777	3.3123	2.8856	2.8019
		Sayyad *et al* [35]	5.6214	3.9775	3.3123	2.8856	2.8021
		Zouatina *et al* [36]	5.6214	3.9776	2.8856	2.8021	1.9533
	0.3	This chapter	5.7267	3.7732	2.7311	1.5952	1.2578
		Sayyad *et al* [35]	5.7266	3.7731	2.7310	1.5952	1.2580
		Zouatina *et al* [36]	—	—	—	—	—

Table 8.4. The variation of the NDNFs of P-FGSBs versus the first three MNs.

		SCSTR					
Type	MNs	1-0-1	2-1-2	2-1-1	1-1-1	2-2-1	1-2-1
NP	1	4.8450	4.7055	4.5881	4.5728	4.4229	4.3576
	2	15.8729	15.0554	14.8552	14.4456	14.1351	13.6862
	3	29.0191	27.0688	26.9349	25.7509	25.4000	24.3015
P	1	4.8879	4.7043	4.5773	4.5484	4.3894	4.3099
	2	15.8845	14.9924	14.7723	14.3734	14.0373	13.6172
	3	28.8722	26.8809	26.7224	25.6232	25.2334	24.2684

Table 8.5. The variation of the NDFNFs of NP/P-FGSBs versus the porosity volume fraction.

	SCSTR					
α	1-0-1	2-1-2	2-1-1	1-1-1	2-2-1	1-2-1
0	4.8450	4.7055	4.5881	4.5728	4.4229	4.3576
0.1	4.8879	4.7043	4.5773	4.5484	4.3894	4.3099
0.2	4.9364	4.7017	4.5637	4.5198	4.3505	4.2544
0.3	4.9897	4.6965	4.5459	4.4857	4.3049	4.1888

Table 8.6. The variation of the NDFNFs of NP/P-FGSBs versus the power-law index.

		SCSTR					
Type	p	1-0-1	2-1-2	2-1-1	1-1-1	2-2-1	1-2-1
NP	0	2.6775	2.6775	2.6775	2.6775	2.6775	2.6775
	1	4.8450	4.7055	4.5881	4.5728	4.4229	4.3576
	2	5.0875	4.9821	4.8620	4.8554	4.6972	4.6302
	5	5.1840	5.1469	5.0358	5.0497	4.8950	4.8338
P	0	2.3556	2.3439	2.3609	2.3447	2.3682	2.3677
	1	4.8879	4.7043	4.5773	4.5484	4.3894	4.3099
	2	5.1546	5.0053	4.8750	4.8538	4.6850	4.6042
	5	5.2589	5.1830	5.0623	5.0616	4.8959	4.8209

Table 8.7. The variation of the NDFNFs of NP/P-FGSBs versus the span-to-depth ratio.

		SCSTR					
Type	L/h	1-0-1	2-1-2	2-1-1	1-1-1	2-2-1	1-2-1
NP	5	4.8450	4.7055	4.5881	4.5728	4.4229	4.3576
	10	5.1956	5.1017	4.9472	4.9886	4.7987	4.7677
	20	5.2984	5.2206	5.0538	5.1148	4.9117	4.8930
	50	5.3285	5.2556	5.0851	5.1523	4.9450	4.9302
P	5	4.8879	4.7043	4.5773	4.5484	4.3894	4.3099
	10	5.2609	5.1098	4.9431	4.9607	4.7606	4.7012
	20	5.3711	5.2318	5.0519	5.0859	4.8721	4.8199
	50	5.4035	5.2679	5.0840	5.1229	4.9050	4.8551

Table 8.8. The variation of the DFNFs of P/NP-FGSBs for different PDPs.

		PDP			
SCSTR	$\alpha = 0$	H	X	O	V
1-0-1	4.8450	4.8879	4.8544	4.8470	4.8654
2-1-2	4.7055	4.7043	4.6973	4.7057	4.7049
2-1-1	4.5881	4.5773	4.5757	4.5887	4.5813
1-1-1	4.5728	4.5484	4.5554	4.5723	4.5610
2-2-1	4.4229	4.3894	4.4018	4.4227	4.4047
1-2-1	4.3576	4.3099	4.3316	4.3568	4.3346

different. Among PDPs, H-PDPs seem to have the strongest effects, while O-PDPs seem to have the least.

8.6 Conclusions

The free vibration analysis of FGSBs consisting of two porous face sheets made of FGMs and a homogenous isotropic metal core was discussed in this chapter. TrSDT (which does not need a shear correction factor) was utilized. The equations of motion were derived using Hamilton's principle, and the resulting equations were analytically solved under simply supported boundary conditions using Navier's method for the free vibration of P/NP-FGSBs. To validate the current formulation, comparisons were made with the results available in the published research literature. Numerical examples were presented to illustrate the influences of the SCSTR, power-law index, span-to-depth ratio, porosity distribution pattern, and porosity volume fraction on the natural frequencies of FGSBs. It was observed that the considered parameters have serious impacts on the natural frequencies of FGSBs. The use of TrSDT not only addressed the current topic well and led to good results but also made the problem easier to deal with. In future research, the solution method described here will be used to study the mechanical behaviors of other kinds of structure made of different materials and with macro/micro dimensions.

References

[1] Reddy J N 2003 *Mechanics of Laminated Composite Plates and Shells Theory and Analysis* (Boca Raton, FL: CRC Press)

[2] Fairbairn W 1849 *An Account of the Construction of the Britannia and Conway Tubular Bridges: With a Complete History of Their Progress from the Conception of the Original Idea, to the Conclusion of the Elaborate Experiments which Determined the Exact Form and Mode of Construction Ultimately Adopted* (London: John Weale and Longman, Brown, Green and Longmans)

[3] Allen H G 1969 *Analysis and Design of Structural Sandwich Panels* (Oxford: Pergamon)

[4] Noor A K, Burton W S and Bert C W 1996 Computational models for sandwich panels and shells *Appl. Mech. Rev.* **49** 155–99

[5] Vinson J R 2001 Sandwich structures *Appl. Mech. Rev.* **54** 201–14

[6] Vinson J R 2005 Sandwich structures: past, present, and future *Sandwich Structures 7: Advancing with Sandwich Structures and Materials* ed O Thomsen, E Bozhevolnaya and A Lyckegaard (Dordrecht: Springer) pp 3–12

[7] AlSaid-Alwan H H S and Avcar M 2020 Analytical solution of free vibration of FG beam utilizing different types of beam theories: a comparative study *Comput. Concr.* **26** 285–92

[8] Koizumi M 1997 FGM activities in Japan *Composites* B **28** 1–4

[9] Kieback B, Neubrand A and Riedel H 2003 Processing techniques for functionally graded materials *Mater. Sci. Eng.* A **362** 81–105

[10] Shen H-S 2009 *Functionally Graded Materials: Nonlinear Analysis of Plates and Shells* (Boca Raton, FL: CRC Press)

[11] Chakraverty S and Pradhan K K 2016 *Vibration of Functionally Graded Beams and Plates* (Amsterdam: Elsevier)

[12] Pradhan K K and Chakraverty S 2019 *Computational Structural Mechanics* (Cambridge, MA: Academic Press)

[13] Pradhan K K and Chakraverty S 2013 Free vibration of Euler and Timoshenko functionally graded beams by Rayleigh–Ritz method *Composites* B **51** 175–84

[14] Jena S K, Chakraverty S, Mahesh V, Harursampath D and Sedighi H M 2022 Free vibration of functionally graded beam embedded in Winkler–Pasternak elastic foundation with geometrical uncertainties using symmetric Gaussian fuzzy number *Eur. Phys. J. Plus* **137** 399

[15] Aydogdu M and Taskin V 2007 Free vibration analysis of functionally graded beams with simply supported edges *Mater. Des.* **28** 1651–6

[16] Avcar M, Hadji L and Civalek O 2021 Natural frequency analysis of sigmoid functionally graded sandwich beams in the framework of high order shear deformation theory *Compos. Struct.* **276** 114564

[17] Mu L and Zhao G P 2016 Fundamental frequency analysis of sandwich beams with functionally graded face and metallic foam core *Shock Vib.* **2016** 3287645

[18] Kahya V and Turan M 2017 Finite element model for vibration and buckling of functionally graded beams based on the first-order shear deformation theory *Composites* B **109** 108–15

[19] Sayyad A S and Ghugal Y M 2019 A sinusoidal beam theory for functionally graded sandwich curved beams *Compos. Struct.* **226** 111246

[20] Hong C Q, Du J C, Liang J, Zhang X H and Han J C 2011 Functionally graded porous ceramics with dense surface layer produced by freeze-casting *Ceram. Int.* **37** 3717–22

[21] Avcar M 2019 Free vibration of imperfect sigmoid and power law functionally graded beams *Steel Compos. Struct.* **30** 603–15

[22] Hadji L, Bernard F and Zouatnia N 2023 Bending and free vibration analysis of porous-functionally-graded (PFG) beams resting on elastic foundations *Fluid Dyn. Mater. Process.* **19** 1043–54

[23] Avcar M, Hadji L and Akan R 2022 The influence of Winkler–Pasternak elastic foundations on the natural frequencies of imperfect functionally graded sandwich beams *Geomech. Eng.* **31** 99–112

[24] Hadji L and Avcar M 2021 Nonlocal free vibration analysis of porous FG nanobeams using hyperbolic shear deformation beam theory *Adv. Nano Res.* **10** 281–93

[25] Hadji L and Avcar M 2021 Free vibration analysis of FG porous sandwich plates under various boundary conditions *J. Appl. Comput. Mech.* **7** 505–19

[26] Akbaş Ş D, Dastjerdi S, Akgöz B and Civalek Ö 2022 Dynamic analysis of functionally graded porous microbeams under moving load *Transp. Porous Media* **142** 209–27

[27] Fan F, Xu Y, Sahmani S and Safaei B 2020 Modified couple stress-based geometrically nonlinear oscillations of porous functionally graded microplates using NURBS-based isogeometric approach *Comput. Meth. Appl. Mech. Eng.* **372** 113400

[28] Vo T P, Thai H T, Nguyen T K, Inam F and Lee J H 2015 A quasi-3D theory for vibration and buckling of functionally graded sandwich beams *Compos. Struct.* **119** 1–12

[29] Delale F and Erdogan F 1983 The crack problem for a nonhomogeneous plane *J. Appl. Mech.* **50** 609–14

[30] Bekkaye T H L, Fahsi B, Bousahla A A, Bourada F, Tounsi A, Benrahou K H, Tounsi A and Al-Zahrani M M 2020 Porosity-dependent mechanical behaviors of FG plate using refined trigonometric shear deformation theory *Comput. Concr.* **26** 439–50

[31] Djedid I K, Benachour A, Houari M S A, Tounsi A and Ameur M 2014 A *n*-order four variable refined theory for bending and free vibration of functionally graded plates *Steel Compos. Struct.* **17** 21–46

[32] Rao S S 2019 *Vibration of Continuous Systems* (New York: Wiley)

[33] Nguyen T K and Nguyen B D 2015 A new higher-order shear deformation theory for static, buckling and free vibration analysis of functionally graded sandwich beams *J. Sandw. Struct. Mater.* **17** 613–31

[34] Vo T P, Thai H T, Nguyen T K, Maheri A and Lee J 2014 Finite element model for vibration and buckling of functionally graded sandwich beams based on a refined shear deformation theory *Eng. Struct.* **64** 12–22

[35] Sayyad A S, Avhad P V and Hadji L 2022 On the static deformation and frequency analysis of functionally graded porous circular beams *Forces Mech.* **7** 100093

[36] Zouatnia N, Hadji L and Kassoul A 2017 An analytical solution for bending and vibration responses of functionally graded beams with porosities *Wind Struct. Int. J.* **25** 329–42

Chapter 9

The torsional vibration of functionally graded nanobeams

Somnath Karmakar and S Chakraverty

This chapter investigates the vibration of functionally graded nanobeams in a magnetic field. Eringen's nonlocal elasticity theory is used to capture the nonlocal effects. The torque effects of an axial magnetic field on the FG nanobeam are given by Maxwell's relation. The material properties are assumed to vary according to the power law in the radial direction. The governing differential equations are solved using the Adomian decomposition method (ADM). Three classical boundary conditions, namely simply supported–simply supported (S–S), clamped–simply supported (C–S), and clamped–clamped (C–C), are considered. The effects of the magnetic field and the elastic foundation parameter on the vibration frequencies are observed for different boundary conditions.

9.1 Introduction

The wide range of application of nanostructures in microelectromechanical and nanoelectromechanical systems (MEMS and NEMS), nanosensors, atomic force microscopy (AFM) and other types of nanosystems and nanodevices has attracted a great deal of interest in the study of their mechanical behavior. Nanomaterials have extraordinary fundamental properties that can be used to make next-generation superstrong metals, ductile ceramics, and wear-free materials. Nanomaterials are used in many sectors such as information technology, solar panels, optics, electronics, medical and health care applications, etc [1–3]. Ignorance of small-scale effects may lead to completely incorrect solutions in the field of nanodesign, thus causing improper designs. In order to capture small-scale effects in the vibration of nanobeams, the nonlocal elasticity theory was introduced by Eringen [4]. Wang *et al* [5] investigated the free vibration of Euler–Bernoulli and Timoshenko nanobeams using an analytical method. The analytic solutions for various beam theories (e.g. the Euler–Bernoulli, Timoshenko, Reddy, and Levinson beam theories) and simply

doi:10.1088/978-0-7503-5301-4ch9

supported edges have been reviewed by Reddy [6]. Aydogdu developed a general nonlocal beam theory and solved the bending, buckling, and vibration problems for nanobeams [7]. Şimşek solved the nonlinear vibration of nanobeams using an analytic method [8].

On the other hand, functionally graded materials (FGMs) that have a gradual variation of their properties are good alternatives for NEMS and MEMS. FGMs were first introduced by Japanese scientists in the 1980s [9]. In recent years, the use of FGMs in high-temperature environments such as nuclear reactors and chemical plants has increased. Analyses of the vibration behaviors of FGM beams have been undertaken by many researchers. Sina *et al* [10] investigated the free vibration of FGM beams using an analytic method. Hein and Feklistova [11] analyzed the vibration of FGM beams using the Euler–Bernoulli beam model under various boundary conditions. The bending and free vibration of FGM beams according to various beam theories were investigated by Thai and Vo [12]. The free vibrations of Euler FGM beams and Timoshenko FGM beams according to the Rayleigh–Ritz technique were discussed by Pradhan and Chakraverty [13]. Jena *et al* [14] implemented a Chebyshev polynomial-based Rayleigh–Ritz method and used Navier's technique for the vibration analysis of functionally graded porous beams. The advantages of using FGMs in microstructures and nanostructures have become well known in recent years. As such, the dynamic behaviors of nano-FG structures are important. In this regard, Şimşek and Yurtcu [15] analyzed the bending and buckling of functionally graded nanobeams using nonlocal theories. The free vibrations of functionally graded bidirectional Euler nanobeams were investigated by Nejad and Hadi [16]. Nanomechanical devices may have to work under different conditions (e.g. thermal environments, magnetic effects, rotational effects, etc); thus, the vibrational behaviors of nano-FGM structures are also changed. Shafiei *et al* [17] investigated the vibration of functionally graded nanotubes in a thermal environment. Abdullah [18] *et al* investigated the effects of thermal stress and magnetic fields on the vibration of nanobeams.

Efficient numerical techniques play an important role in the investigation of functionally graded nanostructures. Different numerical techniques have been implemented by various researchers. Mohammadi and Ghannadpour solved the Timoshenko nanobeam problem using the Rayleigh–Ritz method; they also used Chebyshev polynomials as shape functions [19]. The boundary characteristic orthogonal polynomial-based Rayleigh–Ritz method was introduced by Chakraverty *et al* [20, 21]. The various boundary conditions have been easily implemented using boundary characteristic orthogonal polynomials to investigate the vibration of nanostructural members [22–24]. Jena and Chakraverty solved the Euler–Bernoulli nanobeam problem using the differential transform method [25]. The ADM was implemented by Karmakar and Chakraverty [26] for the thermal vibration of nanobeams.

This chapter addresses the torsional vibration of a functionally graded nanobeam resting on a Winkler elastic foundation model. The governing differential equations are obtained using Hamilton's principle and Eringen's nonlocal theory. The ADM is used to obtain the vibration frequencies under three different classical boundary conditions: S–S, C–S, and C–C. The results obtained are presented via tables and graphs.

9.2 Theory and mathematical formulation

The components of the displacement vector $\vec{u} = (u_1, u_2, u_3)$ for the nanobeam based on the linearized elasticity theory can be written as [27]

$$\begin{aligned} u_1 &= -y\phi \\ u_2 &= x\phi \\ u_3 &= 0, \end{aligned} \tag{9.1}$$

where $\phi = \phi(z, t)$ represents the angle of rotation about the axis.

The longitudinal magnetic field equations may be written as follows using Maxwell's relations [28]:

$$\vec{h} = \nabla \times (\vec{u} \times \vec{H}) \tag{9.2}$$

$$\vec{J} = \nabla \times \vec{h}, \tag{9.3}$$

where $\vec{h}$ and $\vec{J}$ are the distributing vector of the magnetic field and the current density, respectively.

The Lorentz force is defined as [28]

$$\vec{f} = \eta(\vec{J} \times \vec{H}), \tag{9.4}$$

and the Lorentz force resultants due to the longitudinal magnetic field ($\vec{H} = (0, 0, H_z)$) and in the x, y, and z directions are written as [29]

$$f_x = \eta H_z^2\left(\frac{\partial^2 u_1}{\partial x^2} + \frac{\partial^2 u_2}{\partial x \partial y} + \frac{\partial^2 u_1}{\partial z^2}\right) \tag{9.5}$$

$$f_y = \eta H_z^2\left(\frac{\partial^2 u_1}{\partial x \partial y} + \frac{\partial^2 u_2}{\partial y^2} + \frac{\partial^2 u_2}{\partial z^2}\right) \tag{9.6}$$

$$f_z = 0. \tag{9.7}$$

Substituting equation (9.1) into equations (9.5)–(9.7), the Lorentz force can be simplified as follows:

$$f_x = \eta H_z^2\left(-y\frac{\partial^2 \phi}{\partial z^2}\right) \tag{9.8}$$

$$f_y = \eta H_z^2\left(x\frac{\partial^2 \phi}{\partial z^2}\right) \tag{9.9}$$

$$f_z = 0. \tag{9.10}$$

The strain of the nanobeam is given in terms of its displacement by [29]

$$\epsilon_{xz} = \epsilon_{zx} = -\frac{y}{2}\frac{\partial\phi}{\partial z} \tag{9.11}$$

$$\epsilon_{yz} = \epsilon zy = \frac{x}{2}\frac{\partial\phi}{\partial z}. \tag{9.12}$$

The variation of the strain energy density of an isotropic linear elastic material with volume V experiencing an infinitesimal displacement is defined as [29]

$$\begin{aligned}\delta U &= \int_V \left(2\sigma_{xz}\delta\epsilon_{xz} + 2\sigma_{yz}\delta\epsilon_{yz}\right)dV \\ &= T_z\delta\phi\,|_0^L - \int_0^L \frac{\partial T_z}{\partial z}\delta\phi dz,\end{aligned} \tag{9.13}$$

where T_z can be written as [29]

$$T_z = \int_A \left(xt_{yz} - yt_{xz}\right)dA. \tag{9.14}$$

The variation of the kinetic energy of the nanobeam can be written as [29]

$$\begin{aligned}\int_0^t \delta K dt &= \int_0^t \int_V \rho\left[\dot{u}_1\delta\dot{u}_1 + \dot{u}_2\delta\dot{u}_2 + \dot{u}_3\delta\dot{u}_3\right]dVdt \\ &= \int_0^t \int_0^L \rho J_0\dot{\phi}\delta\dot{\phi}dz \\ &= \int_0^L \rho J_0\frac{\partial\phi}{\partial t}\delta\phi dz\,\Big|_0^t - \int_0^t \int_0^L J_0\frac{\partial^2\phi}{\partial t^2}\delta\phi dz dt\end{aligned} \tag{9.15}$$

where J_0 are the polar moments of mass,

$$J_0 = \int_A \rho r^2 dA;\ r^2 = x^2 + y^2;\ \rho = \rho(r). \tag{9.16}$$

The work done by the magnetic field can be written as [29]

$$W_1 = \int_V \left(f_x u_1 + f_y u_2\right)dV = \int_0^L DH_z^2\frac{\partial^2\phi}{\partial z^2}\phi dz, \tag{9.17}$$

where

$$D = \int_A \eta\left(x^2 + y^2\right)dA. \tag{9.18}$$

The work done by the elastic medium (K_T) is given by [29]

$$W_2 = -\frac{1}{2}\int_0^L K_T\phi^2 dz. \tag{9.19}$$

Hence the variation in the total work done can be written as

$$\begin{aligned}\delta W = \delta W_1 + \delta W_2 = \\ 2\int_0^L DH_z^2\frac{\partial^2\phi}{\partial z^2}\delta\phi dz - \int_0^L K_T\phi\delta\phi dz + DH_z^2\phi\frac{\partial\delta\phi}{\partial z}\Big|_0^L \\ - DH_z^2\frac{\partial\phi}{\partial z}\Big|_0^L.\end{aligned} \tag{9.20}$$

Hamilton's variational principle is given by

$$\int_0^t \left(\delta U - \delta K + \delta W\right)dt = 0. \tag{9.21}$$

If we substitute equations (9.13), (9.14), and (9.20) into equation (9.21), the governing equation can be written as [29]

$$-\frac{\partial T_z}{\partial z} + J_0\frac{\partial^2\phi}{\partial t^2} - 2DH_z^2\frac{\partial^2\phi}{\partial z^2} + K_T\phi = 0. \tag{9.22}$$

In the nonlocal strain gradient theory, the stress–strain relationship is defined as [29]

$$\left(1 - \mu^2\frac{\partial^2}{\partial z^2}\right)\sigma_{yz} = 2G\epsilon_{yz} \tag{9.23}$$

$$\left(1 - \mu^2\frac{\partial^2}{\partial z^2}\right)\sigma_{xz} = 2G\epsilon_{xz}. \tag{9.24}$$

where $\mu = e_0\bar{a}$ is the material constant.

By substituting equations (9.23) and (9.24) into equation (9.14), one can obtain [29]

$$\left(1 - \mu^2\frac{\partial^2}{\partial z^2}\right)T_z = I_0\frac{\partial^2\phi}{\partial z^2}, \tag{9.25}$$

where

$$I_0 = \int_A Gr^2 dA, \quad r^2 = x^2 + y^2, \quad G = G(r). \tag{9.26}$$

If we use equation (9.22) in equation (9.25), the governing differential equation can be expressed as

$$\begin{aligned}I_0\frac{\partial^2\phi}{\partial z^2} - J_0\frac{\partial^2\phi}{\partial t^2} + \mu^2 J_0\frac{\partial^4\phi}{\partial z^2\partial t^2} + 2DH_z^2\frac{\partial^2\phi}{\partial z^2} - 2\mu^2 DH_z^2\frac{\partial^4\phi}{\partial z^4} - K_T\phi \\ + \mu^2 K_T\frac{\partial^2\phi}{\partial z^2} = 0.\end{aligned} \tag{9.27}$$

9.2.1 The FGM model

In this section, the functionally graded model is considered. The moduli of rigidity, density, and magnetic permeability of the nanobeam vary in the r direction as follows [29]:

$$G = G_i\left(\frac{r+b}{b}\right)^{n_1} \tag{9.28}$$

$$\rho = \rho_i\left(\frac{r+b}{b}\right)^{n_2} \tag{9.29}$$

$$\eta = \eta_i\left(\frac{r+b}{b}\right)^{n_3}, \tag{9.30}$$

where G_i, ρ_i, and η_i are the moduli of rigidity, density, and magnetic permeability in the inner surface, respectively.

9.3 The solution methodology

In this section, the nondimensional form of the governing differential equation is obtained and a brief discussion of the ADM is presented.

For free vibration, assuming that ϕ varies harmonically with respect to the time variable t, it can be written as:

$$\phi(z, t) = \phi(z)e^{i\omega t}, \tag{9.31}$$

where $\phi(z)$ is the mode shape and ω is the natural frequency.

9.3.1 The nondimensional form

Let us introduce the following nondimensional terms [29]:

$$\bar{z} = \frac{z}{L}, \quad \bar{\mu} = \frac{\mu}{L}, \quad \bar{D} = \frac{2DH_z^2}{I_0}, \quad K_T = \frac{K_T L^2}{I_0}, \quad \Omega^2 = \frac{J_0 L^2 \omega^2}{I_0}. \tag{9.32}$$

If we use the above nondimensional terms in equation (9.27), the nondimensional form of the governing differential equation is

$$-\bar{\mu}^2\bar{D}\frac{d^4\phi}{dz^4} + \left(1 + \bar{D} + \bar{\mu}^2 K_T\right)\frac{d^2\phi}{d\bar{z}^2} - \overline{K_T}\phi = -\Omega^2\left(\phi - \bar{\mu}^2\frac{d^2\phi}{d\bar{z}^2}\right). \tag{9.33}$$

9.3.2 The Adomian decomposition method

Consider the differential equation

$$Fy(x) = g(x), \tag{9.34}$$

where F represents a general nonlinear ordinary differential operator involving both linear and nonlinear parts, and $g(x)$ is a given function of x. The linear terms of Fy are decomposed in the form $Ly + Ry$, where L is an invertible operator which is taken to be the highest order derivative and R is the remainder of the linear operator. Let Ny be the nonlinear terms in Fy.

Equation (9.34) can now be written as

$$Ly + Ry + Ny = g(x). \tag{9.35}$$

Equation (9.35) corresponds to an initial value problem or boundary value problem; by solving this equation for Ly, one can obtain

$$y = \phi + L^{-1}g - L^{-1}Ry - L^{-1}Ny, \tag{9.36}$$

where ϕ is an integrating constant that satisfies the condition $L\phi = 0$.

As in the case of an initial value problem, the operator L^{-1} may be regarded as a definite integration from zero to x.

In order to solve equation (9.36) using the ADM, we decompose y into an infinite sum of convergent series, as follows:

$$y = \sum_{k=0}^{\infty} c_k x^k \tag{9.37}$$

and the nonlinear term $Ny = f(y)$ is decomposed as follows:

$$Ny = f(y) = \sum_{k=0}^{\infty} A_k, \tag{9.38}$$

where A_k are Adomian polynomials, which can be derived as follows [30]:

$$A_0 = f(y_0) \tag{9.39}$$

$$A_1 = y_1 f'(y_0) \tag{9.40}$$

$$A_2 = y_2 f'(y_0) + \frac{y_1^2}{2!} f''(y_0) \tag{9.41}$$

$$A_3 = y_3 f'(y_0) + y_1 y_2 f''(y_0) + \frac{y_1^3}{3!} f'''(y_0); \tag{9.42}$$

other polynomials can also be obtained in a similar way.

Substituting equations (9.35) and (9.37) into equation (9.36) leads to

$$\sum_{k=0}^{\infty} y_k = \phi + L^{-1}g - L^{-1}R\sum_{k=0}^{\infty} y_k - L^{-1}\sum_{k=0}^{\infty} A_k, \tag{9.43}$$

where y_i can be obtained through the use of the following recurrent relations

$$y_0 = \phi + L^{-1}g,\ k = 0 \tag{9.44}$$

$$y_k = -L^{-1}Ry_{k-1} - L^{-1}A_{k-1},\ k \geqslant 1. \tag{9.45}$$

The number that denotes the series summation limit, n, is obtained through the desired convergence and the solutions can only be approximated by a truncated series, i.e.

$$y(x) = \sum_{k=0}^{n} y_k(x).$$

In equation (9.33), the operator $L\phi(\bar{z})$ can be set to $L\phi(\bar{z}) = \phi^4(\bar{z})$, the fourth-order derivative with respect to $\bar{z}$. Hence, equation (9.33) can be rewritten as

$$\phi(x) = \phi_0 - \frac{1}{-\bar{\mu}^2\bar{D}}L^{-1}\left[-\Omega^2\left(\phi - \bar{\mu}^2\frac{d^2\phi}{d\bar{z}^2}\right) - \left(1 + \bar{D} + \bar{\mu}^2K_T\right)\frac{d^2\phi}{d\bar{z}^2} + \overline{K_T}\phi\right], \tag{9.46}$$

where $L^{-1} = \int\int\int\int\cdots d\bar{z}d\bar{z}d\bar{z}d\bar{z}$ and ϕ_0 is the initial approximation given by

$$\phi_0 = \sum_{m=0}^{3}\frac{\bar{z}^m}{m!}. \tag{9.47}$$

9.4 Numerical results and discussion

An efficient Matlab program has been developed to obtain the vibration frequencies of the beam. In this section, the validity and convergence of the presented method are studied, and some new results are provided.

9.4.1 Validation of the presented model

To investigate the validity and reliability of the presented model, the first frequencies of three different scaling parameters under clamped–clamped (C–C) boundary conditions are presented in table 9.1. The results are good in agreement with previous results.

Table 9.1. A comparison between the first frequency parameter and the values reported in [31].

Scaling effects	This chapter	[31]
0	3.1418	3.1418
0.1	29 972	2.9972
0.15	2.8419	2.8419

9.4.2 The convergence of the ADM

A convergence study of the first three frequency parameters under three boundary conditions was carried out for the ADM. Tables 9.2–9.4 present the first three frequency parameters for the boundary conditions S–S, C–S, and C–C. One can observe that $n = 7$, 8, and 10 are sufficient for the convergence of the first, second, and third frequency parameters, respectively.

Table 9.2. The convergence of the first three frequencies for an S–S beam ($\mu = 0.2$, $\bar{D} = 1$, $K_T = 0$).

n	First	Second	Third
5	5.7435	7.401 23	11.2728
6	5.002 24	7.4017	11.0442
7	4.4616	7.4017	11.5537
8	4.4616	7.4017	11.8553
9	4.4616	7.4017	10.2744
10	4.4616	7.4017	10.2742

Table 9.3. The convergence of the first three frequencies for a C–S beam ($\mu = 0.2$, $\bar{D} = 1$, $K_T = 0$).

n	First	Second	Third
5	6.6407	8.0770	11.5350
6	5.7209	8.2720	11.9042
7	5.1126	8.2695	11.5712
8	5.1126	8.2673	11.4492
9	5.1126	8.2676	10.8608
10	5.1126	8.2676	10.8608

Table 9.4. The convergence of the first three frequencies for a C–C beam ($\mu = 0.2$, $\bar{D} = 1$, $K_T = 0$).

n	First	Second	Third
5	6.6407	8.0770	11.5350
6	6.3389	7.4017	11.5735
7	5.1126	8.2695	11.5712
8	6.3085	7.4017	11.9878
9	5.1126	8.2676	10.8608
10	6.3084	7.4017	11.244 08

9.4.3 The effects of the magnetic field coefficient and the elastic media parameter

The magnetic field coefficient $\bar{D}$ and the elastic medium parameter K_T impact the vibration of the beam. Figure 9.1 presents the effect of $\bar{D}$ on the first frequencies of S–S, C–S, and C–C. Here, $\bar{D}$ varies from one to five, $K_T = 0$, and μ is set to 0.2. In addition, figure 9.2 presents the first frequency parameters for S–S, C–S, and C–C beams for $K_T = [0, 5]$, $\bar{D} = 1$, and $\mu = 0.2$.

9.4.4 FGM effects

The effects of variable rigidity, density, and magnetic fields are presented in figures 9.3, 9.4, and 9.5. Figure 9.3 shows that the natural frequency increases when the parameter n_1 increases. However, when n_2 increases, the natural frequency decreases, as presented in figure 9.4. Figure 9.5 address the effects of the n_3 parameter on the frequency. The frequency increases when the magnetic property of the material increases.

9.5 Conclusions

In this chapter, we used nonlocal elasticity theory and Hamilton's principle to derive the governing differential equation of an FG nanobeam supported by an elastic foundation in a magnetic field. The ADM was used to obtain the solution for three

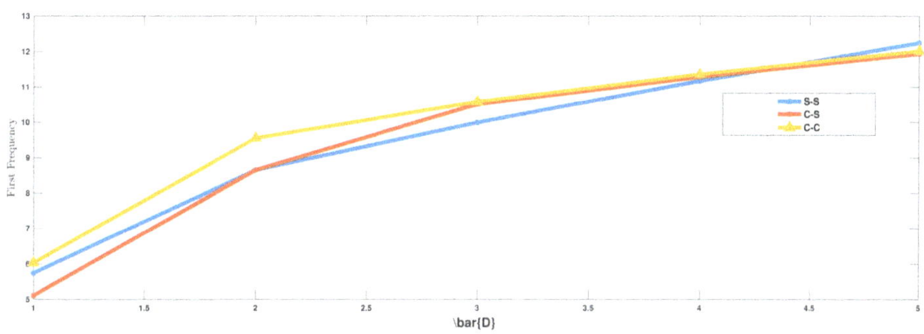

Figure 9.1. The effect of $\bar{D}$ on vibration.

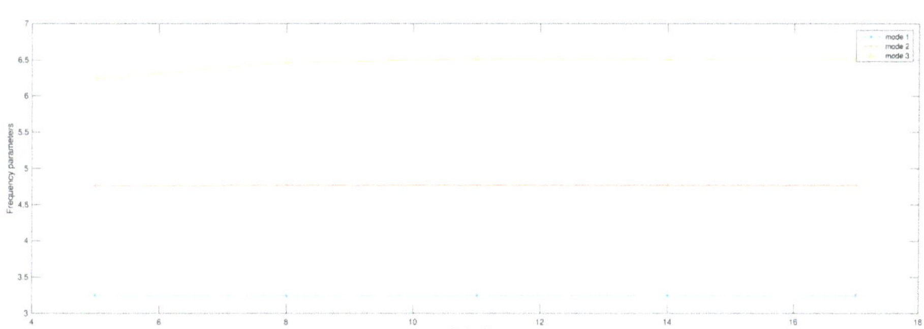

Figure 9.2. The effect of K_T on vibration.

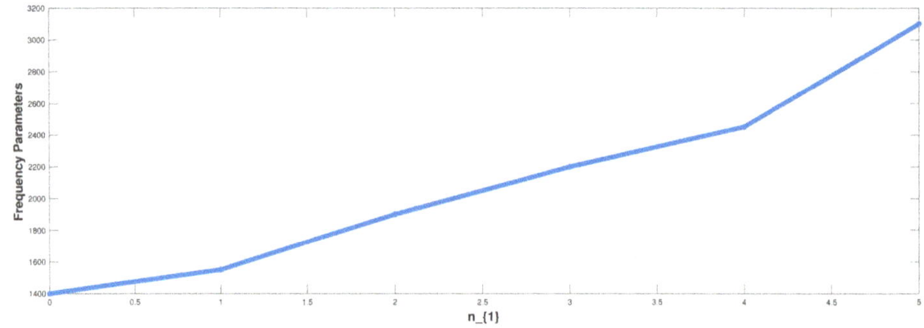

Figure 9.3. The effect of rigidity.

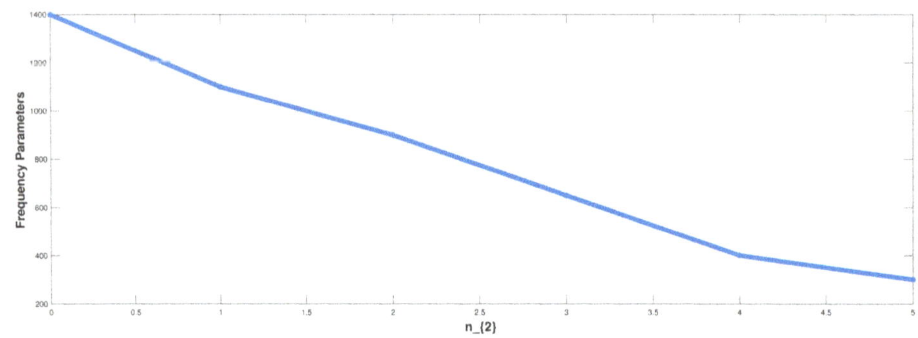

Figure 9.4. The effect of density.

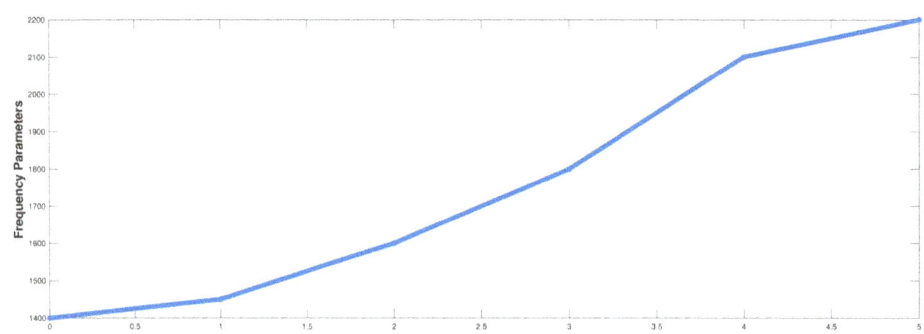

Figure 9.5. The effect of magnetic permeability.

different boundary conditions. The results were compared with results published in the research literature and very good agreement was observed. We conclude the following:

1. An increase in the magnetic parameter increases the frequency.
2. The frequency decreases when the density increases.
3. An increase in the elastic medium parameter K_T increases the frequency.

References

[1] Yezdani U, Khan M G, Kushwah N, Verma A and Khan F 2018 Application of nanotechnology in diagnosis and treatment of various diseases and its future advances in medicinc *J. Pharm. Pharm. Sci.* **7** 1612–33 https://www.wjpps.com/Wjpps_controller/abstract_id/9807
[2] Singh T, Shukla S, Kumar P, Wahla V, Bajpai V K and Rather I A 2017 Application of nanotechnology in food science: perception and overview *Front. Microbiol.* **8** 1501
[3] Vo-Dinh T 2017 *Nanotechnology in Biology and Medicine: Methods, Devices, and Applications* (Boca Raton, FL: CRC Press)
[4] Cemal Eringen A 1972 Nonlocal polar elastic continua *Int. J. Eng. Sci.* **10** 1–16
[5] Wang C M, Zhang Y Y and He X Q 2007 Vibration of nonlocal Timoshenko beams *Nanotechnology* **18** 105401
[6] Reddy J N 2007 Nonlocal theories for bending, buckling and vibration of beams *Int. J. Eng. Sci.* **45** 288–307
[7] Aydogdu M 2009 A general nonlocal beam theory: its application to nanobeam bending, buckling and vibration 2009 *Physica* E **41** 1651–5
[8] Şimşek M 2014 Large amplitude free vibration of nanobeams with various boundary conditions based on the nonlocal elasticity theory *Composites* B **56** 621–8
[9] Uzun B and Yaylı M Ö 2020 Nonlocal vibration analysis of Ti-6Al-4V/ZrO_2 functionally graded nanobeam on elastic matrix *Arab. J. Geosci.* **13** 1–10
[10] Sina S A, Navazi H M and Haddadpour H 2009 An analytical method for free vibration analysis of functionally graded beams *Mater. Des.* **30** 741–7
[11] Hein H and Feklistova L 2011 Free vibrations of non-uniform and axially functionally graded beams using Haar wavelets *Eng. Struct.* **33** 3696–701
[12] Thai H-T and Vo T P 2012 Bending and free vibration of functionally graded beams using various higher-order shear deformation beam theories *Int. J. Mech. Sci.* **62** 57–66
[13] Pradhan K K and Chakraverty S 2013 Free vibration of Euler and Timoshenko functionally graded beams by Rayleigh-Ritz method *Composites* B **51** 175–84
[14] Jena S K, Chakraverty S and Malikan M 2021 Application of shifted Chebyshev polynomial-based Rayleigh-Ritz method and Navier's technique for vibration analysis of a functionally graded porous beam embedded in Kerr foundation *Eng. Comput.* **37** 3569–89
[15] Şimşek M and Yurtcu H H 2013 Analytical solutions for bending and buckling of functionally graded nanobeams based on the nonlocal Timoshenko beam theory *Compos. Struct.* **97** 378–86
[16] Nejad M Z and Hadi A 2016 Non-local analysis of free vibration of bi-directional functionally graded Euler-Bernoulli nano-beams *Int. J. Eng. Sci.* **105** 1–11
[17] Shafiei N and She G-L 2018 On vibration of functionally graded nano-tubes in the thermal environment *Int. J. Eng. Sci.* **133** 84–98
[18] Abdullah S S, Hosseini-Hashemi S, Hussein N A and Nazemnezhad R 2020 Thermal stress and magnetic effects on nonlinear vibration of nanobeams embedded in nonlinear elastic medium *J. Therm. Stresses* **43** 1316–32
[19] Mohammadi B and Ghannadpour S A M 2011 Energy approach vibration analysis of nonlocal Timoshenko beam theory *Procedia Eng.* **10** 1766–71
[20] Bhat R B, Chakraverty S and Stiharu I 1998 Recurrence scheme for the generation of two-dimensional boundary characteristic orthogonal polynomials to study vibration of plates *J. Sound Vib.* **216** 321–7

[21] Chakraverty S, Bhat R B and Stiharu I 1999 Recent research on vibration of structures using boundary characteristic orthogonal polynomials in the Rayleigh-Ritz method *Shock Vibr. Dig.* **31** 187–94

[22] Behera L and Chakraverty S 2014 Free vibration of Euler and Timoshenko nanobeams using boundary characteristic orthogonal polynomials *Appl. Nanosci.* **4** 347–58

[23] Behera L and Chakraverty S 2017 Static analysis of nanobeams using Rayleigh-Ritz method *J. Mech. Mater. Struct.* **12** 603–16

[24] Chakraverty S and Behera L 2016 *Static and Dynamic Problems of Nanobeams and Nanoplates* (Singapore: World Scientific)

[25] Jena S K and Chakraverty S 2018 Free vibration analysis of Euler-Bernoulli nanobeam using differential transform method *Int. J. Comput. Mater. Sci. Eng.* **7** 1850020

[26] Karmakar S and Chakraverty S 2022 Thermal vibration of nonhomogeneous Euler nanobeam resting on Winkler foundation *Eng. Anal. Bound. Elem.* **140** 581–91

[27] Sadd M H 2009 *Elasticity: Theory, Applications, and Numerics* (New York: Academic)

[28] Narendar S, Gupta S S and Gopalakrishnan S 2012 Wave propagation in single-walled carbon nanotube under longitudinal magnetic field using nonlocal Euler-Bernoulli beam theory *Appl. Math. Model.* **36** 4529–38

[29] Zarezadeh E, Hosseini V and Hadi A 2020 Torsional vibration of functionally graded nanorod under magnetic field supported by a generalized torsional foundation based on nonlocal elasticity theory *Mech. Based Des. Struct. Mach.* **48** 480–95

[30] Adomian G 1994 *Solving frontier problems of physics: the decomposition method* 1 (Dordrecht: Kluwer Academic Publishers Group)

[31] Demir Ç and Civalek Ö 2013 Torsional and longitudinal frequency and wave response of microtubules based on the nonlocal continuum and nonlocal discrete models *Appl. Math. Model.* **37** 9355–67

Chapter 10

Determining the accurate free vibration response of thin functionally graded plates using the dynamic stiffness method

Subodh Kumar, Ankit Gupta, V R Shanmukhasundaram, Jay Prakash Srivastava and Prashant Choudhary

In this chapter, the accurate free vibration behavior (including the rotary inertia effect) of thin functionally graded material (FGM) plates is determined using the dynamic stiffness method (DSM). The power-law function is the commonest of the existing property variations of FGM plates. A combination of Kirchhoff–Love plate theory and the concept of the physical neutral surface (PNS) of the FGM plate is used to formulate the equation of motion according to Hamilton's principle. A Levy-type closed-form solution is used to solve the governing differential equation of motion, and then the dynamic stiffness matrix is developed through the application of displacements and force boundary conditions at the two opposite sides of the FGM plates. The Wittrick–Williams algorithm is implemented to solve the dynamic stiffness matrix in order to compute the accurate natural frequencies and mode shapes. The computed DSM frequencies are compared with the published results available in the literature. Consequently, new sets of results are outlined and described for the further design and development of the FGM plates under consideration.

10.1 Introduction

The dynamic analysis of thin-walled FGM structures plays a major role in structural design. Thin FGM plates and plate assembly structures are widely employed in the aerospace, automotive, civil engineering design and marine fields, in addition to many other industries [1]. Thin structural elements such as torsion boxes, the top and bottom skins, the webs of the wing, and the ribs must frequently be integrated to form complete structural units as part of aircraft design [2]. Plate structures subjected to unexpected internal or external excitations may cause excessive

doi:10.1088/978-0-7503-5301-4ch10

vibration or noise in practice. Thus, the study of the vibrational characteristics of thin FGM plates is of great importance to various structural design processes.

The notion of FGMs was first proposed by a Japanese materials scientist in 1984 for the preparation of thermal barrier materials in the Sendai area [3]. FGMs are microscopically inhomogeneous materials in which the material properties continuously vary in one or more desired directions [4, 5]. FGMs are designed in such a way that they are usually mixtures of a metal and a ceramic. Generally, the top surface of an FGM plate is a ceramic-rich material which provides high thermal resistance and eliminates stress singularities, whereas the bottom surface of an FGM plate is a metal-rich material which provides strength and high fracture resistance [6]. FGMs differ from conventional composites in that their material properties vary continuously, whereas the properties of composites change at each laminate interface. FGMs are better than traditional composites because no mismatch of mechanical properties takes place at each laminate interface and no debonding occurs at high temperatures. The smooth gradation of properties in FGMs reduces the thermal stresses, residual stresses, and stress concentrations compared to those found in traditional composites. Due to these potential advantages, FGMs are widely used in different engineering fields [7–9].

This chapter presents a study of how the free vibration behavior of thin functionally graded rectangular plate properties varies with the power law. The study is carries out using the DSM [10]. This chapter also includes the effect of rotatory inertia, the classical plate theory (CPT) of FGM plates, and the concept of a PNS [11–13] for better accuracy. The DSM is a robust and effective alternative to the finite element method (FEM) because it provides more accurate, reliable, and efficient results than conventional FEM in the vibration response. This method is analogous to the conventional FEM. The main distinction between the FEM and the DSM is the use of mass and stiffness matrices. Separate mass and stiffness property matrices are obtained by the discretization of a structural element as a dependence on the assumed shape functions in the FEM, whereas a single-element matrix comprising both mass and stiffness property matrices is computed by exact frequency-dependent shape functions based on the solution of the governing differential equation of the structural element in free vibration in the DSM. The assembly processes are same in these two methods, but the FEM generally leads to a linear eigenvalue solution, whereas the DSM leads to a nonlinear (transcendental) eigen solution via the Wittrick–Williams algorithm [14]. Thus, all assumptions are within the limits of the governing differential equation in the DSM. Therefore, the results computed using the DSM are considered to be exact. The DSM is very accurate and efficient; it has better modeling capabilities than the FEM because it is based on the exact shape functions. The first basic development of the DSM was done by Wittrick and Williams [15] in the early 1970s to investigate the vibration and buckling analyses of isotropic as well as anisotropic plate assemblies through the use of Kirchhoff's plate theory. Thereafter, many authors established the general approach for the treatment of accurate dynamic stiffness matrices by the DSM, but no one described the general approach. For this reason, in 1995, Banerjee [16] proposed a general approach for developing the dynamic stiffness formulation for structures for which which it was perceived that if explicit analytical expressions were used instead of numerical methods in DSM, then a substantial saving of computational time could be achieved. Subsequently, Banerjee and coauthors published multiple research articles [2, 17–21] on the vibration analyses of homogeneous isotropic and composite plates for various plate theories and various sets of boundary conditions using the dynamic stiffness formulation. Among them, Liu and Banerjee [22, 23] investigated the use of a novel spectral dynamic stiffness method to study the free vibration of a rectangular plate under various boundary conditions. Based on Kirchhoff's plate theory, Ghorbel *et al* [24, 25] also devised a DSM for the study of both the bending and the in-plane vibrations of orthotropic plates. Recently, Kumar *et al* [13, 26, 27] studied the free vibration response of uniform and stepped FGM plates using the DSM for Levy-type boundary conditions.

The Wittrick–Williams algorithm was first used in conjunction with the DSM to discover the vibration characteristics of a power law-based functionally graded material (P-FGM) plate while incorporating the rotatory inertia. In this context, the authors of this chapter have reviewed some work that reported the use of the rotatory inertia concept with the CPT in higher-order shear deformation plate theories. Reddy and Phan [28] employed higher-order shear deformation theory to analyze the vibration and buckling behaviors of simply supported elastic plates while considering rotatory inertia and transverse shear strains. Based on the trigonometric shear deformation plate theory, Zenkour [29] analyzed the free vibration of a simply supported thick FGM plate with variation of the power-law property along the thickness of the plate while taking account of rotatory inertia. Due to the limitations and future scope of our current research work (stated above the beginning of this paragraph), we limited ourselves to reviewing works on the vibration of FGM plates based on the incorporation of rotary inertia into the CPT, as this is closely related to our work. In this context, using the framework of Kirchoff's plate theory, Yang and Shen [30] first investigated the vibration response of initially stressed FGM plates resting on Pasternak elastic foundations under impulsive lateral loading and used three different techniques, namely Galerkin's method, modal superposition, and the one-dimensional differential quadrature method (DQM) to obtain the transient and natural frequency responses. Chi and Chung [31] and Chung and Chang [32] analyzed the bending behavior of simply supported FGM plates using power, sigmoidal, and exponential material property variations based on the CPT under [31] transverse loading and thermal loading [32]. Reddy and Kant [33] studied the free vibration analysis of exponential FGM plates with exact three-dimensional elastic solutions by the power series method. Using the framework of Kirchoff's plate theory, Chakraverty and Pradhan [34, 35] examined the free vibration of FGM plates with power-law material property variations resting on elastic foundations [34] under different edge conditions and enhanced their earlier work to account for thermal effects [35] with exponential material property variations. Yin *et al* [36] used nonuniform rational B-spline-based isogeometric analysis with CPT and first-order shear deformation theory with the concept of a PNS to discover the vibration characteristics of thin FGM plates with power-law material property variations. Baferani *et al* [37] discussed the free vibration behavior of thin FGM plates with power-law material property variations using CPT with the concept of a PNS and solved it using a Levy-type method. Li *et al* [38] analytically developed the relationship between the bending, frequencies, and buckling of thin FGM plates and those of corresponding homogenous reference plates using the framework of CPT, allowing anyone to determine the solution of the corresponding problem for the reference homogeneous plate. We were the first to use the DSM approach to study the free vibration characteristics of FGM plates with power, sigmoidal, and exponential property variations and demonstrated that our reported results were the most accurate compared to those available in the research literature. This chapter also includes the effect of rotatory inertia; it uses the CPT with the concept of the PNS of a thin FGM plate for better accuracy.

In this chapter, the free vibration responses of thin FGM plates with power-law property variations are investigated using the DSM with and without consideration of the rotatory inertia. The combination of the CPT with the PNS of the FGM plate is implemented to describe the elementary equation of motion, which is used to study the free vibration response using Hamilton's principle. A Levy-type (closed-form) solution is used to solve the governing differential equation of motion, and then the dynamic stiffness matrix is composed using the amplitudes of the displacements and the force edge conditions at the two opposing ends of the FGM plates. The Wittrick–Williams algorithm is applied as a solution procedure with which to compute the DS matrix used to obtain the exact eigenvalues. The DSM frequencies are compared with those of published research papers and the new set of frequency results obtained using the DSM is reported. These results may be beneficial in the design of FGM plates in cases in which it is necessary to consider rotary inertia in the CPT.

10.2 Mathematical modeling and theoretical formulations

10.2.1 A geometric description of an FGM plate

Consider a P-FGM rectangular plate of length a, width b, and uniform thickness h made from a mixture of ceramic and metal. A schematic representation of the plate in the Cartesian coordinate system (x, y, z) is shown in figure 10.1. The top side of the plate is ceramic, whereas the bottom side is metal.

10.2.2 Material property description of an FGM plate

The power-law model is frequently used to discuss the variation in material properties in the thickness direction. This frequently used model is based on the rule of mixtures and is given by

$$P(z) = (P_c - P_m)\left(\frac{z}{h} + \frac{1}{2}\right)^k + P_m, \tag{10.1}$$

where $P(z)$ denotes the FGM plate's effective material properties, such as its density $\rho(z)$ and Young's modulus $(E(z))$. The pure ceramic and metallic constituents of FGM plates are indicated by the material characteristics P_c and P_m, respectively,

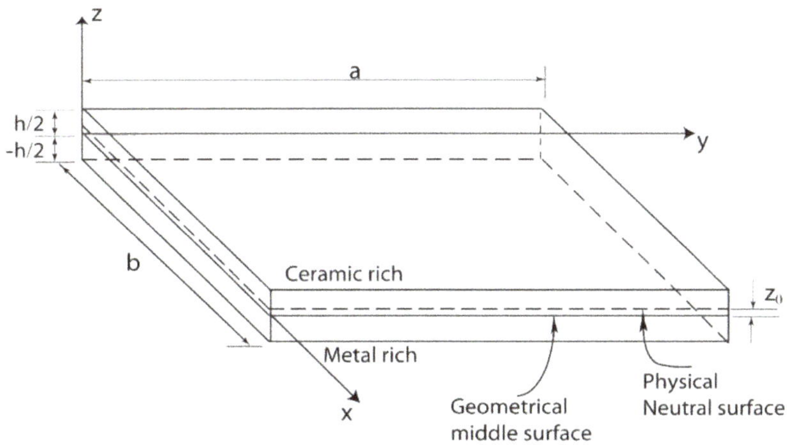

Figure 10.1. A schematic of the P-FGM plate showing the coordinate system, the geometrical middle surface, and the PNS.

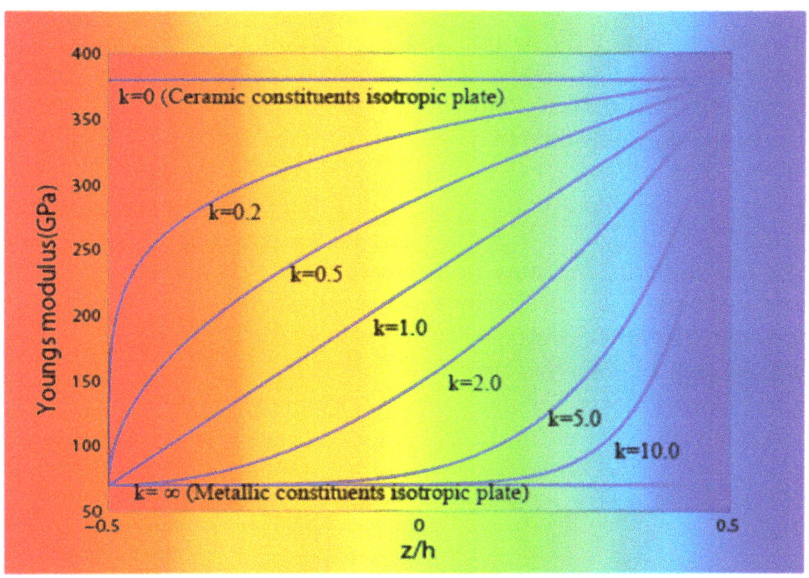

Figure 10.2. A schematic of a P-FGM plate showing the variation in the Young's modulus property in the thickness direction according to the power law.

and k represents the material inhomogeneity index or volume fraction index or power-law exponent and is a non-negative parameter. Figure 10.2 illustrates a typical variation of Young's modulus in the thickness direction of a metal–ceramic FGM plate for various values of k. In this figure, it can be observed that if $k = 0$, then the FGM plate exhibits the properties of a pure ceramic material, whereas it exhibits the properties of a pure metallic material for $k = \infty$. The remaining variations of $k(0 < k < \infty)$ describe the property variations of FGM plates.

10.3 Classical plate theory and the physical neutral surface

The displacement components (in the usual CPT notation) for the bending of the rectangular plate shown in figure 10.1 are given by [2, 13]

$$\begin{aligned} u(x, y, z) &= u_0(x, y) - z\frac{\partial w(x, y)}{\partial x}, \\ v(x, y, z) &= v_0(x, y) - z\frac{\partial w(x, y)}{\partial y}, \\ w(x, y, z) &= w(x, y). \end{aligned} \tag{10.2}$$

Here, the subscript '0' refers to the displacement components defined at the geometric mid-surface of the plate. It should be noted that when one considers the small deflection problem of a thin homogeneous isotropic plate, the mid-surface displacements due to the in-plane motions u_0 and v_0 can be neglected. Thus, the transverse displacement w is considered to be the only unknown in the case of the homogeneous isotropic plate described by equation (10.1) [2, 13]. On the contrary, due to the inhomogeneous nature of FGM plates in the transverse direction, their geometric mid-surfaces have a significant number of in-plane displacements that cannot be ignored. In other words, the PNS of the functionally graded plate does not coincide with its geometrical middle surface due to the variation of material properties in the thickness direction. This leads to the stretching–bending coupling shown in equation (10.1) for FGM plates [2, 13, 26, 27].

However, it is also interesting to note that if the concept of the physical neutral surface is properly utilized in the derivation of the governing equations then this stretching–bending coupling may be eliminated [13, 26, 27]. In this case, a new reference coordinate $z_{ns} = z - z_0$ is introduced at the neutral surface of the plate with z_0 as the distance from the neutral surface to the geometric mid-surface of the FGM plate (see figure 10.1). In this new reference plane defined by z_{ns}, the in-plane displacement components become small and can be neglected. Therefore, the displacement components of functionally graded plates take the following forms:

$$\begin{aligned} u(x, y, z) &= -z_{ns}\frac{\partial w}{\partial x} = -(z - z_0)\frac{\partial w}{\partial x}, \\ v(x, y, z) &= -z_{ns}\frac{\partial w}{\partial y} = -(z - z_0)\frac{\partial w}{\partial y}, \\ w(x, y, z) &= w(x, y). \end{aligned} \tag{10.3}$$

The nonzero linear strains related to the above displacement components are given by

$$\varepsilon_{xx} = -z_{ns}\frac{\partial^2 w}{\partial x^2}; \ \varepsilon_{yy} = -z_{ns}\frac{\partial^2 w}{\partial y^2}; \ \gamma_{xy} = -2z_{ns}\frac{\partial^2 w}{\partial x \partial y}, \tag{10.4}$$

where ε_{xx} and ε_{yy} represent the normal strains in the x- and y-directions, respectively, and γ_{xy} represents the shear strain in the x–y plane. Assuming that the material constituents of the FGM plate follow the generalized Hooke's law, the matrix form of the stress–strain or the constitutive relationships can be written as

$$\begin{bmatrix} \sigma_{xx} \\ \sigma_{yy} \\ \tau_{xy} \end{bmatrix} = \begin{bmatrix} Q_{11} & Q_{12} & 0 \\ Q_{12} & Q_{22} & 0 \\ 0 & 0 & Q_{66} \end{bmatrix} \begin{bmatrix} \varepsilon_{xx} \\ \varepsilon_{yy} \\ \gamma_{xy} \end{bmatrix}, \tag{10.5}$$

where σ_{xx} and σ_{yy} are the normal stresses, τ_{xy} is the shear stress, and the reduced stiffness components Q_{ij} are given by

$$Q_{11} = Q_{22} = \frac{E(z_{ns})}{1 - \nu^2}; \; Q_{12} = Q_{21} = \frac{\nu E(z_{ns})}{1 - \nu^2} \text{ and } Q_{66} = \frac{E(z_{ns})}{2(1 + \nu)}. \tag{10.6}$$

The neutral surface (z_0) is obtained by setting the total axial forces at the cross section, in either the x-direction or the y-direction, equal to zero. Therefore,

$$\sum F_x = \int_{-h/2-z_0}^{h/2-z_0} \sigma_{xx} dA = 0;$$

which leads to

$$z_0 = \frac{\int_{-h/2}^{h/2} E(z) z dz}{\int_{-h/2}^{h/2} E(z) dz} \tag{10.7}$$

$$= \frac{hk(E_c - E_m)}{2(k + 2)(E_c + kE_m)} = \frac{hk(E_{rat} - 1)}{2(k + 2)(E_{rat} + k)}. \tag{10.8}$$

Note that the nondimensional z_0/h depends on both the volume fraction index (k) and the Young's modulus ratio $E_{rat} = E_c/E_m$. Figure 10.3(a) shows the variation of z_0/h with the volume fraction index (k) for four different E_{rat} values. Setting $E_{rat} = 1$ makes the plate homogeneously isotropic, so that the neutral surface coincides with the mid-surface, resulting in $z_0/h = 0$. However, as E_{rat} increases, the neutral surface moves towards the ceramic-rich top surface of the plate due to the relatively higher stiffness of the ceramic material constituent (see table 10.1 for some typical material properties of FGM plates).

10.4 The governing differential equation for FGM plates based on CPT

This subsection presents the development of the governing differential equation of motion which includes the effect of rotary inertia. To derive the governing differential equation, Hamilton's principle is used, as it provides the equation of motion as well as the natural boundary conditions which are essential in the development of the dynamic stiffness matrix. According to Hamilton's principle,

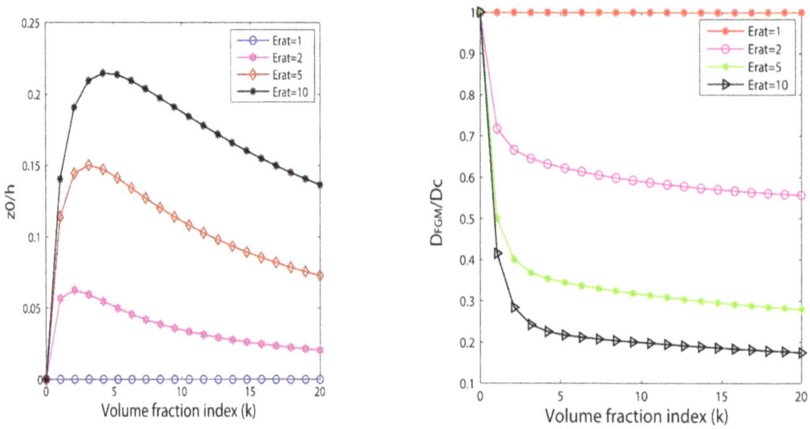

(a) (b)

Figure 10.3. (a) The variation of the nondimensional z_0/h with the volume fraction index (k) for five different E_{rat} values. (b) The variation of the nondimensional $D_{\mathrm{FGM}}/D_{\mathrm{c}}$ with the volume fraction index (k) for five different E_{rat} values.

Table 10.1. The material properties of the constituents of the FGM plate considered in this analysis.

Material constituents	Young's modulus	Density	Poisson's ratio (ν)
Ceramic: Al_2O_3 (Alumina)	$E_c = 380$ GPa	$\rho_c = 3800$ kg m^{-3}	0.3
Metal: Al (Aluminum)	$E_c = 70$ GPa	$\rho_m = 2707$ kg m^{-3}	0.3

$$\delta \int (T - U)dt = 0, \tag{10.9}$$

where T and U represent the kinetic and potential energies of the FGM plate, respectively. The kinetic energy T can be written as

$$T = \frac{1}{2}\int_V \rho\left(\left(\frac{\partial u}{\partial t}\right)^2 + \left(\frac{\partial v}{\partial t}\right)^2 + \left(\frac{\partial w}{\partial t}\right)^2\right)dV \tag{10.10}$$

where ρ is the density of the FGM plate. The potential energy U of the FGM plate is given by

$$U = \frac{1}{2}\int_V \left(\sigma_{xx}\varepsilon_{xx} + \sigma_{yy}\varepsilon_{yy} + \tau_{xy}\gamma_{xy}\right)dV. \tag{10.11}$$

By substituting the displacement components of equation (10.2) into equation (10.9) and the strain and stress relations of equations (10.4) and (10.5) into equation (10.11) and subsequently applying Hamilton's principle, the governing differential equation of motion and the natural boundary conditions are obtained, as follows:

$$D_{\mathrm{FGM}}\left(\frac{\partial^4 w}{\partial x^4} + 2\frac{\partial^4 w}{\partial x^2 \partial y^2} + \frac{\partial^4 w}{\partial y^4}\right) + I_0\frac{\partial^2 w}{\partial t^2} - I_2\left(\frac{\partial^4 w}{\partial x^2 \partial t^2} + \frac{\partial^4 w}{\partial y^2 \partial t^2}\right) = 0 \tag{10.12}$$

The natural boundary conditions[1] are:

$$\begin{aligned} V_x&: \left[-D_{\mathrm{FGM}}\left(\frac{\partial^3 w}{\partial x^3} + (2-\nu)\frac{\partial^3 w}{\partial x \partial y^2}\right) + I_2\frac{\partial^3 w}{\partial x \partial t^2}\right]\delta w, \\ M_{xx}&: -D_{\mathrm{FGM}}\left(\frac{\partial^2 w}{\partial x^2} + \nu\frac{\partial^2 w}{\partial y^2}\right)\delta\phi_y. \end{aligned} \tag{10.13}$$

In the two equations above, D_{FGM} is the effective flexural stiffness and I_0 and I_2 are the transverse inertias of the functionally graded plate. The mathematical expressions for D_{FGM}, I_0, and I_2 are given below.

$$\begin{aligned} D_{\mathrm{FGM}} &= \int_{-h/2-z_0}^{h/2-z_0} z_{ns}^2 Q_{11}(z_{ns})dz_{ns} = \int_{-h/2}^{h/2}(z-z_0)^2 Q_{11}(z)dz \\ &= \frac{h^3}{(1-\nu^2)}\left[\frac{3E_{\mathrm{c}}(k^2+k+2) + E_{\mathrm{m}}(k^3+3k^2+8k)}{12(k+1)(k+2)(k+3)}\right. \\ &\quad \left. - \left(\frac{z_0}{h}\right)\frac{k(E_{\mathrm{c}}-E_{\mathrm{m}})}{(k+1)(k+2)} + \left(\frac{z_0}{h}\right)^2\frac{(E_{\mathrm{c}}+kE_{\mathrm{m}})}{(k+1)}\right] \\ &= \frac{12D_{\mathrm{c}}}{E_{\mathrm{rat}}}\left[\frac{3E_{\mathrm{rat}}(k^2+k+2) + (k^3+3k^2+8k)}{12(k+1)(k+2)(k+3)}\right. \\ &\quad \left. - \left(\frac{z_0}{h}\right)\frac{k(E_{\mathrm{rat}}-1)}{(k+1)(k+2)} + \left(\frac{z_0}{h}\right)^2\frac{(E_{\mathrm{rat}}+k)}{(k+1)}\right] \end{aligned} \tag{10.14}$$

$$I_0 = \int_{-h/2-z_0}^{h/2-z_0}\rho(z_{ns})dz_{ns} = \int_{-h/2}^{h/2}\rho(z)dz = \rho_{\mathrm{c}}\left(\frac{\rho_{\mathrm{rat}}-1}{k+1}+1\right) \tag{10.15}$$

$$\begin{aligned} I_2 &= \int_{-h/2-z_0}^{h/2-z_0} z_{ns}^2\rho(z_{ns})dz_{ns} = \int_{-h/2}^{h/2}(z-z_0)^2\rho(z)dz \\ &= \frac{h^3}{\rho_{\mathrm{rat}}}\left[\frac{3\rho_{\mathrm{rat}}(k^2+k+2) + (k^3+3k^2+8k)}{12(k+1)(k+2)(k+3)}\right. \\ &\quad \left. - \left(\frac{z_0}{h}\right)\frac{k(\rho_{\mathrm{rat}}-1)}{(k+1)(k+2)} + \left(\frac{z_0}{h}\right)^2\frac{(\rho_{\mathrm{rat}}+k)}{(k+1)}\right] \end{aligned} \tag{10.16}$$

[1] Here, only the $x = 0$ and $x = b$ boundaries of the plate are considered, because a Levy-type solution of equation (10.12) is sought, for which the other two sides ($y = 0$ and $y = a$) are assumed to be simply supported [13, 26, 27].

The formula for D_{FGM} shows that it is a function of both the volume fraction index (k) and the ratio of Young's moduli (E_{rat}) of the ceramic and the metal. The variation of the nondimensional $D_{\mathrm{FGM}}/D_{\mathrm{c}}$ with the volume fraction index (k) for five typical E_{rat} values is shown in figure 10.3(b). Here, $D_{\mathrm{c}} = E_{\mathrm{c}}h^3/(12(1 - \nu^2))$ is the bending stiffness of the homogeneous plate with only the ceramic material constituent. Setting $E_{\mathrm{rat}} = 1$ makes the FGM plate a pure ceramic plate, with the result that $D_{\mathrm{FGM}} = D_{\mathrm{c}}$ or $D_{\mathrm{FGM}}/D_{\mathrm{c}} = 1$. However, as can be seen from figure 10.3(b), the effective bending stiffness D_{FGM} decreases for higher values of both k and E_{rat} because the increase of both k and E_{rat} corresponds an increased amount of the metal constituent, which has a relatively lower Young's modulus compared to that of the ceramic constituent.

Another interesting point that arises from the boundary condition equations in equation (10.12) is that the transverse force V_x is connected to w, whereas the bending moment M_{xx} is connected to the rotation $\phi_y = -\partial w/\partial x$. It is important to note that the term I_2 in equations (10.12) and (10.13) incorporates the effect of the rotary inertia on the free vibration of the FGM plate. If I_2 is set to zero, the rotary inertia effect is neglected. The reader may also note that equations (10.12) and (10.13) are the main building blocks in the derivation of the dynamic stiffness matrix of a strip of functionally graded rectangular plate, which leads on to the DSM formulation below.

10.5 The Levy-type closed-form solution

The closed-form or Levy-type solution of the governing differential equation (10.12) which satisfies the natural edge conditions computed in equation (10.13) is required in the following standard form:

$$w(x, y, t) = \sum_{m=1}^{\infty} W_m(x)e^{i\omega t}\sin(\alpha_m y) \text{ with } \alpha_m = m\pi/a; \quad (m = 1,2, \ldots, \infty), \tag{10.17}$$

where ω is the circular frequency or unknown frequency. By substituting equation (10.17) into equation (10.12), the following fourth-order differential equation is obtained:

$$\frac{d^4W_m}{dx^4} - \left(2\alpha_m^2 - \frac{I_2\omega^2}{D_{\mathrm{FGM}}}\right)\frac{d^2W_m}{dx^2} + \left(\alpha_m^4 - \frac{I_0\omega^2}{D_{\mathrm{FGM}}} - \frac{I_2\omega^2\alpha_m^2}{D_{\mathrm{FGM}}}\right)W_m = 0. \tag{10.18}$$

Based on the nature of the four roots of the above equation (10.18), two solutions are possible:

Case I. $\left(\alpha_m^2 - \dfrac{I_2\omega^2}{2D_{\mathrm{FGM}}}\right) \geqslant \sqrt{\left(\dfrac{I_2\omega^2}{2D_{\mathrm{FGM}}}\right)^2 + \dfrac{I_0\omega^2}{D_{\mathrm{FGM}}}}$.

This corresponds to the case in which all four roots are real ($r_{1m}, -r_{1m}, r_{2m}, -r_{2m}$),

$$r_{1m} = \sqrt{\left(\alpha_m^2 - \frac{I_2\omega^2}{2D_{\text{FGM}}}\right) + \sqrt{\left(\frac{I_2\omega^2}{2D_{\text{FGM}}}\right)^2 + \frac{I_0\omega^2}{D_{\text{FGM}}}}},$$
$$r_{2m} = \sqrt{\left(\alpha_m^2 - \frac{I_2\omega^2}{2D_{\text{FGM}}}\right) - \sqrt{\left(\frac{I_2\omega^2}{2D_{\text{FGM}}}\right)^2 + \frac{I_0\omega^2}{D_{\text{FGM}}}}}. \tag{10.19}$$

The solution is

$$W_m(x) = A_m \cosh(r_{1m}x) + B_m \sinh(r_{1m}x) + C_m \cosh(r_{2m}x) + D_m \sinh(r_{2m}x). \tag{10.20}$$

Case II. $\left(\alpha_m^2 - \frac{I_2\omega^2}{2D_{\text{FGM}}}\right) < \sqrt{\left(\frac{I_2\omega^2}{2D_{\text{FGM}}}\right)^2 + \frac{I_0\omega^2}{D_{\text{FGM}}}}.$

This corresponds to two real roots and two imaginary roots (r_{1m}, $-r_{1m}$, r_{2m}, $-r_{2m}$),

$$r_{1m} = \sqrt{\left(\alpha_m^2 - \frac{I_2\omega^2}{2D_{\text{FGM}}}\right) + \sqrt{\left(\frac{I_2\omega^2}{2D_{\text{FGM}}}\right)^2 + \frac{I_0\omega^2}{D_{\text{FGM}}}}},$$
$$r_{2m} = \sqrt{-\left(\alpha_m^2 - \frac{I_2\omega^2}{2D_{\text{FGM}}}\right) + \sqrt{\left(\frac{I_2\omega^2}{2D_{\text{FGM}}}\right)^2 + \frac{I_0\omega^2}{D_{\text{FGM}}}}}. \tag{10.21}$$

The solution is:

$$W_m(x) = A_m \cosh(r_{1m}x) + B_m \sinh(r_{1m}x) + C_m \cos(r_{2m}x) + D_m \sin(r_{2m}x). \tag{10.22}$$

10.6 The development of the dynamic stiffness matrix

We describe the dynamic stiffness (DS) formulation for the first case below. For the sake of brevity, the exact same technique used for the second case is not shown here.

Given the formulations in equation (10.12) and the identified deformation w (in equations (10.13) and (10.19)), the rotation ϕ_y, the shear force V_x, and the bending moment M_{xx} can be obtained, as follows:

$$\phi_{ym} = \Phi_{ym}(x)\sin(\alpha_m y), \tag{10.23}$$

$$\begin{aligned} V_{xm}(x, y) &= v_{xm}(x)\sin(\alpha_m y) \\ &= -D_{\text{FGM}}[\left\{A_m(r_{1m}^3 - (2-\nu)\alpha_m^2 r_{1m} + \left(\frac{I_2\omega^2}{D_{\text{FGM}}}\right) r_{1m}\right\}\sinh(r_{1m}x) + B_m\left\{r_{1m}^3 - (2-\nu)\alpha_m^2 r_{1m}\right. \\ &\left. + \left(\frac{I_2\omega^2}{D_{\text{FGM}}}\right) r_{1m}\right\}\cosh(r_{1m}x) + C_m\left\{r_{2m}^3 + (2-\nu)\alpha_m^2 r_{2m} + \left(\frac{I_2\omega^2}{D_{\text{FGM}}}\right) r_{2m}\right\}\sinh(r_{2m}x) \\ &+ D_m\left\{r_{2m}^3 + (2-\nu)\alpha_m^2 r_{2m} + \left(\frac{I_2\omega^2}{D_{\text{FGM}}}\right) r_{2m}\right\}\cosh(r_{2m}x)]\sin(\alpha_m y) \end{aligned} \tag{10.24}$$

$$
\begin{aligned}
&M_{xxm}(x, y) = M_{xm}(x)\sin(\alpha_m y) \\
&= -D_{\mathrm{FGM}}[A_m(r_{1m}^2 - \nu\alpha_m^2)\cosh(r_{1m}x) + B_m(r_{1m}^2 - \nu\alpha_m^2)\sinh(r_{1m}x) \\
&+ C_m(r_{2m}^2 - \nu\alpha_m^2)\cosh(r_{2m}x) + D_m(r_{2m}^2 - \nu\alpha_m^2)\sinh(r_{2m}x)]\sin(\alpha_m y).
\end{aligned}
\tag{10.25}
$$

The displacement boundary conditions (BCs) are:

$$
\begin{aligned}
x = 0,\ \ W_m = W_1,\ \ \phi_{ym} = \Phi_{y1} \\
x = b,\ \ W_m = W_2,\ \ \phi_{ym} = \Phi_{y2}
\end{aligned}
\tag{10.26}
$$

and the force BCs are:

$$
\begin{aligned}
x = 0,\ \ V_{xm} = -v_1,\ \ M_{xxm} = -M_1 \\
x = b,\ \ V_{xm} = v_2,\ \ M_{xxm} = M_2.
\end{aligned}
\tag{10.27}
$$

In this chapter, the sign conventions used are those given by Kumar and Jana [13]. The substitution of equation (10.26) into equations (10.20) and (10.21) yields the following matrix relationship for the displacement BCs:

$$
\begin{bmatrix} W_1 \\ \Phi_1 \\ W_2 \\ \Phi_2 \end{bmatrix} = \begin{bmatrix} 1 & 0 & 1 & 0 \\ 0 & -r_{1m} & 0 & -r_{2m} \\ C_{h1} & S_{h1} & C_{h2} & S_{h2} \\ -r_{1m}S_{h1} & -r_{1m}C_{h1} & -r_{2m}S_{h2} & -r_{2m}C_{h2} \end{bmatrix} \begin{bmatrix} A_m \\ B_m \\ C_m \\ D_m \end{bmatrix}
\tag{10.28}
$$

$$
\text{i. e.}\quad \delta = AC,
\tag{10.29}
$$

where

$$
\begin{aligned}
&C_{hi} = \cosh(r_i b),\ S_{hi} = \sinh(r_i b),\ C_i = \cos(r_{im} b) \text{ and } S_i = \sin(r_{im} b) \\
&\quad \text{with } i = 1, 2.
\end{aligned}
\tag{10.30}
$$

Substituting equation (10.27) into equations (10.24) and (10.23) allows the resulting matrix equation for the force BCs to be computed:

$$
\begin{bmatrix} v_1 \\ M_1 \\ v_2 \\ M_2 \end{bmatrix} = \begin{bmatrix} 0 & R_1 & 0 & R_2 \\ L_1 & 0 & L_2 & 0 \\ -R_1S_{h1} & -R_1C_{h1} & -R_2S_{h2} & -R_2C_{h2} \\ -L_1C_{h1} & -L_1S_{h1} & -L_2C_{h2} & -L_2S_{h2} \end{bmatrix} \begin{bmatrix} A_m \\ B_m \\ C_m \\ D_m \end{bmatrix}
$$

$$
\text{i.e}\ \ F = RC,
\tag{10.31}
$$

where

$$
R_i = D_{\mathrm{FGM}}[r_{im}^3 - \alpha^2 r_{im}(2 - \nu)] + r_{im}I_2\omega^2,\ L_i = D_{\mathrm{FGM}}(r_{im}^3 - \alpha^2\nu) \text{ with } i = 1, 2.
\tag{10.32}
$$

We now use equations (10.29) and (10.31) and eliminate the constant vector C, thereby obtaining the dynamic stiffness matrix K for the FGM plate element:

$$F = K\delta, \tag{10.33}$$

where

$$K = RA^{-1}. \tag{10.34}$$

The abovementioned relation in equation (10.34) signifies that the symmetric 4 × 4 DS matrix features and comprises six independent terms s_{vv}, s_{vm}, s_{mm}, f_{vv}, f_{vm}, and f_{mm} that represent the effects on the shear force and bending moment of a unit displacement implemented on the 'same' (s) nodal line and on the 'far' (f) nodal line [2, 13]. Therefore, the dynamic stiffness (K) may be represented in the following matrix form:

$$K = \begin{bmatrix} s_{vv} & s_{vm} & f_{vv} & f_{vm} \\ s_{vm} & s_{mm} & -f_{vm} & f_{mm} \\ f_{vv} & -f_{vm} & s_{vv} & -s_{vm} \\ f_{vm} & f_{mm} & -s_{vm} & s_{mm} \end{bmatrix}. \tag{10.35}$$

The explicit mathematical relationships between these six independent elements of the DS matrix K are represented by

$$\left.\begin{aligned} S_{vv} &= (r_{2m}R_1 + r_{1m}R_2)(r_{2m}C_{h1}S_2 + r_{1m}C_2S_{h1})/\Delta, \\ S_{vm} &= -r_{2m}\{R_1(C_2^2 - C_2C_{h1} + S_2^2) - R_2S_2S_{h1}\} \\ &\quad - r_{1m}\{R_1S_2S_{h1} - R_2((C_2 - C_{h1})C_{h1} + S_{h1}^2)\}/\Delta, \\ S_{mm} &= (L_1 - L_2)(r_{1m}C_{h1}S_2 - r_{2m}C_2S_{h1})/\Delta\,, \\ f_{vv} &= (r_{2m}R_1 - r_{1m}R_2)(r_{2m}S_2 + r_{1m}S_{h1})/\Delta\,, \\ f_{vm} &= (C_{h2} - C_{h1})(r_{2m}R_1 - r_{1m}R_2)/\Delta\,, \\ f_{mm} &= (L_1 - L_2)(r_{2m}S_{h1} - r_{1m}S_{h2})/\Delta\,, \end{aligned}\right\}, \tag{10.36}$$

where r_{im}, S_{hi}, C_{hi}, C_i, S_i, L_i, R_i (with i = 1,2) are defined in equations (10.30) and (10.32), respectively, and the parameter Δ is defined as:

$$\Delta = S_2S_{h1}(r_{1m}^2 - r_{2m}^2) + r_{1m}r_{2m}\{(C_2 - C_{h1})^2 + S_2^2 - S_{h1}^2\}. \tag{10.37}$$

10.7 The computation of the natural frequencies using the Wittrick–Williams algorithm

The above 4 × 4 frequency-dependent symmetric DS matrix K of equation (10.34) now can be used to determine the exact natural frequencies and mode shapes through an assembly process of FGM plate elements, which is almost similar to the FEM, except that line nodes are used in the DSM instead of the point nodes of the FEM. A schematic illustration of the assembly process can be found in figure 10.4,

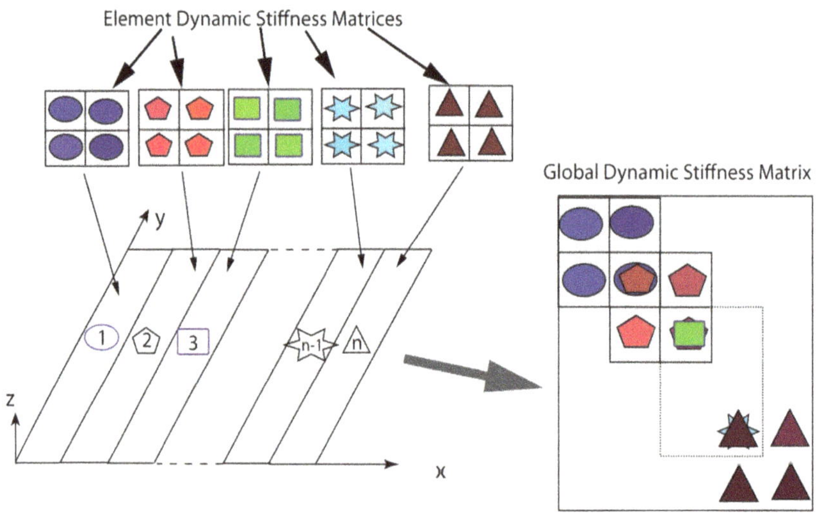

Figure 10.4. A schematic of the global DS matrix construction procedure used for FGM plates.

in which it can clearly be observed that there are line nodes (as opposed to the point nodes of the FEM) for each strip in the DS plate elements. As in the case of the FEM, the complete assembled dynamic stiffness matrix seems to be banded. After the assembly process is complete, a penalty method is utilized to address all possible Levy-type edge conditions by adding a very large stiffness to the appropriate term of the leading diagonal of the assembled matrix in order to arrest the particular degree of freedom at the edge of the plate.

The Wittrick–Williams approach, which is a reliable, computationally efficient, and accurate algorithm, is now used to calculate the natural frequencies of the FGM plate using the global dynamic stiffness matrix with the necessary constraints. This algorithm is well suited to the DSM; it utilizes the Sturm sequence property of the global DS matrix and ensures that no natural frequencies of the analyzed structures are missing. The details of the Wittrick–Williams algorithm have been discussed in numerous papers. Thus, it is not shown here for the sake of brevity. In brief, it fundamentally offers the number of eigenvalues of a structure which remain less than a randomly chosen trial frequency in a computationally efficient manner, since it works by taking consecutive trail frequencies via the bisection method to any desired accuracy. The DSM results discussed in the following section were produced using the Wittrick–Williams algorithm.

10.8 Results and discussion

The methodology discussed above was coded as a MATLAB program to obtain the eigenvalues (natural frequencies) and mode shapes of thin FGM plates while considering the rotary inertia effect. Various numerical investigations are discussed based on the use of this program. Some of the important DSM frequencies and mode shapes are highlighted in this section. The influences of various geometrical parameters and P-FGM properties on the frequency results of FGM plates are also discussed. The mechanical properties of the constituent materials of the FGM plates used in this chapter are those shown in table 10.1. However, some tables consider different material properties for validation purposes, and the same are mentioned accordingly.

In order to efficiently describe the results, several acronyms are used in this section. To describe the edge conditions of the Levy type, the symbols 'S,' 'C,' and 'F' are used to denote 'simply supported,' 'clamped,' and 'free' edges, respectively. As an example, SCSF denotes the following edge conditions at the four edges: simply supported (S) at $y = 0$, clamped (C) at $x = 0$, simply supported (S) at $y = a$, and free (F) at $x = b$. The symbol 'RI' denotes 'rotary inertia.' The modes of a given frequency are denoted by the letters 'm' and 'n,' where 'm' stands for the number of half sine waves in the 'y' direction and 'n' stands for the nth-lowest natural frequency for a given value of 'm.'

The first part of this section compares the DSM frequency results with values found in the research literature for isotropic and P-FGM plates while taking account of rotating inertia in the CPT. The comparison demonstrates that the DSM findings are very precise. The DSM results for a square FGM plate and a rectangular FGM plate, which can be used to modify the material parameters of P-FGMs, are then given in various tables. Finally, some graphs and tables are displayed to illustrate how different FGM plate design parameters affect the natural frequency of the plate.

10.8.1 Comparisons with published results

The DSM frequency results are now compared with the results for P-FGM plates available in the published literature. Tables 10.2(a) and 10.2(b) compare the DSM frequency results for P-FGM plates with the results in the articles published by Hashemi and coauthors [38, 39], Baferani *et al* [40], and Thai and Choi [41]. Table 10.2(a) shows the nondimensional frequencies $\left(\tilde{\omega} = \omega h\sqrt{\rho_c/E_c}\right)$ for five different values of the volume fraction index, namely (k = 0, 0.5, 1, 4, 10) for a square P-FGM plate under simply supported (SSSS) edge conditions. Table 10.2(b) represents the nondimensional frequencies $\left(\tilde{\omega} = \omega h\sqrt{\rho_m/E_m}\right)$ for five different values of the volume fraction index, namely (k = 0, 0.5, 1, 2, 5) for a square P-FGM plate

Table 10.2(a). A comparison of nondimensional frequencies $\left(\tilde{\omega} = \omega h\sqrt{\rho_c/E_c}\right)$ for an SSSS P-FGM plate with $h/a = 0.05$.

	Volume fraction index (k)				
Reference	0	0.5	1	4	10
[38]	0.0148	0.0128	0.0115	0.0101	0.0096
[39]	0.0148	0.0125	0.0113	0.0098	0.0094
Present	0.0149	0.0126	0.0114	0.0099	0.0095

Table 10.2(b). A comparison of nondimensional frequencies $\left(\widehat{\omega} = \omega h\sqrt{\rho_m / E_m}\right)$ for six Levy-type P-FGM plates with $h/a = 0.05$.

Levy-type edge conditions	Reference	Volume fraction index (k)				
		0	0.5	1	2	5
SSSS	[40]	0.0291	0.0249	0.0227	0.0209	0.0197
	[41]	0.0291	0.0247	0.0222	0.0202	0.0191
	Present	0.0293	0.0248	0.0224	0.0203	0.0193
SCSC	[40]	0.0421	0.0359	0.0324	0.0295	0.0278
	[41]	0.0424	0.0359	0.0324	0.0294	0.0278
	Present	0.0430	0.0364	0.0328	0.0298	0.0282
SFSF	[40]	0.0142	0.0121	0.0109	0.0099	0.0094
	[41]	0.0143	0.0121	0.0109	0.0099	0.0094
	Present	0.0143	0.0121	0.0109	0.0099	0.0094
SCSF	[40]	0.0187	0.0159	0.0143	0.0130	0.0123
	[41]	0.0188	0.0159	0.0143	0.0130	0.0123
	Present	0.0189	0.0160	0.0144	0.0131	0.0124
SCSS	[40]	0.0342	0.0296	0.0268	0.0245	0.0231
	[41]	0.0348	0.0295	0.0266	0.0241	0.0228
	Present	0.0351	0.0297	0.0268	0.0243	0.0231
SSSF	[40]	0.0172	0.0146	0.0133	0.0121	0.0114
	[41]	0.0173	0.0146	0.0132	0.0120	0.0114
	Present	0.0174	0.0147	0.0133	0.0121	0.0114

for all possible combinations of Levy-type edge conditions. The reader may be interested to note that each of these studies used a different methodology to calculate these natural frequencies. As a result, this comparison demonstrates that the DSM results are in excellent agreement with the results reported using other methodologies.

Note that all these articles used different methodologies for the computation of these natural frequencies. Thus, this comparison shows that the DSM results are in excellent agreement with the published results obtained using several different methodologies.

10.8.2 Frequency results obtained using the DSM for square FGM plates

New sets of DSM results for P-FGM plates are presented in this paragraph as well as the two subsequent subsections, which can be regarded as significant scientific contributions to this work. For the sake of uniformity and clarity, the reporting of these DSM data uses a single nondimensional frequency formula $\omega^* = \omega a^2\sqrt{\rho_c h / D_c}$ with $D_c = E_c h^3/(12(1 - \nu^2))$ for reasons of simplicity and ease of comprehension. A square FGM plate with a power-law material property variation and each of the six Levy-type boundary criteria is treated in this subsection.

In these following tables, frequency results outside brackets were computed while considering the effect of rotary inertia, whereas the results within brackets were computed without considering the effect of rotary inertia.

Two special cases of the volume fraction index arise for P-FGM plates. The volume fraction index $k = 0$ corresponds to a homogenous isotopic plate with a ceramic constituent (Al_2O_3), whereas $k = \infty$ corresponds to a homogenous isotopic plate with a metallic constituent (Al). In MATLAB computations, $k = \infty$ is unattainable. Thus, a sufficiently large value of $k = 10^{10}$ is used in place of $k = \infty$ for the computation, which provides the required precision for the frequency results.

Table 10.3 presents the first six frequencies obtained from the DSM for a square P-FGM plate for eight different volume fraction indices (k). From table 10.3, it can clearly be observed that the frequency values decrease with an increase in the volume fraction index (k). This reduction in frequency can be explained by noting that increasing k introduces more metallic constituents into the P-FGM structure, thereby reducing the relative stiffness of the plate. It can also be seen that the DSM frequency results for FGM plates and the homogeneous isotropic plate ($k = 0$, i.e. a special case of the FGM plate) obtained without taking rotary inertia into account (shown within brackets) match the frequencies reported by Leissa (in which the rotary inertia was also not considered) to up to four decimal places [42].

Figure 10.5 is a plot of the thickness-to-side ratio and the fundamental frequencies of the square FGM plate for the case in which all sides are simply supported and the volume fraction index $k = 2$. The plot depicts the effect of rotary inertia on the side-to-thickness ratio plate parameter of the SSSS square FGM plate. The plot shows if rotary inertia is not taken into account, then there is no effect on the frequency results as a result of varying the thickness of the plate. This is is the case because the thickness parameter is not involved in the governing differential equation and the boundary condition of the plate. However, if rotary inertia is considered, the frequency results increase for a decrease in the thickness of the plate. This means that thicker plates have lower nondimensional frequencies than thinner plates.

10.8.3 Frequency results obtained using the DSM for rectangular FGM plates

This subsection reports the DSM natural frequencies for rectangular plates with P-FGM material property changes. Consideration is given to all six Levy-type edge conditions. The fundamental frequency of a rectangular P-FGM plate for each of the six boundary conditions is shown in table 10.4 for eight volume fraction indices and five aspect ratios. The results shown within brackets are the results obtained without considering rotary inertia. Table 10.4 shows that as k increases, the fundamental frequency of a rectangular P-FGM plate decreases for a given aspect ratio.

Tables 10.5(a) and 10.5(b) represent all the possible combinations of Levy-type boundary conditions of thin FGM plates that have seven volume fraction indices, namely $k = 0$, 0.5, 1.0, 2.0, 5.0, 10, and ∞ for five thickness-to-side ratios, namely

Table 10.3. The nondimensional natural frequencies $\left(\omega^* = \omega a^2 \sqrt{\rho_c h / D_c}\right)$ of a square P-FGM plate for all Levy-type edge conditions with $h/a = 0.05$.

BSs	*m n*	$k=0$	$k=0.2$	$k=0.5$	$k=1.0$	$k=2.0$	$k=5.0$	$k=10.0$	$k=\infty$
SCSC	1 1	28.8850	26.8052	24.4574	22.0351	20.0291	18.9883	18.3905	14.6885
		(28.9508)[a]	(26.8659)	(24.5141)	(22.0894)	(20.0832)	(19.0419)	(18.4395)	(14.7535)
	2 1	54.4543	50.5343	46.1050	41.5308	37.7383	35.7714	34.6523	27.6909
		(54.7431)	(50.8007)	(46.3537)	(41.7688)	(37.9752)	(36.0063)	(34.8673)	(27.8377)
	1 2	68.9292	63.9674	58.3601	52.5684	47.7658	45.2751	43.8601	35.0516
		(69.3270)	(64.3344)	(58.7026)	(52.8963)	(48.0921)	(45.5986)	(44.1561)	(35.2539)
	2 2	93.7651	87.0170	79.3840	71.4925	64.9415	61.5454	59.6338	47.6811
		(94.5853)	(87.7737)	(80.0900)	(72.1683)	(65.6137)	(62.2118)	(60.2437)	(48.0982)
	3 1	101.1678	93.8878	85.6490	77.1270	70.0484	66.3796	64.3247	51.4455
		(102.2162)	(94.8548)	(86.5515)	(77.9907)	(70.9072)	(67.2308)	(65.1039)	(51,9786)
	1 3	127.6593	118.4736	108.0753	97.3165	88.3771	83.7444	81.1567	64.9168
		(129.095)	(119.7981)	(109.3111)	(98.4991)	(89.5530)	(84.9100)	(82.2238)	(65.6472)
SCSS	1 1	23.5946	21.8957	19.9779	17.9994	16.3609	15.5108	15.0224	11.9982
		(23.6463)	(21.9434)	(20.0225)	(18.0421)	(16.4034)	(15.5529)	(15.0609)	(10.0245)
	2 1	51.4057	47.7051	43.5239	39.2060	35.6261	33.7694	32.7127	26.1406
		(51.6743)	(47.9529)	(43.7552)	(39.4274)	(35.8464)	(33.9879)	(32.9127)	(26.2772)
	1 2	58.3274	54.1287	49.3842	44.4841	40.4213	38.3142	37.1160	29.6605
		(58.6464)	(54.4229)	(49.6588)	(44.7471)	(40.6829)	(38.5736)	(37.3534)	(29.8226)
	2 2	85.4098	79.2629	72.3105	65.1232	59.1574	56.0646	54.3222	43.4323
		86.1345	79.9315	72.9343	65.7204	59.7514	56.6535	(54.4229)	(43.8008)
	3 1	99.2476	92.1058	84.0235	75.6635	68.7197	65.1207	63.1045	50.4690
		(100.2690)	(93.0481)	(84.9027)	(76.5050)	(69.5565)	(65.9502)	(63.8638)	(50.9888)
	1 3	110.0228	103.9619	94.8383	85.3999	77.5590	73.4952	71.2219	56.9654
		(113.2280)	(105.0738)	(95.8757)	(86.3927)	(78.5461)	(74.4737)	(72.1177)	(57.5783)
SSSS	1 1	19.6987	18.2803	16.6793	15.0276	13.6598	10.9502	10.5423	10.0171
		(19.7392)	(18.3177)	(16.7142)	(15.0610)	(13.6931)	(10.9831)	(10.5724)	(10.0377)
	1 2	49.0963	45.5620	41.5687	37.4449	34.0262	32.2530	31.2436	24.9663
		(49.3480)	(45.7942)	(41.7854)	(37.6524)	(34.2326)	(32.4578)	(31.4310)	(25.0943)
	2 1	49.0963	45.5620	41.5687	37.4449	34.0262	32.2530	31.2436	24.9663
		(49.3480)	(45.7942)	(41.7854)	(37.6524)	(34.2326)	(32.4578)	(31.4310)	(25.0943)
	2 2	78.3153	72.6789	66.3044	59.7153	54.2465	51.4112	49.8125	39.8246
		(78.9568)	(73.2707)	(66.8566)	(60.2438)	(54.7722)	(51.9324)	(50.2896)	(40.1508)
	1 3	97.6968	90.6665	82.7106	74.4816	67.6467	64.1041	62.1191	49.6804
		(98.6960)	(91.5883)	(83.5708)	(75.3048)	(68.4653)	(64.9156)	(62.862)	(50.1885)
	3 1	97.6968	90.6665	82.7106	74.4816	67.6467	64.1041	62.1191	49.6804
		(98.6960)	(91.5883)	(83.5708)	(75.3048)	(68.4653)	(64.9156)	(62.8620)	(50.1885)
SCSF	1 1	12.6694	11.7571	10.7275	9.6656	8.7864	8.3302	8.0675	6.4426
		(12.6874)	(11.7737)	(10.743)	(9.6805)	(8.8012)	(8.3449)	(8.0809)	(6.4517)
	1 2	32.9276	30.5571	27.8795	25.1152	22.8244	21.6361	20.9576	16.7442
		(33.0651)	(30.6839)	(27.9978)	(25.2286)	(22.9372)	(21.7480)	(21.0600)	(16.8141)
	2 1	41.5163	38.5275	35.1513	31.6655	28.7763	27.2777	26.4228	21.1117
		(41.7019)	(38.6987)	(35.3110)	(31.8184)	(28.9286)	(27.4287)	(26.5610)	(21.2061)

	2 2	62.5807	58.0763	52.9840	47.7224	43.3573	41.0939	39.8127	31.8233
		(63.0148)	(58.4767)	(53.3577)	(48.0801)	(43.7133)	(41.4469)	(40.1357)	(32.0441)
	1 3	71.7530	66.5891	60.7477	54.7080	49.6938	47.0945	45.6325	36.4876
		(72.3976)	(67.1838)	(61.3026)	(55.2392)	(50.2221)	(47.6183)	(46.1118)	(36.8153)
	3 1	89.7505	83.2917	75.9841	68.4270	62.1517	58.8988	57.0726	45.6396
		(90.6114)	(84.0860)	(76.7251)	(69.1363)	(62.8570)	(59.5981)	(57.7127)	(46.0774)
SSSF	1 1	11.6690	10.8287	9.8805	8.9025	8.0928	7.6726	7.4306	5.9339
		(11.6846)	(10.8430)	(9.8938)	(8.9152)	(8.1055)	(7.6853)	(7.4422)	(5.9418)
	1 2	27.6537	25.6628	23.4143	21.0934	19.1703	18.1727	17.6023	14.0624
		(27.7563)	(25.7574)	(23.5026)	(21.1780)	(19.2545)	(18.2562)	(17.6787)	(14.1146)
	2 1	41.0152	38.0624	34.7270	31.2833	28.4291	26.9487	26.1041	20.8569
		(41.1967)	(38.2299)	(34.8833)	(31.4330)	(28.5781)	(27.0964)	(26.2392)	(20.9492)
	2 2	58.6760	54.4525	49.6783	44.7458	40.6542	38.5325	37.3304	29.8377
		(59.0655)	(54.8119)	(50.0137)	(45.0668)	(40.9737)	(38.8493)	(37.6203)	(30.0358)
	1 3	61.3500	56.9346	51.9408	46.7787	42.4941	40.2727	39.0208	31.1975
		(61.8606)	(57.4057)	(52.3804)	(47.1995)	(42.9126)	(40.6877)	(39.4006)	(31.4571)
	3 1	89.4390	83.0026	75.7204	68.1897	61.9363	58.6948	56.8748	45.4812
		(90.2941)	(83.7915)	(76.4564)	(68.8942)	(62.6369)	(59.3894)	(57.5106)	(45.9160)
SFSF	1 1	9.6214	8.9285	8.1467	7.3405	6.6730	6.3267	6.1270	4.8926
		(9.6314)	(8.9378)	(8.1554)	(7.3487)	(6.6813)	(6.3349)	(6.1345)	(4.8977)
	1 2	16.0985	14.9394	13.6309	10.2809	11.1629	10.5829	10.2496	8.1864
		(16.1348)	(14.9728)	(13.6621)	(10.3108)	(11.1927)	(10.6124)	(10.2767)	(8.2048)
	1 3	36.5064	33.8785	30.9086	27.8408	25.2967	23.9773	23.2283	18.5641
		(36.7256)	(34.0808)	(31.0974)	(28.0215)	(25.4765)	(24.1556)	(23.3915)	(18.6756)
	2 1	38.7841	35.9919	32.8382	29.5823	26.8841	25.4845	24.6853	19.7224
		(38.9450)	(35.7228)	(32.5956)	(29.3716)	(26.7039)	(25.3194)	(24.5184)	(19.8041)
	2 2	46.4909	43.1442	39.3626	35.4573	32.2194	30.5401	29.5847	23.6414
		(46.7381)	(43.3722)	(39.5754)	(35.6610)	(32.4222)	(30.7412)	(29.7687)	(23.7671)
	2 3	70.1204	65.0740	59.3657	53.4639	48.5644	46.0245	44.5953	35.6574
		(70.7401)	(65.6457)	(59.8991)	(53.9745)	(49.0723)	(46.5281)	(45.0561)	(35.9725)

[a] Without consideration of rotary inertia. *** Frequency results within brackets exactly match those of Leissa [42].

5, 10, 20, 25, and 50 and three aspect ratios, viz. 0.5, 1, and 2. It can be seen from tables 10.4, 10.5(a), and 10.5(b) that the nondimensional frequencies increase with an increase in aspect ratio for all possible combinations of Levy-type edge conditions except for the SFSF edge condition. The increase in frequency observed for five sets of Levy-type boundary conditions is due to the increase in the stiffness of the plate when one edge of the plate becomes smaller. In contrast, the free edge of the plate becomes larger for the SFSF edge condition; as a consequence, the stiffness of the plate is reduced, which causes the decrease in the frequency of the plate.

It is important to note from tables 10.4, 10.5(a), and 10.5(b) that the first nondimensional frequencies are increased by increasing the a/h ratio, irrespective of edge conditions and aspect ratios. When rotary inertia is taken into account, the nondimensional frequency

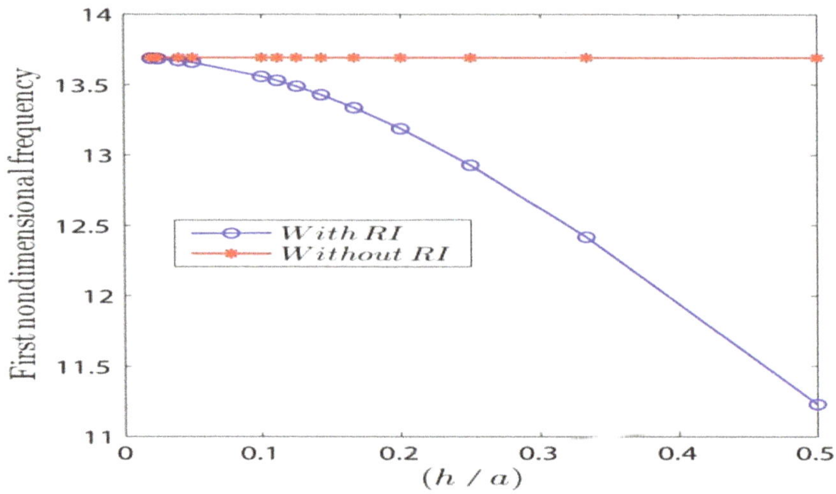

Figure 10.5. The variation between the thickness-to-side ratio and the first nondimensional frequency of an SSSS square FGM plate for $k = 2$.

tends to decrease as the plate thickness increases. This means that the effect of the thickness ratio is less significant for thinner plates and more significant for thicker plates. However, if rotary inertia is not taken into account, the nondimensional frequencies remain unaltered despite any variation in the plate thickness (as shown in figure 10.3).

From tables 10.3–10.5(b) and figure 10.6, it is clear that the thickness of the FGM plate and the edge boundary conditions determine the difference in the frequency results if rotary inertia is taken into consideration for both square and rectangular plates. Compared to free edges (e.g. the SFSF edge condition), the effect of rotary inertia is more pronounced when the edges are clamped (such as in the SCSC edge condition). It can be observed from tables 10.5(a) and 10.5(b) that as the thickness of an FGM plate increases, the influence of material inhomogeneity on the nondimensional frequency becomes more considerable. It can be also observed from tables 10.4, 10.5(a), and 10.5(b) that the effect of the boundary condition is more significant for the higher values of the aspect ratio.

10.8.4 The effects of E_{rat} and ρ_{rat} on the DSM results

10.8.4.1 The effect of E_{rat} with constant ρ_{rat}

The effects of the Young's modulus ratio $E_{\text{rat}} = E_c/E_m$ and the density ratio $\rho_{\text{rat}} = \rho_c/\rho_m$ on the nondimensional frequencies $\left(\omega^* = \omega a^2\sqrt{\rho_c h/D_c}\right)$ with $D_c = E_c h^3/(12(1 - \nu^2))$ of P-FGM plates are highlighted in this subsection. Tables 10.6(a) and 10.6(b) cover the DSM fundamental frequency results of all the possible Levy-type edge conditions of a rectangular P-FGM plate for $k = 2$ for five sets of E_{rat}, namely 1, 5, 10, 20, and 50 and four side-to-thickness ratios, namely 5, 10, 20, and 50 evaluated for three aspect ratios, namely 0.5, 1.0, and 2.0 at constant $\rho_{\text{rat}} = 1$. It is important to note that the fundamental frequencies in

Table 10.4. Nondimensional fundamental frequencies $\left(\omega^* = \omega a^2\sqrt{\rho_c h/D_c}\right)$ with $D_c = E_c h^3/(12(1-\nu^2))$ for all Levy-type edge conditions of a rectangular **P-FGM** plate with and without consideration of rotary inertia (RI) for $h/a = 0.05$.

BCS	a/b	$k = 0$	$k = 0.2$	$k = 0.5$	$k = 1.0$	$k = 2.0$	$k = 5.0$	$k = 10.0$	$k = \infty$
SCSC	0.5	13.6136	10.6336	11.5263	10.3827	9.4346	8.9428	8.6631	6.9227
		(13.6858)[a]	(10.7002)	(11.5884)	(10.4422)	(9.4938)	(9.0016)	(8.7168)	(6.9594)
	1.5	56.2511	52.2007	47.6292	42.9134	39.0089	36.9830	35.8172	28.6046
		(56.3481)	(52.2902)	(47.7127)	(42.9935)	(39.0886)	(37.0620)	(35.8895)	(28.6539)
	2	95.1170	88.2679	80.5382	72.5650	65.9640	62.5387	60.5667	48.3685
		(95.2625)	(88.4021)	(80.6634)	(72.6850)	(66.0835)	(62.6572)	(60.6751)	(48.4425)
SCSS	0.5	12.8514	11.9263	10.8810	9.8015	8.9065	8.4423	8.1782	6.5352
		(12.3370)	(11.4485)	(10.4463)	(9.4131)	(8.5582)	(8.1144)	(7.8577)	(6.5693)
	1.5	42.4582	39.4009	35.9505	32.3912	29.4443	27.9152	27.0352	21.5907
		(42.5278)	(39.4651)	(36.0104)	(32.4486)	(29.5015)	(27.9719)	(27.0870)	(21.6261)
	2	69.2269	64.2421	58.6164	52.8138	48.0099	45.5171	44.0815	35.2030
		(69.3270)	(64.3344)	(58.7026)	(52.8963)	(48.0921)	(45.5986)	(44.1561)	(35.2393)
SSSS	0.5	12.2741	11.3905	10.3922	9.3612	8.5065	8.0633	7.6752	6.2416
		(12.3370)	(11.4485)	(10.4463)	(9.4131)	(8.5582)	(8.1144)	(7.8577)	(6.2736)
	1.5	32.0287	29.7224	27.1196	24.4349	22.2122	21.0589	20.3947	16.2871
		(32.0762)	(29.7662)	(27.1605)	(24.4741)	(22.2512)	(21.0976)	(20.4301)	(16.3113)
	2	49.2847	45.7358	41.7309	37.6002	34.1807	32.4062	31.3838	25.0621
		(49.3480)	(45.7942)	(41.7854)	(37.6524)	(34.2327)	(32.4578)	(31.4310)	(25.0943)
SCSF	0.5	10.3791	9.6319	8.7878	7.9164	7.1941	6.8194	6.6057	5.2779
		(10.4255)	(9.6747)	(8.8278)	(7.9546)	(7.2321)	(6.8572)	(6.6403)	(5.3015)
	1.5	16.8130	15.5973	14.2317	10.8233	11.6576	11.0526	10.7076	8.5470
		(16.8225)	(15.6110)	(14.2440)	(10.8350)	(11.6699)	(11.0647)	(10.7147)	(8.5545)
	2	22.7995	21.1577	19.3052	17.3950	15.8140	14.9935	14.5198	11.5939
		(22.8155)	(21.1724)	(19.3190)	(17.4081)	(15.8271)	(15.0065)	(14.5317)	(11.6020)
SSSF	0.5	10.2538	9.5156	8.6817	7.8208	7.1073	6.7372	6.5260	5.2142
		(10.2992)	(9.5575)	(8.7208)	(7.8582)	(7.1445)	(6.7741)	(6.5598)	(5.2373)
	1.5	13.7007	10.7141	11.6009	10.4529	9.5028	9.0097	8.7252	6.9670
		(13.7111)	(10.7237)	(11.6099)	(10.4615)	(9.5113)	(9.0182)	(8.7329)	(6.9723)
	2	16.1257	14.9644	13.6543	10.3033	11.1852	10.2699	10.2699	8.2002
		(16.1348)	(14.9728)	(13.6621)	(10.3108)	(11.1927)	(10.6124)	(10.2766)	(8.2048)
SFSF	0.5	9.6960	8.9980	8.2095	7.3956	6.7210	6.3711	6.1713	4.9306
		(9.7362)	(9.0351)	(8.2442)	(7.4287)	(6.7540)	(6.4038)	(6.2013)	(4.9616)
	1.5	9.5538	8.8658	8.0896	7.2892	6.6269	6.2831	6.0845	4.8582
		(9.5582)	(8.8699)	(8.0934)	(7.2929)	(6.6305)	(6.2867)	(6.0879)	(4.8476)
	2	9.5100	8.8251	8.0525	7.2559	6.5968	6.2546	6.0569	4.8360
		(9.5125)	(8.8274)	(8.0547)	(7.2580)	(6.5988)	(6.2567)	(6.0587)	(4.8709)

[a] Without consideration of rotary inertia.

Table 10.5(a). The first nondimensional frequency $\left(\omega^* = \omega a^2\sqrt{\rho_c h/D_c}\right)$ with $D_c = E_c h^3/(12(1-\nu^2))$ for the SCSC-, SCSS-, and SSSS-type edge conditions of a rectangular **P-FGM** plate with the inclusion of rotary inertia.

BCs	*a*/*b*	*h*/*a*	$k = 0$	$k = 0.5$	$k = 1.0$	$k = 2.0$	$k = 5.0$	$k = 10.0$	$k = \infty$
SCSC	**0.5**	**20**	13.6136	11.5263	10.3827	9.4346	8.9428	8.6631	6.9227
		25	13.6394	11.5485	10.4040	9.4558	8.9639	8.6823	6.9359
		40	13.6676	11.5728	10.4272	9.4789	8.9868	8.7033	6.9502
		50	13.6741	11.5784	10.4326	9.4842	8.9921	8.7081	6.9535
		100	13.6829	11.5859	10.4398	9.4914	8.9992	8.7146	6.9580
	1	**20**	28.8850	24.4574	22.0351	20.0291	18.9883	18.3905	14.6885
		25	28.9086	24.4778	22.0546	20.0485	19.0075	18.4081	14.7005
		40	28.9343	24.4999	22.0758	20.0696	19.0285	18.4272	14.7136
		50	28.9403	24.5050	22.0807	20.0745	19.0333	18.4316	14.7166
		100	28.9482	24.5118	22.0872	20.0810	19.0398	14.7206	14.7206
	2	**20**	95.1170	80.5382	72.5650	65.9640	62.5387	60.5667	48.3685
		25	95.1693	80.5832	72.6082	66.0069	62.5813	60.6056	48.3951
		40	95.2261	80.6321	72.6550	66.0535	62.6275	60.6479	48.4240
		50	95.2392	80.6434	72.6658	66.0643	62.6382	60.6577	48.4307
		100	95.2567	80.6584	72.6802	66.0787	62.6525	60.6707	48.4396
SCSS	**0.5**	**20**	12.8514	10.8810	9.8015	8.9065	8.4423	8.1782	6.5352
		25	12.8755	10.9017	9.8213	8.9262	8.4619	8.1961	6.5474
		40	12.9017	10.9243	9.8429	8.9477	8.4832	8.2156	6.5607
		50	12.9078	10.9295	9.8479	8.9527	8.4882	8.2201	6.5638
		100	12.9159	10.9365	9.8546	8.9594	8.4948	8.2261	6.5679
	1	**20**	23.5946	19.9779	17.9994	16.3609	15.5108	15.0224	11.9982
		25	23.6132	19.9939	18.0147	16.3762	15.5259	15.0362	10.0077
		40	23.6333	20.0113	18.0314	16.3928	15.5424	15.0513	10.0179
		50	23.6380	20.0154	18.0352	16.3966	15.5462	15.0547	10.0203
		100	23.6442	20.0207	18.0404	16.4017	15.5513	15.0594	10.0235
	2	**20**	69.2269	58.6164	52.8138	48.0099	45.5171	44.0815	35.2030
		25	69.2629	58.6474	52.8435	48.0394	45.5464	44.1083	35.2213
		40	69.3020	58.6810	52.8757	48.0715	45.5782	44.1374	35.2412
		50	69.3110	58.6888	52.8831	48.0789	45.5856	44.1442	35.2458
		100	69.3230	58.6991	52.8930	48.0888	45.5954	44.1531	35.2519
SSSS	**0.5**	**20**	12.2741	10.3922	9.3612	8.5065	8.0633	7.8109	6.2416
		25	12.2966	10.4116	9.3798	8.5250	8.0816	7.8277	6.2530
		40	12.3212	10.4327	9.4001	8.5452	8.1016	7.8460	6.2655
		50	12.3269	10.4376	9.4047	8.5498	8.1062	7.8502	6.2684
		100	12.3345	10.4442	9.4110	8.5561	8.1124	7.8559	6.2723
	1	**20**	19.6987	16.6793	15.0276	13.6598	10.9502	10.5423	10.0171
		25	19.7133	16.6918	15.0396	13.6718	10.9620	10.5531	10.0245
		40	19.7291	16.7054	15.0526	13.6847	10.9749	10.5648	10.0326
		50	19.7327	16.7086	15.0556	13.6877	10.9778	10.5676	10.0344
		100	19.7376	16.7128	15.0596	13.6917	10.5712	10.0369	10.0369

	2	**20**	49.2847	41.7309	37.6002	34.1807	32.4062	31.3838	25.0621
		25	49.3322	41.7718	37.6393	34.2196	32.4449	31.4192	25.0862
		40	49.3075	41.7505	37.6190	34.1994	32.4248	31.4008	25.0737
		50	49.3379	41.7767	37.6440	34.2243	32.4495	31.4234	25.0891
		100	49.3455	41.7832	37.6503	34.2306	32.4557	31.4291	25.0930

Table 10.5(b). The first nondimensional frequency $\left(\omega^* = \omega a^2\sqrt{\rho_c h/D_c}\right)$ with $D_c = E_c h^3/(12(1-\nu^2))$ for the SCSF-, SFSS-, and SFSF-type edge conditions of a rectangular **P-FGM** plate with the inclusion of rotary inertia.

BCs	*a*/*b*	*a*/*h*	$k = 0$	$k = 0.5$	$k = 1.0$	$k = 2.0$	$k = 5.0$	$k = 10.0$	$k = \infty$
SCSF	**0.5**	**20**	10.3791	8.7878	7.9164	7.1941	6.8194	6.6057	5.2779
		25	10.3957	8.8021	7.9301	7.2077	6.8329	6.6181	5.2864
		40	10.4138	8.8177	7.9450	7.2226	6.8477	6.6316	5.2956
		50	10.4180	8.8213	7.9485	7.2260	6.8511	6.6347	5.2977
		100	10.4236	8.8262	7.9531	7.2306	6.8557	6.6389	5.3006
	1	**20**	12.6694	10.7275	9.6656	8.7864	8.3302	8.0675	6.4426
		25	12.6758	10.7331	9.6709	8.7917	8.3355	8.0723	6.4459
		40	12.6829	10.7391	9.6767	8.7975	8.3412	8.0775	6.4494
		50	12.6845	10.7405	9.6780	8.7988	8.3425	8.0787	6.4503
		100	12.6866	10.7424	9.6798	8.8006	8.3443	8.0804	6.4514
	2	**20**	22.7995	19.3052	17.3950	15.8140	14.9935	14.5198	11.5939
		25	22.8052	19.3102	17.3997	15.8187	14.9981	14.5241	11.5968
		40	22.8115	19.3155	17.4048	15.8238	15.0032	14.5288	11.6000
		50	22.8129	19.3168	17.4060	15.8250	15.0044	14.5298	11.6007
		100	22.8148	19.3184	17.4076	15.8265	15.0059	14.5313	11.6017
SFSS	**0.5**	**20**	10.2538	8.6817	7.8208	7.1073	6.7372	6.5260	5.2142
		25	10.2701	8.6957	7.8342	7.1206	6.7504	6.5381	5.2225
		40	10.2878	8.7110	7.8488	7.1352	6.7648	6.5513	5.2315
		50	10.2919	8.7145	7.8522	7.1385	6.7681	6.5544	5.2336
		100	10.2973	8.7192	7.8567	7.1430	6.7726	6.5584	5.2364
	1	**20**	11.6690	9.8805	8.9025	8.0928	7.6726	7.4306	5.9339
		25	11.6746	9.8853	8.9071	8.0974	7.6772	7.4348	5.9367
		40	11.6806	9.8905	8.9121	8.1024	7.6821	7.4393	5.9398
		50	11.6820	9.8917	8.9132	8.1035	7.6833	7.4403	5.9405
		100	11.6839	9.8933	8.9148	8.1050	7.6848	7.4417	5.9415
	2	**20**	16.1257	13.6543	10.3033	11.1852	10.6050	10.2699	8.2002
		25	16.1290	13.6571	10.3060	11.1879	10.6076	10.2723	8.2018
		40	16.1325	13.6602	10.3089	11.1908	10.6105	10.2750	8.2036
		50	16.1333	13.6609	10.3096	11.1915	10.6112	10.2756	8.2041
		100	16.1344	13.6618	10.3105	11.1924	10.6121	10.2764	8.2046

(Continued)

Table 10.5(b). (*Continued*)

BCs	*a*/*b*	*a*/*h*	$k = 0$	$k = 0.5$	$k = 1.0$	$k = 2.0$	$k = 5.0$	$k = 10.0$	$k = \infty$
SFSF	**0.5**	**20**	9.6960	8.2095	7.3956	6.7210	6.3711	6.1713	4.9306
		25	9.7105	8.2220	7.4075	6.7328	6.3828	6.1820	4.9379
		40	9.7261	8.2355	7.4204	6.7457	6.3956	6.1937	4.9459
		50	9.7298	8.2386	7.4234	6.7487	6.3986	6.1964	4.9477
		100	9.7346	8.2428	7.4274	6.7527	6.4025	6.2000	4.9502
	1	**20**	9.6214	8.1467	7.3405	6.6730	6.3267	6.1270	6.1270
		25	9.6250	8.1498	7.3434	6.6760	6.3296	6.1297	4.8945
		40	9.6289	8.1532	7.3467	6.6792	6.3328	6.1326	4.8964
		50	9.6298	8.1540	7.3474	6.6800	6.3336	6.1333	4.8969
		100	9.6310	8.1550	7.3484	6.6810	6.3345	6.1342	4.8975
	2	**20**	9.5100	8.0525	7.2559	6.5968	6.2546	6.0569	4.8360
		25	9.5109	8.0533	7.2567	6.5975	6.2554	6.0575	4.8364
		40	9.5118	8.0541	7.2575	6.5983	6.2562	6.0583	4.8369
		50	9.5121	8.0543	7.2577	6.5985	6.2563	6.0584	4.8370
		100	9.5124	8.0546	7.2579	6.5987	6.2566	6.0587	4.8372

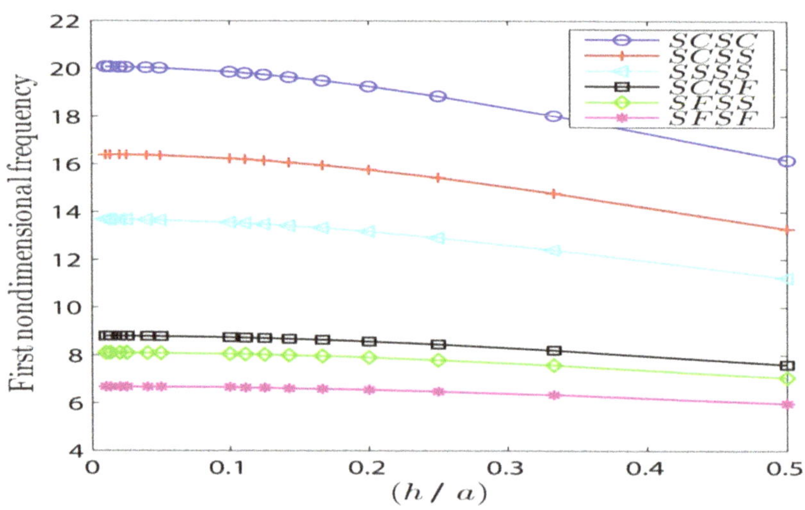

Figure10.6. The variation between the thickness-to-side ratio and the first nondimensional frequency of an FGM square plate for $k = 2$ for all Levy-type boundary conditions.

tables 10.6(a) and 10.6(b) decrease with increasing values of E_{rat}, regardless of the aspect ratio and boundary conditions. This happens because the flexural rigidity/stiffness of the plate decreases as E_{rat} increases.

10.8.4.2 The effect of ρ_{rat} *with constant* E_{rat}

Tables 10.7(a) and 10.7(b) depict the DSM frequency results for all the possible Levy-type boundary conditions of a rectangular P-FGM plate for $k = 2$ for five

Table 10.6(a). The first nondimensional frequencies for SCSC-, SCSS-, and SSSS-type edge conditions of a rectangular **P-FGM** plate for different values of E_{rat} at a constant $\rho_{rat} = 1$.

			E_{rat}				
BCs	*a*/*b*	*a*/*h*	1	5	10	20	50
SCSC	**0.5**	20	13.6136	8.6601	7.3307	6.4369	5.7737
		25	13.6394	8.6805	7.3504	6.4559	5.7919
		50	13.6741	8.7080	7.3770	6.4815	5.8164
		100	13.6829	8.7149	7.3836	6.4879	5.8226
	1.0	20	28.8850	18.3883	15.5736	13.6805	10.2749
		25	28.9086	18.4070	15.5917	13.6980	10.2916
		50	28.9403	18.4321	15.6160	13.7213	10.3140
		100	28.9482	18.4384	15.6220	13.7272	10.3196
	2.0	20	95.1170	60.5628	51.2993	45.0681	40.4406
		25	95.1693	60.6042	60.6042	45.1067	40.4776
		50	95.2392	60.6596	51.3930	45.1583	40.5271
		100	95.2567	60.6735	51.4064	45.1712	40.5395
SCSS	**0.5**	20	12.8514	8.1754	6.9205	6.0768	5.4508
		25	12.8755	8.1944	6.9388	6.0945	5.4677
		50	12.9078	8.2200	6.9635	6.1183	5.4905
		100	12.9159	8.2264	6.9698	6.1243	5.4962
	1.0	20	23.5946	15.0207	10.7217	11.1754	10.0272
		25	23.6132	15.0354	10.7359	11.1891	10.0404
		50	23.6380	15.0551	10.7550	11.2075	10.0580
		100	23.6442	15.0600	10.7598	11.2121	10.0624
	2.0	20	69.2269	44.0790	37.3374	32.8025	29.4347
		25	69.2629	44.1075	37.3650	32.8290	29.4601
		50	69.3110	44.1456	37.4018	32.8645	29.4941
		100	69.3230	44.1551	37.4110	32.8734	29.5027
SSSS	**0.5**	20	12.2741	7.8083	6.6098	5.8041	5.2062
		25	12.2966	7.8262	6.6271	5.8207	5.2221
		50	12.3269	7.8501	6.6502	5.8430	5.2434
		100	12.3345	7.8561	6.6560	5.8486	5.2488
	1.0	20	19.6987	10.5410	10.6218	9.3309	8.3724
		25	19.7133	10.5525	10.6329	9.3416	8.3826
		50	19.7327	10.5679	10.6478	9.3560	8.3964
		100	19.7376	10.5717	10.6515	9.3596	8.3998
	2.0	20	49.2847	31.3824	26.5834	23.3552	20.9577
		25	49.3075	31.4004	26.6008	23.3720	20.9738
		50	49.3379	31.4245	26.6242	23.3944	20.9953
		100	49.3455	31.4306	26.6300	23.4000	21.0007

Table 10.6(b). The first nondimensional frequencies for SCSF-, SFSS-, and SFSF-type edge conditions of a rectangular **P-FGM** plate for different values of E_{rat} at a constant $\rho_{rat} = 1$.

			E_{rat}				
BCs	*a*/*b*	*a*/*h*	1	5	10	20	50
SCSF	**0.5**	20	10.3791	6.6039	5.5909	4.9098	4.4043
		25	10.3957	6.6170	5.6036	4.9220	4.4160
		50	10.4180	6.6347	5.6207	4.9385	4.4318
		100	10.4236	6.6391	5.6249	4.9426	4.4358
	1.0	20	12.6694	18.5430	20.0278	20.9176	21.5121
		25	12.6758	18.5536	20.0395	20.9302	21.5251
		50	12.6845	18.5677	20.0552	20.9469	21.5426
		100	12.6866	18.5712	20.0592	20.9511	21.5470
	2.0	20	22.7995	33.3724	36.0454	37.6476	38.7179
		25	22.8052	33.3818	36.0558	37.6587	38.7295
		50	22.8129	33.3942	36.0698	37.6735	38.7450
		100	22.8148	33.3974	36.0732	37.6773	38.7488
SFSS	**0.5**	20	10.2538	6.5242	5.5235	4.8507	4.3513
		25	10.2701	6.5371	5.5359	4.8626	4.3627
		50	10.2919	6.5543	5.5526	4.8787	4.3781
		100	10.2973	6.5587	5.5568	4.8827	4.3820
	1.0	20	11.6690	7.4302	6.2939	5.5296	4.9619
		25	11.6746	6.2982	5.5337	5.5337	4.9659
		50	11.6820	7.4406	6.3040	5.5392	4.9712
		100	11.6839	7.4421	6.3054	5.5406	4.9725
	2.0	20	16.1257	10.2700	8.7006	7.6448	6.8605
		25	16.1290	10.2725	8.7031	7.6472	6.8628
		50	16.1333	10.2760	8.7064	7.6504	6.8659
		100	16.1344	10.2769	8.7073	7.6512	6.8667
SFSF	**0.5**	20	9.6960	6.1697	5.2236	4.5875	4.1153
		25	9.7105	6.1811	5.2347	4.5981	4.1255
		50	9.7298	6.1964	5.2495	4.6124	4.1392
		100	9.7346	6.2003	5.2532	4.6159	4.1426
	1.0	20	9.6214	6.1268	5.1901	4.5600	4.0920
		25	9.6250	6.1297	5.1929	4.5627	4.0946
		50	9.6298	6.1335	5.1966	4.5662	4.0980
		100	9.6310	6.1345	5.1975	4.5671	4.0988
	2.0	20	9.5100	6.0571	5.1318	4.5092	4.0467
		25	9.5109	6.0578	5.1324	4.5098	4.0474
		50	9.5121	6.0587	5.1333	4.5107	4.0482
		100	9.5124	6.0590	5.1336	4.5109	4.0484

Table 10.7(a). The first nondimensional frequencies for SCSC-, SCSS-, and SSSS-type edge conditions of a rectangular **P-FGM** plate for different values of ρ_{rat} at a constant $E_{rat} = 1$.

			ρ_{rat}				
BCs	*a*/*b*	*a*/*h*	1	5	10	20	50
SCSC	**0.5**	20	13.6136	8.6601	7.3307	6.4369	5.7737
		25	13.6394	8.6805	7.3504	6.4559	5.7919
		50	13.6741	8.7080	7.3770	6.4815	5.8164
		100	13.6829	8.7149	7.3836	6.4879	5.8226
	1.0	20	28.8850	18.3883	15.5736	13.6805	10.2749
		25	28.9086	18.4070	15.5917	13.6980	10.2916
		50	28.9403	18.4321	15.6160	13.7213	10.3140
		100	28.9482	18.4384	15.6220	13.7272	10.3196
	2.0	20	95.1170	60.5628	51.2993	45.0681	40.4406
		25	95.1693	60.6042	60.6042	45.1067	40.4776
		50	95.2392	60.6596	51.3930	45.1583	40.5271
		100	95.2567	60.6735	51.4064	45.1712	40.5395
SCSS	**0.5**	20	12.8514	8.1754	6.9205	6.0768	5.4508
		25	12.8755	8.1944	6.9388	6.0945	5.4677
		50	12.9078	8.2200	6.9635	6.1183	5.4905
		100	12.9159	8.2264	6.9698	6.1243	5.4962
	1.0	20	23.5946	15.0207	10.7217	11.1754	10.0272
		25	23.6132	15.0354	10.7359	11.1891	10.0404
		50	23.6380	15.0551	10.7550	11.2075	10.0580
		100	23.6442	15.0600	10.7598	11.2121	10.0624
	2.0	20	69.2269	44.0790	37.3374	32.8025	29.4347
		25	69.2629	44.1075	37.3650	32.8290	29.4601
		50	69.3110	44.1456	37.4018	32.8645	29.4941
		100	69.3230	44.1551	37.4110	32.8734	29.5027
SSSS	**0.5**	20	12.2741	7.8083	6.6098	5.8041	5.2062
		25	12.2966	7.8262	6.6271	5.8207	5.2221
		50	12.3269	7.8501	6.6502	5.8430	5.2434
		100	12.3345	7.8561	6.6560	5.8486	5.2488
	1.0	20	19.6987	10.5410	10.6218	9.3309	8.3724
		25	19.7133	10.5525	10.6329	9.3416	8.3826
		50	19.7327	10.5679	10.6478	9.3560	8.3964
		100	19.7376	10.5717	10.6515	9.3596	8.3998
	2.0	20	49.2847	31.3824	26.5834	23.3552	20.9577
		25	49.3075	31.4004	26.6008	23.3720	20.9738
		50	49.3379	31.4245	26.6242	23.3944	20.9953
		100	49.3455	31.4306	26.6300	23.4000	21.0007

Table 10.7(b). The first nondimensional frequencies for SCSF-, SFSS-, and SFSF-type edge conditions of a rectangular **P-FGM** plate for different values of ρ_{rat} at a constant $E_{\text{rat}} = 1$.

			ρ_{rat}				
BCs	*a*/*b*	*a*/*h*	1	5	10	20	50
SCSF	**0.5**	20	10.3791	15.1857	16.3999	17.1274	17.6133
		25	10.3957	15.2128	16.4300	17.1595	17.6468
		50	10.4180	15.2492	16.4706	17.2027	17.6917
		100	10.4236	15.2583	16.4807	17.2135	17.7030
	1.0	20	12.6694	18.5430	20.0278	20.9176	21.5121
		25	12.6758	18.5536	20.0395	20.9302	21.5251
		50	12.6845	18.5677	20.0552	20.9469	21.5426
		100	12.6866	18.5712	20.0592	20.9511	21.5470
	2.0	20	22.7995	33.3724	36.0454	37.6476	38.7179
		25	22.8052	33.3818	36.0558	37.6587	38.7295
		50	22.8129	33.3942	36.0698	37.6735	38.7450
		100	22.8148	33.3974	36.0732	37.6773	38.7488
SFSS	**0.5**	20	10.2538	15.0025	16.2020	16.9207	17.4008
		25	10.2701	15.0290	16.2315	16.9522	17.4336
		50	10.2919	15.0645	16.2711	16.9944	17.4775
		100	10.2973	15.0735	16.2811	17.0050	17.4886
	1.0	20	11.6690	17.0791	18.4466	19.2663	19.8138
		25	11.6746	17.0882	18.4568	19.2771	19.8251
		50	11.6820	17.1003	18.4703	19.2916	19.8402
		100	11.6839	17.1034	18.4737	19.2952	19.8439
	2.0	20	16.1257	23.6041	25.4948	26.6281	27.3852
		25	16.1290	23.6094	25.5008	26.6344	27.3918
		50	16.1333	23.6165	25.5087	26.6429	27.4006
		100	16.1344	23.6183	25.5107	26.6450	27.4028
SFSF	**0.5**	20	9.6960	14.1869	15.3213	16.0011	16.4551
		25	9.7105	14.2103	15.3475	16.0290	16.4842
		50	9.7298	14.2418	15.3826	16.0663	16.5231
		100	9.7346	14.2498	15.3914	16.0757	16.5329
	1.0	20	9.6214	14.0826	15.2103	15.8863	16.3379
		25	9.6250	14.0884	15.2169	15.8933	16.3451
		50	9.6298	14.0963	15.2256	15.9026	16.3549
		100	9.6310	14.0983	15.2278	15.9049	16.3573
	2.0	20	9.5100	13.9208	15.0360	15.7045	16.1511
		25	9.5109	13.9222	15.0376	15.7063	16.1529
		50	9.5121	13.9242	15.0398	15.7085	16.1553
		100	9.5124	13.9247	15.0403	15.7091	16.1559

values of ρ_{rat}, namely 1, 5, 10, 20, and 50 and four values of the side-to-thickness ratio, namely 5, 10, 20, and 50 evaluated for three aspect ratios, namely 0.5, 1.0, and 2.0 at a constant $E_{rat}= 1$. It can be observed from tables 10.7(a) and 10.7(b) that irrespective of the boundary conditions and aspect ratios, the fundamental frequencies increase with increasing ρ_{rat}. This is due to the reduction in the mass of the FGM plate.

10.9 Conclusions

In this chapter, the Wittrick–Williams algorithm was used for the accurate free vibration analysis of thin functionally graded plates. The analysis incorporated rotary inertia and variation of the power-law properties of the Levy-type boundary conditions. Hamilton's principle was used to find the equation of motion and the natural boundary condition by considering the CPT in conjunction with the physical neutral surface. A Levy-type closed-form solution was used to solve the equation of motion and a generic boundary condition was used to develop the dynamic stiffness matrix. The expressions for every frequency-dependent element of the dynamic stiffness matrix were given in detail. The Wittrick–Williams algorithm was used to solve the overall dynamic stiffness matrix in order to obtain the accurate natural frequencies and mode shapes of FGM plates. The entire process of these computations was implemented in the MATLAB software in order to obtain highly accurate natural frequencies.

The natural frequency results computed using the DSM formulation were compared with results published in the research literature and excellent agreement was observed. In a comparison of the DSM results with $k = 0$ for the homogeneous ceramic constituent isotropic plate, an exact match was found with the results of Reddy and Phan [28], Boscolo and Banerjee [2], and Liu and Banerjee [22, 23], in which rotary inertia was considered. This shows that these findings represent the most accurate results, which can thus be used as benchmark results for comparison and validation purposes in the future.

In addition, the effects on the DSM frequency results of FGM plate parameters such as the side-to-thickness ratio, the aspect ratio, the edge conditions, the volume fraction index and the material property were discussed in detail with their respective physical interpretations using several tables and graphs. Finally, it was concluded that these parametric studies can make an important contribution to FGM plate design.

References and further reading

[1] Swaminathan K and Sangeetha D M 2017 Thermal analysis of FGM plates—a critical review of various modeling techniques and solution methods *Compos. Struct.* **160** 43–60

[2] Boscolo M and Banerjee J R 2011 Dynamic stiffness elements and their applications for plates using first order shear deformation theory *Comput. Struct.* **89** 395–410

[3] Koizumi M 1997 FGM activities in Japan *Composites* B **28** 1–4

[4] Qian L F and Batra R C 2005 Design of bidirectional functionally graded plate for optimal natural frequencies *J. Sound Vib.* **280** 415–24

[5] Liu D Y, Wang C Y and Chen W Q 2010 Free vibration of FGM plates with in-plane material inhomogeneity *Compos. Struct.* **92** 1047–51

[6] Birman V and Byrd L W 2007 Modeling and analysis of functionally graded materials and structures *Appl. Mech. Rev.* **60** 195

[7] I S and Miyamoto Y 1996 Functionally graded materials *4th Int. Symp. on Functionally Graded Materials* p 803

[8] Suresh S and Mortensen A 1997 Functionally graded metals and metal–ceramic composites: part 2. Thermomechanical behaviour *Int. Mater. Rev.* **42** 85–116

[9] Saurabh R, Subodh K and Ankit G 2022 A comprehensive review of design and development of advanced tailored material on sustainability aspects *Sustainable Advanced Manufacturing and Materials Processing* (Boca Raton, FL: CRC Press) pp 159–85

[10] Leung A Y T 2012 *Dynamic Stiffness and Substructures* (London: Springer) p 242

[11] Zhang D G and Zhou Y H 2008 A theoretical analysis of FGM thin plates based on physical neutral surface *Comput. Mater. Sci.* **44** 716–20

[12] Abrate S 2008 Functionally graded plates behave like homogeneous plates *Composites* B **39** 151–8

[13] Kumar S, Ranjan V and Jana P 2018 Free vibration analysis of thin functionally graded rectangular plates using the dynamic stiffness method *Compos. Struct.* **197** 39–53

[14] Wittrick W H and Williams F W 1971 A general algorithm for computing natural frequencies of elastic structures *Q. J. Mech. Appl. Math.* **24** 263–84

[15] Wittrick W H and Williams F W 1974 Buckling and vibration of anisotropic or isotropic plate assemblies under combined loadings *Int. J. Mech. Sci.* **16** 209–39

[16] Banerjee J R 1997 Dynamic stiffness formulation for structural elements: a general approach *Comput. Struct.* **63** 101–3

[17] Boscolo M and Banerjee J R 2011 Dynamic stiffness method for exact inplane free vibration analysis of plates and plate assemblies *J. Sound Vib.* **330** 2928–36

[18] Boscolo M and Banerjee J R 2012 Dynamic stiffness formulation for composite Mindlin plates for exact modal analysis of structures. Part I: theory *Comput. Struct.* **96–7** 61–73

[19] Boscolo M and Banerjee J R 2012 Dynamic stiffness formulation for composite Mindlin plates for exact modal analysis of structures. Part II: results and applications *Comput. Struct.* **96–7** 74–83

[20] Boscolo M and Banerjee J R 2014 Layer-wise dynamic stiffness solution for free vibration analysis of laminated composite plates *J. Sound Vib.* **333** 200–27

[21] Papkov S O and Banerjee J R 2015 A new method for free vibration and buckling analysis of rectangular orthotropic plates *J. Sound Vib.* **339** 342–58

[22] Liu X and Banerjee J R 2015 An exact spectral-dynamic stiffness method for free flexural vibration analysis of orthotropic composite plate assemblies—part I: theory *Compos. Struct.* **132** 1274–87

[23] Liu X and Banerjee J R 2016 Free vibration analysis for plates with arbitrary boundary conditions using a novel spectral-dynamic stiffness method *Comput. Struct.* **164** 108–26

[24] Ghorbel O, Casimir J B, Hammami L, Tawfiq I and Haddar M 2016 In-plane dynamic stiffness matrix for a free orthotropic plate *J. Sound Vib.* **364** 234–46

[25] Ghorbel O, Casimir J B, Hammami L, Tawfiq I and Haddar M 2015 Dynamic stiffness formulation for free orthotropic plates *J. Sound Vib.* **346** 361–75

[26] Kumar S and Jana P S 2019 Application of dynamic stiffness method for accurate free vibration analysis of sigmoid and exponential functionally graded rectangular plates *Int. J. Mech. Sci.* **163** 105105

[27] Kumar S and Jana P S 2022 Accurate solution for free vibration behaviour of stepped FGM plates implementing the dynamic stiffness method *Structures* **45** 1971–89

[28] Reddy J N and Phan N D 1985 Stability and vibration of isotropic, orthotropic and laminated plates according to a higher-order shear deformation theory *J. Sound Vib.* **98** 157–70

[29] Zenkour A M 2005 On vibration of functionally graded plates according to a refined trigonometric plate theory *Int. J. Struct. Stab. Dyn.* **5** 279–97

[30] Yang J and Shen H-S 2001 Dynamic response of initially stressed functionally graded rectangular thin plates *Composite Structures* **54** 497–508

[31] Chi S H and Chung Y L 2006 Mechanical behavior of functionally graded material plates under transverse load—part I: analysis *Int. J. Solids Struct.* **43** 3657–74

[32] Chung Y L and Chang H X 2008 Mechanical behavior of rectangular plates with functionally graded coefficient of thermal expansion subjected to thermal loading *J. Therm. Stresses* **31** 368–88

[33] Reddy K S K and Kant T 2014 Three-dimensional elasticity solution for free vibrations of exponentially graded plates *J. Eng. Mech.* **140** 04014047

[34] Chakraverty S and Pradhan K K 2014 Free vibration of exponential functionally graded rectangular plates in thermal environment with general boundary conditions *Aerosp. Sci. Technol.* **36** 132–56

[35] Chakraverty S and Pradhan K K 2014 Free vibration of functionally graded thin rectangular plates resting on Winkler elastic foundation with general boundary conditions using Rayleigh–Ritz method *Int. J. Appl. Mech.* **06** 1450043

[36] Yin S, Yu T and Liu P 2013 Free vibration analyses of FGM thin plates by isogeometric analysis based on classical plate theory and physical neutral surface *Adv. Mech. Eng.* **5** 634584

[37] Hasani Baferani A, Saidi A R and Jomehzadeh E 2011 An exact solution for free vibration of thin functionally graded rectangular plates *Proc. Inst. Mech. Eng.* C **225** 526–36

[38] Li S-R, Wang X and Batra R C 2015 Correspondence relations between deflection, buckling load, and frequencies of thin functionally graded material plates and those of corresponding homogeneous plates *J. Appl. Mech.* **82** 111006

[39] Hosseini-Hashemi S, Rokni Damavandi Taher H, Akhavan H and Omidi M 2010 Free vibration of functionally graded rectangular plates using first-order shear deformation plate theory *Appl. Math. Model.* **34** 1276–91

[40] Hosseini-Hashemi S, Fadaee M and Seyed Rasoul A 2011 Study on the free vibration of thick functionally graded rectangular plates according to a new exact closed-form procedure *Compos. Struct.* **93** 722–35

[41] Baferani A, Hasani A R, Saidi and Ehteshami H 2011 Accurate solution for free vibration analysis of functionally graded thick rectangular plates resting on elastic foundation *Compos. Struct.* **93** 1842–53

[42] Arthur W L 1973 The free vibration of rectangular plates *J. Sound Vib.* **31** 257–93

[43] Thai H-T, Park M and Choi D-H 2013 A simple refined theory for bending, buckling, and vibration of thick plates resting on elastic foundation *Int. J. Mech. Sci.* **73** 40–52

Chapter 11

Effect of functionally graded magneto-electro-elastic facings on the damped nonlinear transient response of a sandwich plate with agglomerated CNT core

Vinyas Mahesh, Vishwas Mahesh, Arjun Siddharth Mangalasseri, Dineshkumar Harursampath and Sathiskumar A Ponnusami

This chapter numerically investigates the controlled nonlinear transient response of sandwich plates made of functionally graded magneto-electro-elastic (MEE) face sheets and an agglomerated carbon nanotube (ACNT) core using the finite element (FE) method. The damping of the plate is accomplished through an active constrained layer damping (ACLD) treatment. Using the control feedback loop technique, a voltage proportional to the displacement/velocity is fed back to the piezoelectric (PE) layer of the ACLD patch. The viscoelastic (VE) layer in the ACLD patch contributes to passive damping. The carbon nanotubes (CNTs) are either partially or completely agglomerated and modeled using the Eshelby–Mori–Tanaka approach. The equations of motion of a plate with an ACLD patch are obtained using layer-wise shear deformation theory (LSDT) and Hamilton's principle. A synergy of three-field (magnetic, electric, and elastic) coupling, agglomeration, and electromagnetic (EM) circuit effects is established, which makes this work complex. This method represents a ready tool for the realistic design of smart structures and systems. Numerical examples are solved to assess the influence of CNT distributions, the functional gradation of MEE composites, electromagnetic circuits, and coupling. This work serves as a novel benchmark solution for further research into the active control of agglomerated CNT structures.

11.1 Introduction

The excellent material properties of CNTs, such as specific strength and stiffness, represent an enhancement compared to those of conventional carbon-fiber

doi:10.1088/978-0-7503-5301-4ch11

composites [1, 2]. This makes them a promising candidate for numerous engineering applications, including aerospace, marine, and automobile engineering, when reinforced with a polymer or metal matrix. The effective material properties of composites reinforced with CNTs depend entirely on their dispersion and orientation [3, 4]. The CNTs' large aspect ratio, in addition to their minimum diameter and radial stiffness, makes it difficult to achieve uniform dispersion throughout the matrix. This leads to agglomeration, which drastically influences the effective material properties, including the mechanical and multifunctional properties [5].

Hence, several researchers have focused on assessing the various responses of structures made of agglomerated CNTs. Using FEs, the influence of CNT agglomeration on the frequency response of composite plates was studied by Craveiro and Loja [6]. Syah *et al* [7] demonstrated the influence of agglomerated CNTs on the dynamic response of sandwich shells through numerical simulation. Garcia-Macias and Castro-Triguero [8] probed the frequency response of polymer plates reinforced with agglomerated CNTs. They stressed the integrated effect of CNTs' waviness and agglomeration. Moradi-Dastjerdi *et al* [9] attempted to study the effects of agglomeration on the stability characteristics of CNTs/piezoelectric sandwich plates. Bisheh *et al* [10] investigated the application of CNTs/piezoelectric shells for structural health monitoring (SHM) based on wave propagation techniques and assessed the effect of agglomeration on the overall structural response. Daghigh *et al* [11] investigated the effects of elastic foundations and CNT agglomeration on the bending and buckling of CNT-reinforced nanoplates using Eringen's nonlocal elasticity theory. Using semi-analytical methods, Fakhar *et al* [12] investigated the effects of agglomeration on the fluid–structure interaction of CNT-reinforced pipes. Ghasemi *et al* [13] computationally assessed the vibration response of agglomerated CNT-reinforced hybrid shells using Kirchoff–Love's first approximation. Arani and Zamani [14] investigated the bending response of beams resting on elastic foundations; their study included agglomeration effects. Hajmohammad *et al* [15] studied the coupled effect of agglomerated CNTs and auxeticity on the dynamic response of blast-loaded hybrid plates. Moradi-Dastjerdi and Behdinan [16] probed the natural frequencies of smart plates composed of agglomerated CNTs and piezoelectric phases. Singh *et al* [17] presented a semi-analytical model for the prediction of the nonlinear dynamic and buckling responses of composite plates affected by the agglomeration of CNTs. The free vibrations of an agglomerated CNT composite panel with constant and variable thickness were studied by Shahmohammadi *et al* [18] using the isogeometric finite strip method. Yadav *et al* [19] examined the effect of various agglomeration patterns of CNTs on the nonlinear damped response of a three-phase composite shell.

Recently, the triple-interaction capability exhibited by the new form of smart materials called MEE composites has profoundly influenced various structural applications. Such composites can assist in various multifunctional activities ranging from sensing [20, 21] to energy harvesting [22–25]. They also exhibit inherent vibration control abilities [26–31], which makes them best suited for structural health monitoring. Researchers have proposed various computational models with which to investigate the different structural responses of MEE beams, plates, and shells [32]. However, since they offer a good trade-off between the computational effectiveness and the effort involved, FE methods have been found useful in investigating MEE structural responses [33–36]. Meanwhile, in contrast to the basic displacement models, the coupled field interactions of MEE structures are well captured by the higher-order shear deformation theory (HSDT), which is also more reliable [37–39].

11.2 Problem description

This chapter evaluates the damped response of a smart sandwich plate with a CNT core and functionally graded MEE (FGMEE) facings using the FE framework. The core of the plate is assumed to be agglomerated with CNTs (ACNT). To this end, partial and complete agglomeration states are considered. Further, the facings are functionally graded with 'B'-rich and 'F'-rich gradation patterns. The material properties of the FGMEE layer are altered using the gradient index '*r*,' which is discussed in subsequent sections. A schematic of the smart ACNT/FGMEE plate with ACLD patches is shown in figure 11.1(a), and the representative volume element (RVE) of the ACNT core is depicted in figure 11.1(b). The material properties of the MEE face sheets and PE layer used in this study are listed in table 11.1.

11.3 Materials and methods

The effective material properties of ACNTs are derived using the two-parameter micromechanics model [40]. The overall volume fraction of CNT reinforcement in the RVE (V^r) is the summation of its volume fraction dispersed in the matrix (V^r_m) and agglomerated (V^r_{incl}). It can be mathematically expressed as,

$$V^r = V^r_{\text{incl}} + V^r_m \tag{11.1}$$

Using the parameters μ and η, it can be shown that

$$\mu = \frac{V_{\text{incl}}}{V}; \quad \eta = \frac{V^r_{\text{incl}}}{V^r}; \quad f_r = \frac{V^r}{V}. \tag{11.2}$$

If we consider the effect of agglomeration, the effective shear and bulk moduli both inside and outside the region of the inclusions can be expressed as [40]:

$$K_{\text{in}} = K_m + \frac{(\delta_r - 3K_m\alpha_r)f_r\eta}{3(\mu - f_r\eta(1 - \alpha_r))}$$

$$K_{\text{out}} = K_m + \frac{(\delta_r - 3K_m\alpha_r)f_r(1 - \eta)}{3[1 - \mu - f(1 - \eta) + (1 - \eta)f_r\alpha_r]}$$

$$G_{\text{in}} = G_m + \frac{(\eta_r - 3G_m\beta_r)f_r\eta}{2(\mu - f_r\eta(1 - \beta_r))}$$

$$G_{\text{out}} = G_m + \frac{(\eta_r - 2G_m\beta_r)f_r(1 - \eta)}{2[1 - \mu - f_r(1 - \eta) + f_r(1 - \eta)\beta_r]}, \tag{11.3}$$

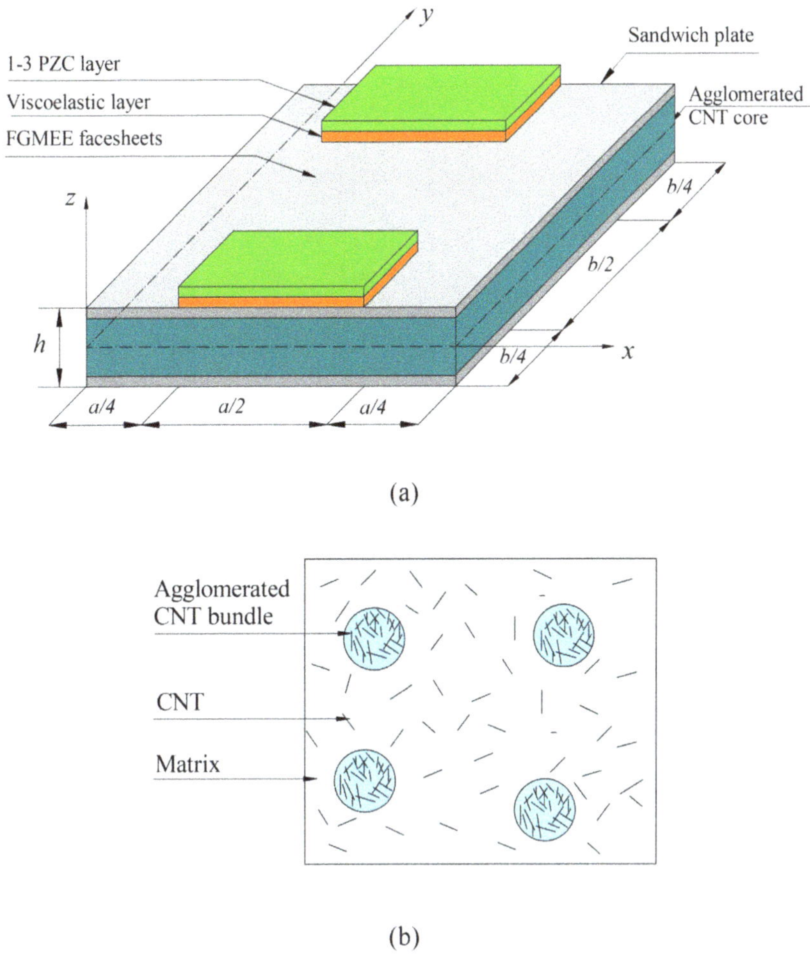

Figure 11.1. A schematic representation of (a) the geometry of a sandwich ACNT/FGMEE plate and (b) the agglomerated CNT RVE.

where

$$\alpha_r = \frac{k_r - l_r + 3(K_m + G_m)}{3(K_m + k_r)}$$

$$\beta_r = \frac{1}{5}\left\{\frac{4G_m + 2k_r + l_r}{3(G_m + k_r)} + \frac{4G_m}{G_m + p_r} + \frac{2[G_m(3K_m + G_m) + G_m(3K_m + 7G_m)]}{G_m(3K_m + G_m) + m_r(3K_m + 7G_m)}\right\}$$

$$\delta_r = \frac{1}{3}\left[n_r + 2l_r + \frac{(2k_r + l_r)(3K_m + G_m - l_r)}{G_m + k_r}\right]$$

Table 11.1. Material properties of the MEE material and the piezoelectric patch of the ACLD [28].

Material property	Material constants	MEE (0.5 V_f)	PE patch
Elastic constants (GPa)	$Q_{11} = Q_{22}$	220	9.29
	Q_{12}	120	6.18
	$Q_{13} = Q_{23}$	120	6.05
	Q_{33}	215	35.44
	$Q_{44} = Q_{55}$	45	1.58
	Q_{66}	50	1.54
Piezoelectric constants (C m^{-2})	$e_{31} = e_{32}$	−3.5	−0.1902
	e_{33}	9.0	18.4107
	$e_{15} = e_{24}$	0	0.004
Dielectric constants (10^{-9} C^2 N^{-1} m^{-2})	$\varepsilon_{11} = \varepsilon_{22}$	0.85	—
	ε_{33}	6.3	—
Magnetic permeability (10^{-4} N s^2 C^{-2})	$\mu_{11} = \mu_{22}$	−2.0	—
	μ_{33}	0.9	—
Piezomagnetic constants (N A^{-1} m^{-1})	$q_{31} = q_{32}$	350	—
	q_{33}	320	—
	$q_{15} = q_{24}$	200	
Magneto-electric constants (10^{-12} N s V^{-1} C^{-1})	$m_{11} = m_{22}$	5.5	
	m_{33}	2600	
Density (kg m^{-3})	ρ	5550	5090

$$\eta_r = \frac{1}{5}\left[\frac{2}{3}(n_r - l_r) + \frac{8G_m p_r}{G_m + p_r} + \frac{8m_r G_m(3K_m + 4G_m)}{3K_m\left(G_m + p_r\right) + G_m(7m_r + G_m)} + \frac{2(k_r - l_r)(2G_m + l_r)}{3(G_m + k_r)}\right]. \tag{11.4}$$

Meanwhile, the entire composite's effective bulk and shear moduli can be represented using equations (11.3) and (11.4), as follows:

$$K = K_{\text{out}}\left[1 + \frac{\mu\left(\dfrac{K_{\text{in}}}{K_{\text{out}}} - 1\right)}{1 + \alpha(1 - \mu)\left(\dfrac{K_{\text{in}}}{K_{\text{out}}} - 1\right)}\right]$$

$$G = G_{\text{out}}\left[1 + \frac{\mu\left(\dfrac{G_{\text{in}}}{G_{\text{out}}} - 1\right)}{1 + \beta(1 - \mu)\left(\dfrac{G_{\text{in}}}{G_{\text{out}}} - 1\right)}\right].$$

Here,

$$\nu_{\text{out}} = \frac{(3K_{\text{out}} - 2G_{\text{out}})}{2(3K_{\text{out}} + 2G_{\text{out}})}; \quad \alpha = \frac{1 + \nu_{\text{out}}}{3(1 - \nu_{\text{out}})}; \quad \beta = \frac{2(4 - 5\nu_{\text{out}})}{15(1 - \nu_{\text{out}})}. \tag{11.5}$$

With agglomerated CNTs, the composite's elastic modulus and Poisson ratio are given by:

$$E = \frac{9KG}{3K + G}; \ \nu = \frac{3K - 2G}{6K + 2G}. \tag{11.6}$$

The different forms of CNT distribution considered in this study are listed below:

uniform distribution (UD) $f_r = f_{r*}$

symmetric functionally graded (SFG) $f_r = \left(\frac{4\,|\,z\,|}{h}\right) f_{r*}$

unsymmetrical functionally graded (USFG) $f_r = \left(1 - \frac{2z}{h}\right) f_{r*}$,

where

$$f_{r*} = \frac{w_r}{w_r + \left(\frac{\rho_r}{\rho_m}\right) - \left(\frac{\rho_r}{\rho_m}\right) w_r}. \tag{11.7}$$

The constitutive relations of the FGMEE material can be shown as [29]:

$$\{\sigma\} = [Q]\{\varepsilon\} - [e]\{E\} - [q]\{H\}$$

$$\{D\} = [e]^T\{\varepsilon\} + [\eta]\{E\} + [m]\{H\}$$

$$\{B\} = [q]^T\{\varepsilon\} + [m]\{E\} + [\mu]\{H\}. \tag{11.8}$$

Here,

$$V_{fg}(z) = V_l + (V_u - V_l)\left\{\left(\frac{z}{h}\right) + \left(\frac{1}{2}\right)\right\}^r,$$

which is the generalized representation of the FG material coefficient, such as $[e]$, $[h]$, $[f]$, $[q]$, $[C]$, $[\mu]$, $[\alpha]$, $[p]$ or $[\tau]$. Further, the material properties corresponding to the bottom and top layers are generalized by the notations V_l and V_u, respectively. Meanwhile, the gradient index of the FGMEE material is denoted by r.

The constitutive equations for the PE layer of the ACLD patch can be explicitly represented as follows [27]:

$$\{\sigma_b^p\} = \left[Q_b^p\right]\{\varepsilon_b^p\} - \left[Q_{bs}^p\right]\{\varepsilon_s^p\} - \{\bar{e}_b^p\}E_z$$

$$\{\sigma_s^p\} = \left[Q_{bs}^p\right]^T\{\varepsilon_b^p\} + \left[Q_s^p\right]\{\varepsilon_s^p\} - \{\bar{e}_b^p\}E_z$$

$$D_z^p = \{\bar{e}_b^p\}^T\{\varepsilon_b^p\} + \{\bar{e}_s^p\}^T\{\varepsilon_s^p\} + \in_{33}^p E_z^p. \tag{11.9}$$

Using the Golla-Hughes McTavish (GHM) approach, the viscoelastic layer of the ACLD patch can be modeled as follows [29]:

$$(\sigma_s)_v = \int_0^t G(t-\tau)\frac{\partial(\varepsilon_s^k)}{\partial\tau}\partial\tau. \tag{11.10}$$

Further, developing the FE formulation for an ACNT/FGMEE plate using the LSDT and an eight-noded isoparametric element, in a similar manner to that illustrated in reference [28], the final equations of motion can be written as follows:

$$[M^*]\{\ddot{X}_t\} + [C_d^*]\{\dot{X}_t\}[K^*]\{X_t\} = \{F^*\}, \tag{11.11}$$

where $[M^*]$, $[C_d^*]$, $[K^*]$ are the equivalent mass, damping, and stiffness matrices, respectively. The reader is encouraged to review Mahesh [28] for a detailed derivation of the FE formulation.

11.4 Results and discussion

In this section, the damped behavior of ACNT/FGMEE plates is numerically investigated using FE methods. For more insight into the formulation, the reader is referred to [28]. The parameters adopted for the analysis are as follows: control gain $K_d = 600$; central ACLD patch; vertical orientation of the PE fibers in the ACLD patch; simply supported boundary condition; $a = b = 1.0$ m; $a/h = 100$; $h_v = 50.8$ µm; $h_p = 250$ µm; mechanical pulse load $q_0 = 2$ kN m^{-2}.

The electromagnetic constraints used for the numerical simulations were [28]:

$$\text{Closed-circuit: } \phi = \psi = 0 \left(z = \pm\frac{h}{2}\right)$$

$$\text{Open-circuit: } D_z = B_z = 0 \left(z = \pm\frac{h}{2}\right). \tag{11.12}$$

The formulation was verified, and the results are shown in tables 11.2 and 11.3. It can be inferred from these results that the proposed formulation accurately replicates the coupled interaction and agglomeration effects.

Figures 11.2 and 11.3 illustrate the ACNT/FGMEE sandwich plate's attenuated response in the partial and complete agglomeration cases, respectively. In addition, comparative studies to assess the influence of the CNT arrangement, the MEE functional gradation, and gradient index on the damped response are also highlighted in these figures. It can be noticed in these figures that the smart sandwich plate with the SFG type of CNT arrangement is the most convenient to control using an ACLD treatment. This may be attributed to the higher stiffness of the plate. Meanwhile, the USFG arrangement requires a greater effort to dampen the vibration amplitude. The clear contrast between figures 11.2(a) and (b) suggests that for the 'B'-rich bottom, the vibration amplitude diminishes for higher values of the gradation index, whereas for the 'F'-rich bottom, the trend is reversed. This is because a higher gradient index enhances the percentage of the PE phase, which in

Table 11.2. A comparison of the natural frequencies of MEE plates (h = 0.3 m; a = b=1 m).

Stacking sequence	Mode no.	Non-dimensional frequency		% Error
		Chen *et al* [41]	This chapter	
BFB	1	1.3434	1.3482	0.3601
	2	2.2199	2.2279	0.3597
	3	2.2199	2.2279	0.3597
FBF	1	1.4463	1.4346	−0.8109
	2	2.3602	2.3390	−0.8979
	3	2.3602	2.3392	−0.8905

Table 11.3. A verification of the fundamental natural frequencies of completely agglomerated plates.

	η = 1.00; μ = 0.25		η = 1.00; μ = 0.5		η = 1.00; μ = 0.75	
CNT distribution	Craveiro and Loja [6]	Present	Craveiro and Loja [6]	Present	Craveiro and Loja [6]	Present
UD	7.2119	7.2011	9.8010	9.7893	8.5212	8.4438
SFG	7.2801	7.2573	10.6382	10.4593	8.8357	8.8033
USFG	6.9518	6.9321	7.9533	7.9434	8.9882	8.9454

turn reduces the stiffness of the plate. One more prominent observation in this study is that compared to the complete agglomeration case, the vibration amplitude is smaller for the ACNT core with the partial agglomeration of CNTs. Complete agglomeration weakens the material and facilitates more vibration.

The influence of electromagnetic circuits, such as closed and open circuits, on the damped response is studied in figure 11.4, which includes both partial and complete agglomeration cases. Due to the higher energy dissipation capability of open-circuit conditions, the vibration amplitudes are smaller than those of closed-circuit conditions. In addition, it can be seen that the influence of EM circuits is more predominant in the case of partial agglomeration than for complete agglomeration. Further, the percentage differences in the vibration amplitudes of ACNT/FGMEE plates with different gradation patterns and gradient indices are shown in figure 11.5.

The analysis was extended to investigate the effect of coupling on the controlled response of ACNT/FGMEE plates. Here, the coupling effect is assessed by comparing the transient response of the plate, including complete coupling between all three fields, with that of the response of the plate obtained under the influence of the elastic field alone. As can be seen in figure 11.6, multifield coupled interactions enhance the stiffness and reduce the vibration amplitude. Further, the influence of coupling is found to be predominant when the ACNT core is partially agglomerated. Analogously, the influence of the gradation pattern and index associated with the coupling is illustrated in figure 11.7 in terms of the percentage difference between the vibration amplitudes. Likewise, considering r = 0.5, 'B'-rich gradation and the

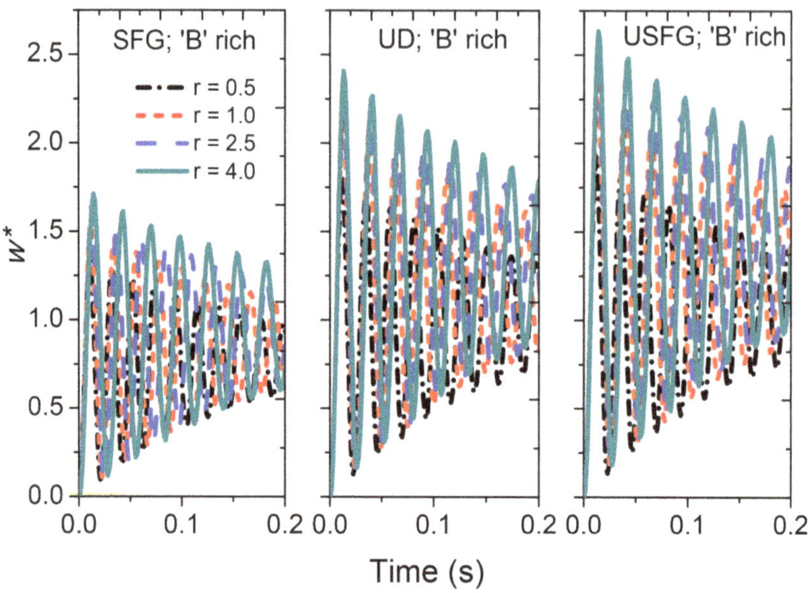

(a)

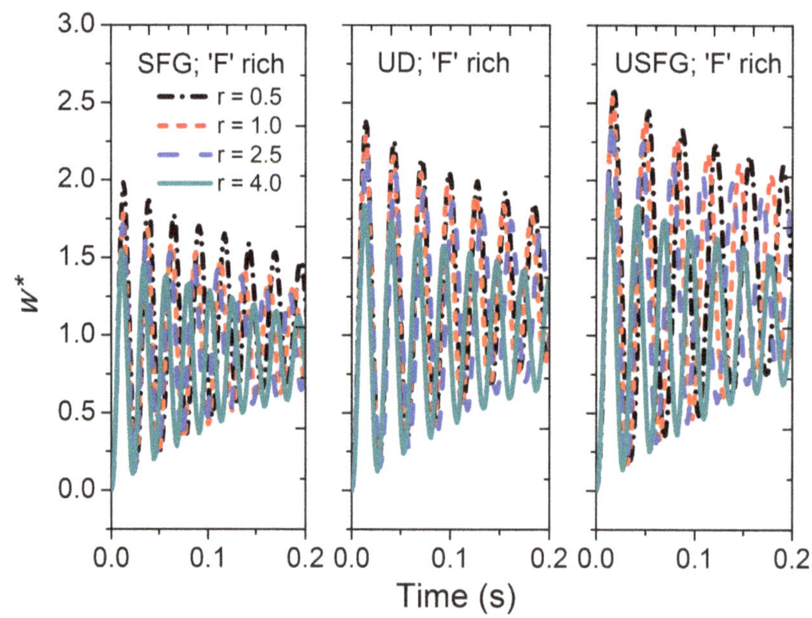

(b)

Figure 11.2. The controlled vibration responses of ACNT/FGMEE plates with a partially agglomerated core and (a) 'B'-rich and (b) 'F'-rich FGMEE facings.

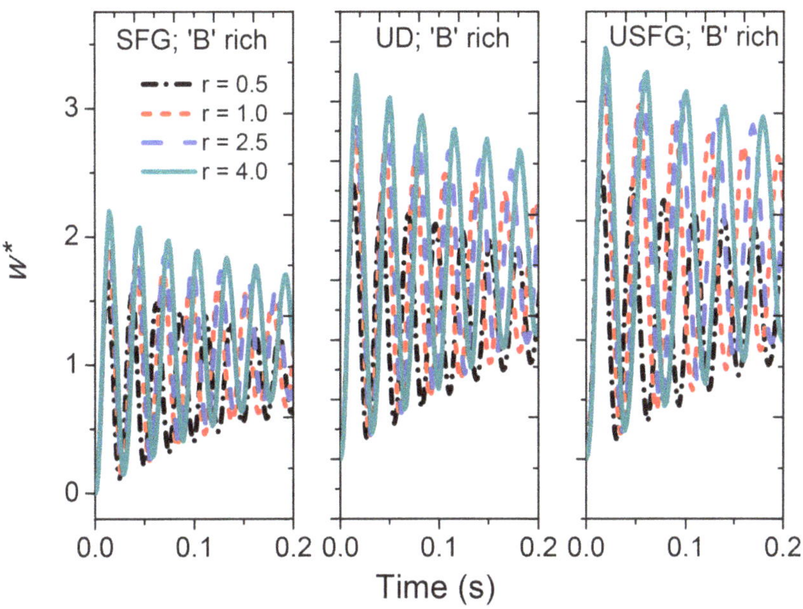

(a)

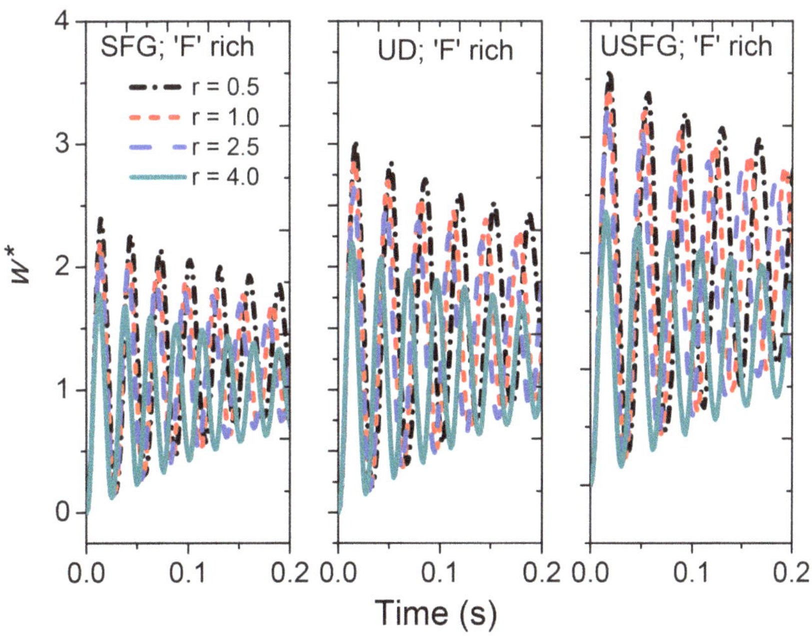

(b)

Figure 11.3. The controlled vibration responses of ACNT/FGMEE plates with a completely agglomerated core and (a) 'B'-rich and (b) 'F'-rich FGMEE facings.

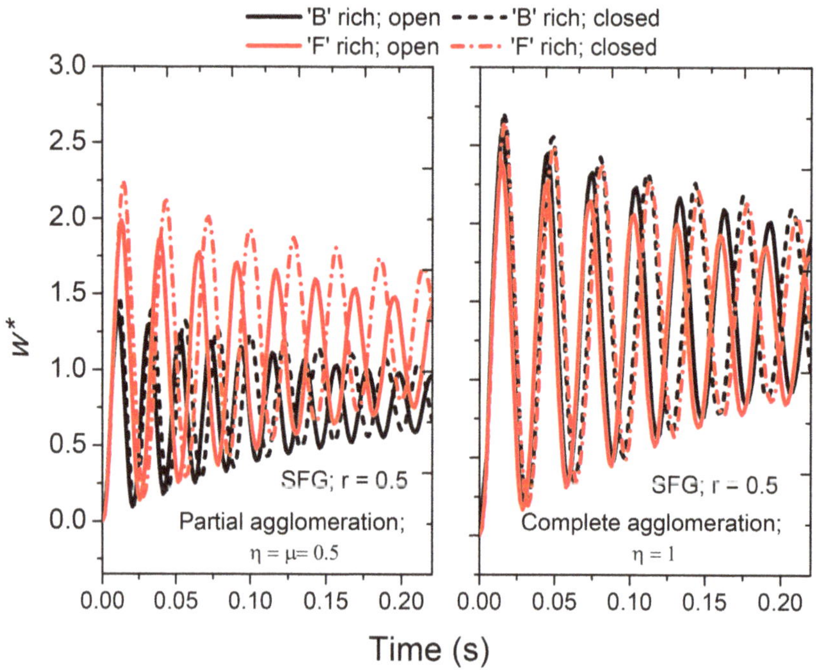

Figure 11.4. The effect of EM circuits on the controlled vibration response of ACNT/FGMEE plates.

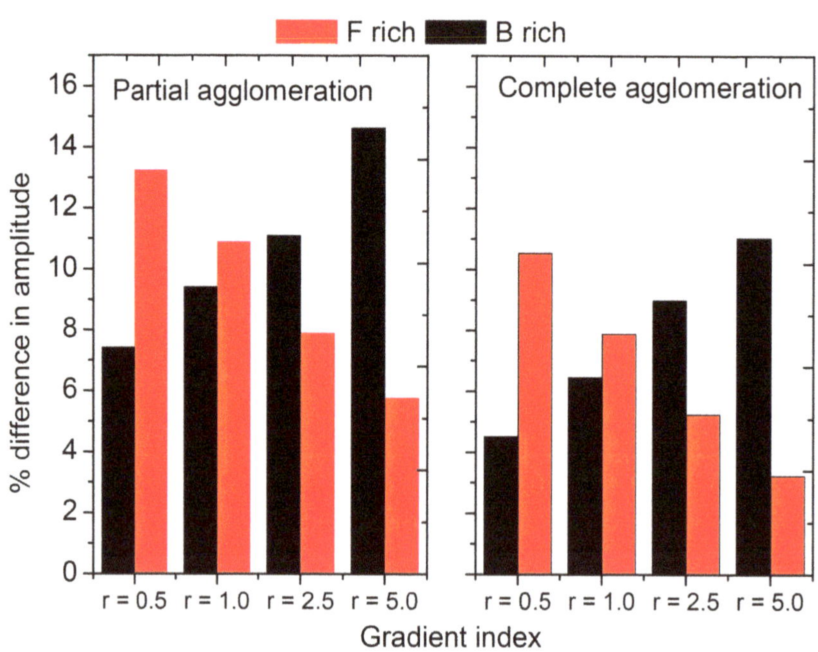

Figure 11.5. The effects of an EM circuit associated with the gradation index on the controlled vibration responses of ACNT/FGMEE plates.

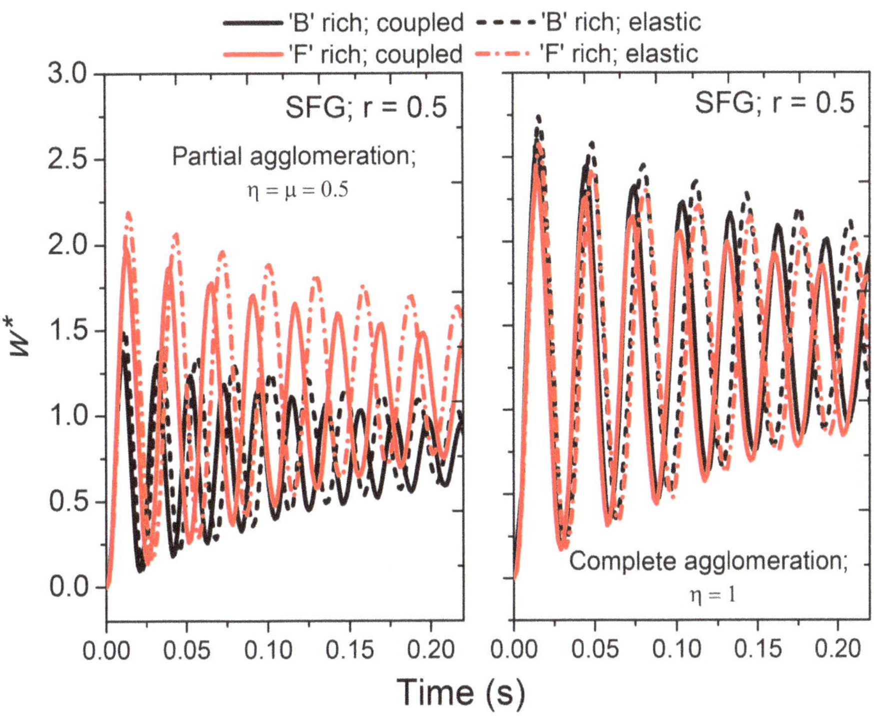

Figure 11.6. The effect of coupling fields on the controlled vibration responses of ACNT/FGMEE plates.

partial agglomeration state, the influence of different CNT arrangements associated with the variation of vibration amplitudes of the plates with different EM circuits and coupling fields is elucidated in figure 11.8. As can be seen, the SFG arrangement has a predominant effect associated with both EM circuits and coupling fields.

11.5 Conclusions

This work investigates the damped transient response of a smart sandwich plate with an agglomerated CNT core and functionally graded magneto-electro-elastic facings treated with ACLD patches. To this end, finite element (FE) methods were incorporated. The viscoelastic patch of ACLD was modeled using the GHM method. The effective properties of the agglomerated CNT core were determined through the micromechanics approach. The numerical evaluations suggest that the 'B'-rich bottom has a superior effect on the vibration attenuation characteristics. Further, a higher gradient index improves vibration damping when the MEE facings are 'B' rich, but this trend is reversed for the 'F'-rich bottom. The effect of ACLD treatment on vibration damping is superior when the CNT core is partially agglomerated, as opposed to a complete agglomeration state. Meanwhile, the externally applied EM circuit predominantly influences the damping properties. Due to enhanced energy dissipation, The open-circuit condition facilitates better vibration controllability in smart plates.

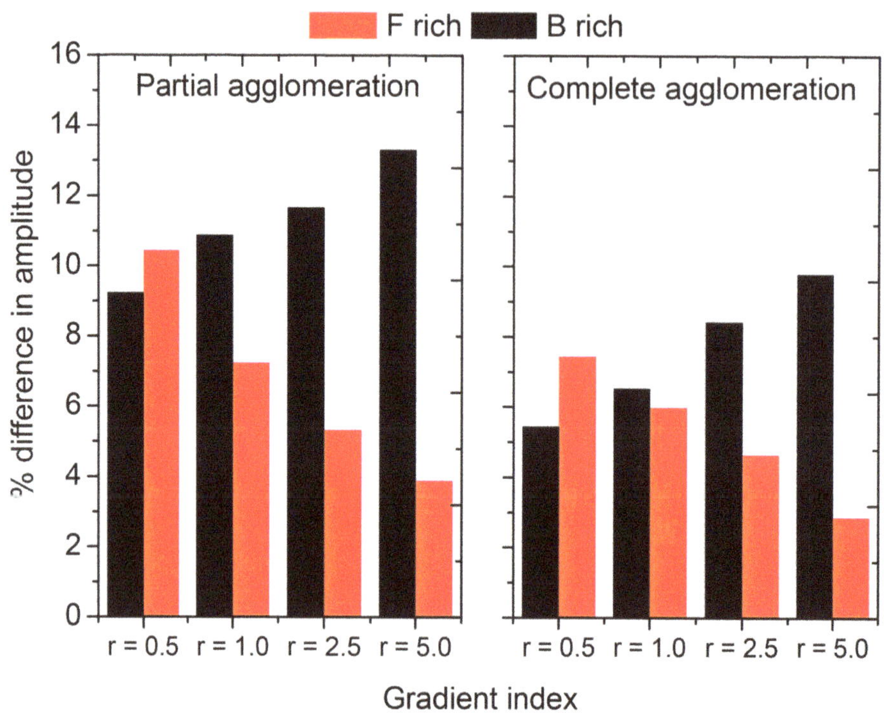

Figure 11.7. The effect of coupling fields associated with the gradation index on the controlled vibration responses of ACNT/FGMEE plates.

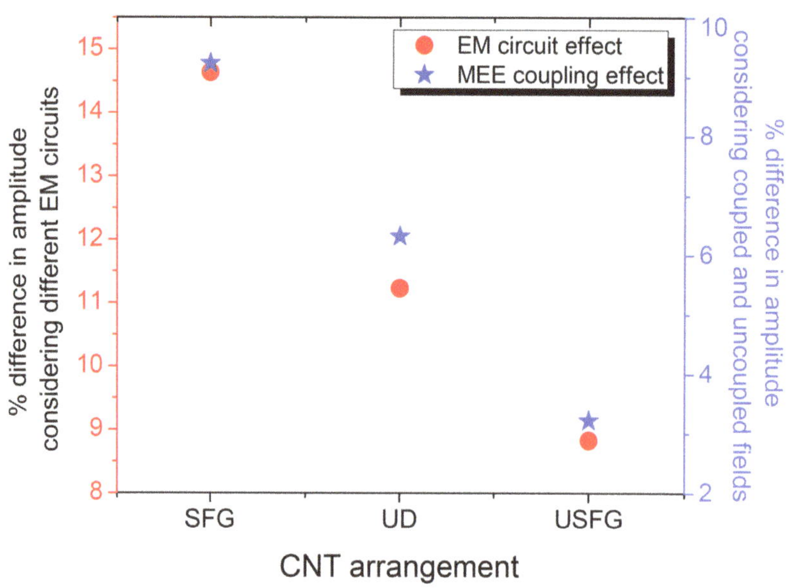

Figure 11.8. The effect of the CNT arrangement associated with EM circuits and coupling fields on the controlled vibration responses of ACNT/FGMEE plates.

Acknowledgments

The financial support of The Royal Society, London, through Newton International Fellowship (NIF\R1\212432) is sincerely acknowledged by the authors Vinyas Mahesh and Sathiskumar A Ponnusami.

References

[1] Qian D, Wagner G J, Liu W K, Yu M-F and Ruoff R S 2002 Mechanics of carbon nanotubes *Appl. Mech. Rev.* **55** 495–533

[2] Haggenmueller R, Gommans H H, Rinzler A G, Fischer J E and Winey K I 2000 Aligned singlewall carbon nanotubes in composites by melt processing methods *Chem. Phys. Lett.* **330** 219–25

[3] Qian D, Dickey E C, Andrews R and Rantell T 2000 Load transfer and deformation mechanisms in carbon nanotube–polystyrene composites *Appl. Phys. Lett.* **76** 2868–70

[4] Shaffer M S P and Windle A H 1999 Fabrication and characterisation of carbon nanotube/poly(vinyl alcohol) composites *Adv. Mater.* **11** 937–41

[5] Vigolo B, Penicaud A, Coulon C, Sauder C, Pailler R and Journet C *et al* 2000 Macroscopic fibres and ribbons of oriented carbon nanotubes *Science* **290** 1331–4

[6] Craveiro D S and Loja M A R 2020 A study on the effect of carbon nanotubes' distribution and agglomeration in the free vibration of nanocomposite plates *C* **6** 79

[7] Syah R, Ramdan D, Elveny M, Mohandes M, Khan A, Nouri A and Albadarin A B 2021 Computational simulation of critical speed and dynamic analysis of agglomerated CNTs/fiber/polymer/metal laminates rotating cylindrical shell *Arab. J. Chem.* **14** 103358

[8] García-Macías E and Castro-Triguero R 2018 Coupled effect of CNT waviness and agglomeration: a case study of vibrational analysis of CNT/polymer skew plates *Compos. Struct.* **193** 87–102

[9] Moradi-Dastjerdi R, Behdinan K, Safaei B and Qin Z 2020 Buckling behavior of porous CNT-reinforced plates integrated between active piezoelectric layers *Eng. Struct.* **222** 111141

[10] Bisheh H, Rabczuk T and Wu N 2020 Effects of nanotube agglomeration on wave dynamics of carbon nanotube-reinforced piezocomposite cylindrical shells *Composites* B **187** 107739

[11] Daghigh H, Daghigh V, Milani A, Tannant D, Lacy T E Jr and Reddy J N 2020 Nonlocal bending and buckling of agglomerated CNT-reinforced composite nanoplates *Composites* B **183** 107716

[12] Fakhar M H, Fakhar A and Tabatabaei H 2020 Mathematical modeling of pipes reinforced by agglomerated CNTs conveying turbulent nanofluid and application of semi-analytical method for studying the instable Nusselt number and fluid velocity *J. Comput. Appl. Math.* **378** 112945

[13] Ghasemi A R, Mohandes M, Dimitri R and Tornabene F 2019 Agglomeration effects on the vibrations of CNTs/fiber/polymer/metal hybrid laminates cylindrical shell *Composites* B **167** 700–16

[14] Ghorbanpour Arani A and Zamani M H 2018 Bending analysis of agglomerated carbon nanotube-reinforced beam resting on two parameters modified Vlasov model foundation *Indian J. Phys.* **92** 767–77

[15] Hajmohammad M H, Kolahchi R, Zarei M S and Nouri A H 2019 Dynamic response of auxetic honeycomb plates integrated with agglomerated CNT-reinforced face sheets subjected to blast load based on visco-sinusoidal theory *Int. J. Mech. Sci.* **153** 391–401

[16] Moradi-Dastjerdi R and Behdinan K 2021 Free vibration response of smart sandwich plates with porous CNT-reinforced and piezoelectric layers *Appl. Math. Model.* **96** 66–79

[17] Singh V, Vescovini R, Kumar R, Patel S N and Watts G 2022 Nonlinear vibration and instability of a randomly distributed CNT-reinforced composite plate subjected to localised in-plane parametric excitation *Appl. Math. Model.* **101** 453–80

[18] Shahmohammadi M A, Azhari M, Salehipour H and Civalek Ö 2021 A novel composite model for vibration of thin-walled layered composite panels incorporating the agglomeration of CNTs *Aerosp. Sci. Technol.* **116** 106897

[19] Yadav A, Amabili M, Panda S K, Dey T and Kumar R 2021 Nonlinear damped vibrations of three-phase CNT-FRC circular cylindrical shell *Compos. Struct.* **255** 112939

[20] Mahesh V and Harursampath D 2020 Nonlinear vibration of functionally graded magneto-electro-elastic higher order plates reinforced by CNTs using FEM *Eng. Comput.* **38** 1–23

[21] Vinyas M, Harursampath D and Thoi T N 2021 A higher order coupled frequency characteristics study of smart magneto-electro-elastic composite plates with cut-outs using finite element methods *Def. Technol.* **17** 100–18

[22] Vinyas M 2021 Porosity effect on the energy harvesting behaviour of functionally graded magneto-electro-elastic/fiber reinforced composite beam *Eur. Phys. J. Plus* **137** 48

[23] Siddharth Mangalasseri A, Mahesh V, Mahesh V, Ponnusami S A and Harursampath D 2022 Investigation on the interphase effects on the energy harvesting characteristics of three phase magneto-electro-elastic cantilever beam *Mech. Adv. Mater. Struct.* 1–13

[24] Mangalasseri A S, Mahesh V, Mukunda S, Mahesh V, Ponnusami S A and Harursampath D 2022 Vibration-based energy harvesting characteristics of functionally graded magneto-electro-elastic beam structures using lumped parameter model *J. Vib. Eng. Technol.* **10** 1705–20

[25] Mahesh V and Mangalasseri A S 2022 Agglomeration effects of CNTs on the energy harvesting performance of multifield interactive magneto-electro-elastic/nanocomposite unimorph smart beam *Mech. Based Des. Struct. Mach.* 1–27

[26] Vinyas M 2020 Interphase effect on the controlled frequency response of three-phase smart magneto-electro-elastic plates embedded with active constrained layer damping: FE study *Mater. Res. Express* **6** 125707

[27] Mahesh V and Kattimani S 2019 Finite element simulation of controlled frequency response of skew multiphase magneto-electro-elastic plates *J. Intell. Mater. Syst. Struct.* **30** 1757–71

[28] Mahesh V 2022 Active control of nonlinear coupled transient vibrations of multifunctional sandwich plates with agglomerated FG-CNTs core/magneto-electro-elastic facesheets *Thin-Walled Struct.* **179** 109547

[29] Mahesh V 2022 Nonlinear damping of auxetic sandwich plates with functionally graded magneto-electro-elastic facings under multiphysics loads and electromagnetic circuits *Compos. Struct.* **290** 115523

[30] Vinyas M 2019 Vibration control of skew magneto-electro-elastic plates using active constrained layer damping *Compos. Struct.* **208** 600–17

[31] Ly D K, Mahesh V, Thongchom C and Nguyen-Thoi T 2023 Hybrid control of laminated FG-CNTRC shell structures using an advanced smoothed finite element approach based on zig-zag theory *Thin-Walled Struct.* **184** 110463

[32] Vinyas M 2021 Computational analysis of smart magneto-electro-elastic materials and structures: review and classification *Arch. Comput. Meth. Eng.* **28** 1205–48

[33] Vinyas M, Vishwas M, Harursampath D and Abouelregal A E 2021 Simulation-based assessment of coupled frequency response of magneto electro elastic auxetic multifunctional structures subjected to various electromagnetic circuits *Proc. Inst. Mech. Eng.* **236** 228–96
[34] Vinyas M and Kattimani S C 2017 Static analysis of stepped functionally graded magneto-electro-elastic plates in thermal environment: a finite element study *Compos. Struct.* **178** 63–86
[35] Vinyas M and Kattimani S C 2017 Static behavior of thermally loaded multilayered magneto-electro-elastic beam *Struct. Eng. Mech.* **63** 481–95
[36] Vinyas M and Kattimani S C 2017 Static studies of stepped functionally graded magneto-electro-elastic beam subjected to different thermal loads *Compos. Struct.* **163** 216–37
[37] Vinyas M, Sunny K K, Harursampath D, Nguyen-Thoi T and Loja M A R 2019 Influence of interphase on the multi-physics coupled frequency of three-phase smart magneto-electro-elastic composite plates *Compos. Struct.* **226** 111254
[38] Vinyas M, Nischith G, Loja M A R, Ebrahimi F and Duc N D 2019 Numerical analysis of the vibration response of skew magneto-electro-elastic plates based on the higher-order shear deformation theory *Compos. Struct.* **214** 132–42
[39] Vinyas M 2019 A higher-order free vibration analysis of carbon nanotube-reinforced magneto-electro-elastic plates using finite element methods *Composites* B **158** 286–301
[40] Shi D L, Feng X Q, Huang Y Y, Hwang K C and Gao H 2004 The effect of nanotube waviness and agglomeration on the elastic property of carbon nanotube-reinforced composites *J. Eng. Mater. Technol.* **126** 250–7
[41] Chen J, Chen H, Pan E and Heyliger P R 2007 Modal analysis of magneto-electro-elastic plates using the state-vector approach *J. Sound Vib.* **304** 722–34

IOP Publishing

Functionally Graded Structures
Modelling and computation of static and dynamical problems
Snehashish Chakraverty, Subrat Kumar Jena and Ömer Civalek

Chapter 12

Lamb-type waves in functionally graded orthotropic piezoelectric plates

Sudarshan Dhua, Arindam Nath and Arpita Maji

This chapter illustrates the propagation behavior of Lamb-type waves in functionally graded (FG) orthotropic piezoelectric plates. The functionally grading in the plate is due to spatial variations in the material constants. Here, the power series technique is used to solve a dynamic system of equations. The dispersion equation is deduced using appropriate boundary conditions. Numerical analysis and graphical representation are used to illustrate the effects of functionally grading and different material parameters. The results of this study can be helpful for the design and performance improvement of transducers made using FG materials and for non-destructive ultrasonic evaluation.

12.1 Introduction

Smart materials such as piezoelectric materials have attracted attention and play a significant role in automation engineering in the modern era due to the advancement of technology. As a result of their increased sensitivity and efficiency, piezoelectric materials with higher anisotropy also play a significant role. As a result, ongoing efforts are being made to develop more sophisticated smart materials in order to meet demands in various fields, such as aeronautics, medical equipment, acoustic devices, and so forth. Therefore, a few researchers have developed and improved piezoelectric properties through the use of higher anisotropy by introducing piezoelectric orthotropic composite materials. Shindo *et al* [1] calculated the transient dynamic stress intensity factor and dynamic energy release rate for a cracked piezoelectric orthotropic ceramic using Laplace and Fourier transformations.

A class of advanced micro- and nanoscale structures, known as small-scale FG structures, exhibits progressive change in its material characteristics in at least one

doi:10.1088/978-0-7503-5301-4ch12

direction [2]. Many microelectromechanical systems (MEMS) and nanoelectromechanical systems (NEMS) are projected to operate better overall because of the adjustable material qualities of FG small-scale structures. Mechanical models of small-scale FG structures have been the subjects of several scientific investigations in recent years. In addition, more research is being done to examine the mechanical behaviors of small-scale FG piezoelectric (FGP) devices, including their static bending, free vibration [3], and thermo-electro-mechanical responses [4]. Recently, Nath *et al* [5] studied the effect of an FG substrate on shear waves in a piezo-electro-magnetic composite using a Green's function technique.

In several numerical solutions [6, 7], an inhomogeneous medium has been divided into a multilayer model, and the finite element technique has been used to analyze each layer using assumed homogeneous material properties to study wave propagation in inhomogeneous media. The Wentzel–Kramers–Brillouin (WKB) approach [8], the special functions method [9], the perturbation technique [10], and the power series technique [11] are a few examples of analytical solutions that have been studied. The WKB approach has regularly been used to study horizontal shear waves in inhomogeneous layered structures. When variations in the material parameters follow certain patterns, special functions, such as the cylindrical [12] and Airy functions [13], are used to solve conventional equations. Under electrical and magnetic open boundary conditions, Wu *et al* [14] used the Legendre orthogonal polynomial series expansion to investigate the wave propagation issue in an FG magneto-electro-elastic plate. The propagation problem can also be solved using the perturbation technique [10], since there is just one parameter that changes a little and smoothly. In layered FG piezoelectric material structures, shear waves have been solved using the power series approach because of the high precision and versatility of this technique. Although the wave propagation properties of typical piezoelectric structures have been extensively investigated, the published papers on the properties of waves propagating in orthotropic piezoelectric structures are relatively limited.

This chapter considers Lamb waves in an orthotropic piezoelectric plate that is FG, with material properties that change continuously in the thickness direction. The power series approach is used to solve coupled second-order governing equations. This chapter also describes and graphically illustrates how the dispersion curve is affected by changes in the gradient parameter. This theoretical foundation can provide guidance for the characterization of graded material coefficients through the experimental measurement of Lamb waves.

12.2 Statement and solution to the problem

Consider a functionally graded orthotropic piezoelectric (FGOP) plate as, shown in figure 12.1. The thickness of the plate is denoted by h. The upper and lower surfaces are traction free. It is assumed that the Cartesian coordinate system (x,y,z) is selected so that the poling direction lies in the positive direction of the z-axis and that the z-axis is positive downwards. Without loss of generality, it is also assumed that the Lamb wave moves in the positive x-axis direction.

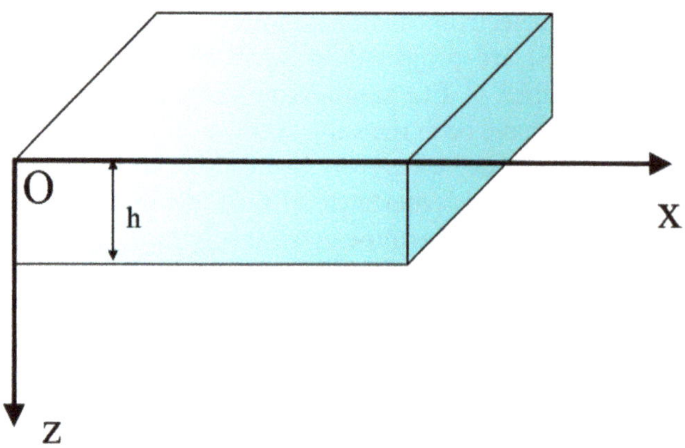

Figure 12.1. A schematic representation of the problem.

The constitutive equation for a piezoelectric elastic solid can be expressed as follows:

$$\left.\begin{aligned} \sigma_{ij} &= C_{ijkl}S_{kl} - e_{kij}E_k \\ D_i &= e_{ikl}S_{kl} + \varepsilon_{il}E_l \end{aligned}\right\}, \tag{12.1}$$

where σ_{ij} is the stress tensor, S_{kl} is the strain tensor, E_k is the electric field, and D_i is the electric displacement; C_{ijkl}, e_{kij}, and ε_{il} are the elastic constants, the piezoelectric constants, and the dielectric constants, respectively.

For the functionally graded orthotropic piezoelectric medium, the constitutive equations may be expressed as [15]

$$\left.\begin{aligned} \sigma_{11} &= C_{11}(z)S_{11} + C_{12}(z)S_{22} + C_{13}(z)S_{33} - e_{31}(z)E_3 \\ \sigma_{22} &= C_{12}(z)S_{11} + C_{22}(z)S_{22} + C_{23}(z)S_{33} - e_{32}(z)E_3 \\ \sigma_{33} &= C_{13}(z)S_{11} + C_{23}(z)S_{22} + C_{33}(z)S_{33} - e_{33}(z)E_3 \\ \sigma_{23} &= 2C_{44}(z)S_{23} - e_{24}(z)E_2,\ \sigma_{13} = 2C_{55}(z)S_{13} - e_{15}(z)E_1 \\ \sigma_{12} &= 2C_{66}(z)S_{12} \\ D_1 &= 2e_{15}(z)S_{13} + \varepsilon_{11}(z)E_1,\quad D_2 = 2e_{24}(z)S_{23} + \varepsilon_{22}(z)E_2 \\ D_3 &= e_{31}(z)S_{11} + e_{32}(z)S_{22} + e_{33}(z)S_{33} + \varepsilon_{33}(z)E_3 \end{aligned}\right\}, \tag{12.2}$$

where $S_{ij} = \frac{1}{2}(u_{i,j} + u_{j,i})$ is the strain tensor and $E_i = -\phi_{,i}$ is the electric field.

The material parameters of the FGOP plate vary in the thickness direction, so the material parameters take the form of a power series [16]:

$$\begin{gathered} C_{ij}(z) = \sum_{n=0}^{\infty} C_{ij}^{n}\left(\frac{z}{h}\right)^n,\quad e_{ij}(z) = \sum_{n=0}^{\infty} e_{ij}^{n}\left(\frac{z}{h}\right)^n,\quad \varepsilon_{ij}(z) = \sum_{n=0}^{\infty} \varepsilon_{ij}^{n}\left(\frac{z}{h}\right)^n, \\ \rho(z) = \sum_{n=0}^{\infty} \rho^{n}\left(\frac{z}{h}\right)^n, \end{gathered} \tag{12.3}$$

where ρ is the mass density of the medium.

For a Lamb wave propagating in the plate in the positive direction of the x-axis, the mechanical displacement component is

$$u = u(x, z, t), \quad v = 0, \quad w = w(x, z, t) \tag{12.4}$$

and the electrical potential is

$$\phi = \phi(x, z, t). \tag{12.5}$$

The governing equations without the body force for a piezoelectric medium are

$$\left.\begin{aligned} \sigma_{ij,j} &= \rho \ddot{u}_i \\ D_{i,i} &= 0 \end{aligned}\right\}. \tag{12.6}$$

In view of equations (12.1), (12.2), (12.4), and (12.5), the governing equation (12.6) for a functionally graded orthotropic piezoelectric medium can be written as follows:

$$\begin{aligned} &C_{11}(z)\frac{\partial^2 u}{\partial x^2} + (C_{13}(z) + C_{55}(z))\frac{\partial^2 w}{\partial x \partial z} + (e_{13}(z) + e_{15}(z))\frac{\partial^2 \phi}{\partial x \partial z} \\ &+ C_{55}(z)\frac{\partial^2 u}{\partial z^2} + C'_{55}(z)(\frac{\partial u}{\partial z} + \frac{\partial w}{\partial x}) + e'_{15}(z)\frac{\partial \phi}{\partial x} = \rho(z)\frac{\partial^2 u}{\partial t^2} \end{aligned} \tag{12.7}$$

$$\begin{aligned} &C_{55}(z)\frac{\partial^2 w}{\partial x^2} + (C_{13}(z) + C_{55}(z))\frac{\partial^2 u}{\partial x \partial z} + C_{33}(z)\frac{\partial^2 w}{\partial z^2} + e_{15}(z)\frac{\partial^2 \phi}{\partial x^2} \\ &+ e_{33}(z)\frac{\partial^2 \phi}{\partial z^2} + C'_{13}(z)\frac{\partial u}{\partial x} + C'_{33}(z)\frac{\partial w}{\partial z} + e'_{33}(z)\frac{\partial \phi}{\partial z} = \rho(z)\frac{\partial^2 w}{\partial t^2} \end{aligned} \tag{12.8}$$

$$\begin{aligned} &(e_{15}(z) + e_{13}(z))\frac{\partial^2 u}{\partial x \partial z} + e_{15}(z)\frac{\partial^2 w}{\partial x^2} + e_{33}(z)\frac{\partial^2 w}{\partial z^2} - \varepsilon_{11}(z)\frac{\partial^2 \phi}{\partial x^2} \\ &- \varepsilon_{33}(z)\frac{\partial^2 \phi}{\partial z^2} + e'_{13}(z)\frac{\partial u}{\partial x} + e'_{33}(z)\frac{\partial w}{\partial z} - \varepsilon'_{33}(z)\frac{\partial \phi}{\partial z} = 0. \end{aligned} \tag{12.9}$$

For Lamb wave propagation in a functionally graded piezoelectric material in the direction of the x-axis, the electric potential and displacement components can be given in the following forms:

$$\{u, w, \phi\}(x, z, t) = \left\{Z_u(z), Z_w(z), Z_\phi(z)\right\}e^{ik(x-ct)}. \tag{12.10}$$

If we substitute equation (12.10) into equations (12.7)–(12.9), considering the z-polarized orthotropic piezoelectric material, the governing equations are as follows:

$$\begin{aligned} &c_{55}\ D^2 Z_u + c'_{55}\ DZ_u + (\rho c^2 - c_{11})k^2 Z_u + (c_{13} + c_{55})ik\ DZ_w \\ &+ (e_{13} + e_{15})ik\ DZ_\phi + ik(c'_{55}Z_w + e'_{15}Z_\phi) = 0 \end{aligned} \tag{12.11}$$

$$\begin{aligned} &c_{33}\ D^2 Z_w + c'_{33}\ DZ_w + (\rho c^2 - c_{55})k^2 Z_w + (c_{55} + c_{13})ik\ DZ_u \\ &- e_{15}k^2 Z_\phi + e_{33}D^2 Z_\phi + c'_{13}ikZ_u + e'_{33}DZ_\phi = 0 \end{aligned} \tag{12.12}$$

$$\begin{aligned}&\varepsilon_{33}\ D^2Z_\phi+\varepsilon'_{33}\ DZ_\phi-\varepsilon_{11}k^2Z_\phi-e_{33}\ D^2Z_w-(e_{15}+e_{31})ik\ DZ_u\\&+e_{15}k^2Z_w+e'_{13}\ ikZ_u-e'_{33}\ DZ_w=0.\end{aligned}\tag{12.13}$$

Taking equation (12.3) into account, the solution of equations (12.11)–(12.13) may be considered to be as follows [16]:

$$Z_u=\sum_{n=0}^{\infty}s_n\left(\frac{z}{h}\right)^n,\quad Z_w=i\sum_{n=0}^{\infty}t_n\left(\frac{z}{h}\right)^n,\quad Z_\phi=i\sum_{n=0}^{\infty}r_n\left(\frac{z}{h}\right)^n,\tag{12.14}$$

where the coefficients s_n, r_n and t_n can be obtained from the recursion relations.

By substituting equations (12.3) and (12.14) into equations (12.11)–(12.13), obtaining the recursive relations for s_n, r_n and t_n, and equating the coefficients of $\left(\frac{z}{h}\right)^n$ to zero, we obtain the following:

$$\sum_{i=0}^{\infty}G_{1i}=0,\quad\sum_{i=0}^{\infty}G_{2i}=0,\quad\sum_{i=0}^{\infty}G_{3i}=0,\tag{12.15}$$

where

$$\begin{aligned}G_{1i}=&(n-i+1)[(n-i+2)c_{55}^{i}s_{n-i+2}+(i+1)c_{55}^{i+1}{}'s_{n-i+1}]\\&-(kh)(i+1)[c_{55}^{i+1}{}'t_{n-i}+e_{15}^{i+1}{}'r_{n-i}]\\&-(n-i+1)(kh)[(c_{13}^{i}+c_{55}^{i})t_{n-i}+(e_{15}^{i}+e_{13}^{i})r_{n-i+1}]+k^2h^2(c^2\rho^i-c_{11}^{i})s_{n-i}\end{aligned}$$

$$\begin{aligned}G_{2i}=&(n-i+1)[(n-i+2)(c_{33}^{i}t_{n-i+2}+e_{33}^{i}r_{n-i+2})+(i+1)(c_{33}^{i+1}{}'t_{n-i+1}+e_{33}^{i+1}{}'r_{n-i+1})]\\&+(kh)[(n-i+1)(c_{13}^{i}+c_{55}^{i})s_{n-i+1}+(i+1)c_{13}^{i+1}{}'s_i]\\&+(kh)^2[(\rho^ic^2-c_{55}^{i})t_{n-i}-e_{15}^{i}r_{n-i}]\end{aligned}$$

$$\begin{aligned}G_{3i}=&(n-i+1)[(n-i+2)(\varepsilon_{33}^{i}r_{n-i+2}-e_{33}^{i}t_{n-i+2})+(i+1)(\varepsilon_{33}^{i+1}{}'r_{n-i+1}-e_{33}^{i+1}{}'t_{n-i+1})]\\&+(kh)^2[e_{15}^{i}t_{n-i}-\varepsilon_{11}^{i}r_{n-i}]\\&-(kh)[(e_{15}^{i}+e_{13}^{i})(n-i+1)s_{n-i+1}+(i+1)e_{13}^{i+1}{}'s_{n-i}].\end{aligned}$$

The unknown decoupling coefficients s_0, s_1, t_0, t_1, r_0, and r_1 are determined by the following matrix:

$$[s_{0j},\ s_{1j},\ t_{0j},\ t_{1j},\ r_{0j},\ r_{1j}]=I\tag{12.16}$$

For $n\geqslant 2$, the unknown coefficients s_i, r_i, and t_i can be obtained by solving the recursive relations (12.15). Therefore, to obtain the values of s_2, r_2, and t_2, the linear functions involving the coefficients s_0, s_1, r_0, r_1, t_0, and t_1 are solved successively.

Equation (12.14) can now be rewritten as

$$Z_u=\sum_{j=1}^{n}c_j\sum_{n=0}^{\infty}s_{nj}\left(\frac{z}{h}\right)^n,\quad Z_w=\sum_{j=1}^{n}c_j\sum_{n=0}^{\infty}t_{nj}\left(\frac{z}{h}\right)^n,\quad Z_\phi=\sum_{j=1}^{n}c_j\sum_{n=0}^{\infty}r_{nj}\left(\frac{z}{h}\right)^n,\tag{12.17}$$

where c_j are arbitrary constants to be determined and the coefficients s_{nj}, t_{nj}, and r_{nj} used in equation (12.17) can be evaluated with the help of recursive relation (12.15) and equation (12.16).

Here, in the view of equation (12.3), the functionally gradient of the material constant m is taken to be $m(z) = m^0\left(1 + \alpha\left(\frac{z}{h}\right)\right)$, where α is the functionally grading parameter.

Due to the insignificant contribution of the higher terms in equation (12.1), the terms $n > 2$ are disregarded. Therefore, the solution of equations (12.11)–(12.13) can be written as

$$Z_u = c_1 + c_2\left(\frac{z}{h} + s_2\left(\frac{z}{h}\right)^2\right), \quad Z_w = c_3 + c_4\left(\frac{z}{h} + t_2\left(\frac{z}{h}\right)^2\right),$$
$$Z_\phi = c_5 + c_6\left(\frac{z}{h} + r_2\left(\frac{z}{h}\right)^2\right), \tag{12.18}$$

where s_2, r_2, and t_2 are provided in appendix A.

Therefore, the mechanical displacements and potential functions of the functionally graded piezoelectric plate are as follows:

$$u = \left\{c_1 + c_2\left(\frac{z}{h} + s_2\left(\frac{z}{h}\right)^2\right)\right\}\exp[ik(x - ct)] \tag{12.19}$$

$$v = \left\{c_3 + c_4\left(\frac{z}{h} + t_2\left(\frac{z}{h}\right)^2\right)\right\}\exp[ik(x - ct)] \tag{12.20}$$

$$\phi = \left\{c_5 + c_6\left(\frac{z}{h} + r_2\left(\frac{z}{h}\right)^2\right)\right\}\exp[ik(x - ct)]. \tag{12.21}$$

12.3 Boundary conditions

i. The free surface is traction free, i.e. at $z = 0$ and $z = h$, we have

$$\sigma_{33}(x, 0) = 0, \quad \sigma_{33}(x, h) = 0 \tag{12.22}$$

$$\sigma_{13}(x, 0) = 0, \quad \sigma_{13}(x, h) = 0. \tag{12.23}$$

ii. The electric boundary conditions for the free surface are considered to be electrically short, i.e. for $z = 0$ and $z = h$, we have

$$\phi(x, 0) = 0, \quad \phi(x, h) = 0. \tag{12.24}$$

12.4 The dispersion relation

By substituting equations (12.19)–(12.21) into the boundary conditions (12.22)–(12.24), we can write the dispersion relation for the electrically short case as

$$\left| \left[M_{ij}^{S} \right] \right| = 0, \tag{12.25}$$

where the elements of matrix $[M_{ij}^{S}]$ are defined in appendix A.

12.5 Numerical results and discussion

For the numerical computation, the linear type of functionally graded material is taken into account. The effects on Lamb wave propagation of the functionally grading parameter α are shown for two different kinds of material. The following data were considered for calculation:

Data 1 for material 1: polyvinylidene fluoride (PVDF) [17]

$$c_{11} = 238.24 \times 10^{9}\ \mathrm{Nm^{-2}},\quad c_{33} = 10.64 \times 10^{9}\ \mathrm{Nm^{-2}},\quad c_{13} = 2.19 \times 10^{9}\ \mathrm{Nm^{-2}}$$

$$c_{55} = 4.4 \times 10^{9}\ \mathrm{Nm^{-2}},\quad e_{15} = -0.135\ \mathrm{Cm^{-2}},\quad e_{13} = -0.13\ \mathrm{Cm^{-2}},$$

$$e_{33} = -0.276\ \mathrm{Cm^{-2}}$$

$$\varepsilon_{11} = 0.110\ 677 \times 10^{-9}\ \mathrm{C^2N^{-1}m^{-2}},\quad \varepsilon_{33} = 0.106\ 073 \times 10^{-9}\ \mathrm{C^2N^{-1}m^{-2}},$$

$$\rho = 7500\ \mathrm{kgm^{-3}}$$

Data 2 for material 2: polyvinylidene fluoride (PVDF) [15]

$$c_{11} = 3.61 \times 10^{9}\ \mathrm{Nm^{-2}},\quad c_{33} = 1.63 \times 10^{9}\ \mathrm{Nm^{-2}},\quad c_{13} = 1.42 \times 10^{9}\ \mathrm{Nm^{-2}}$$

$$c_{55} = 0.59 \times 10^{9}\ \mathrm{Nm^{-2}},\quad e_{15} = -15.93 \times 10^{-3}\ \mathrm{Cm^{-2}},$$

$$e_{13} = 32.075 \times 10^{-3}\ \mathrm{Cm^{-2}},\quad e_{33} = -21.19 \times 10^{-3}\ \mathrm{Cm^{-2}}$$

$$\varepsilon_{11} = 0.053\ 985 \times 10^{-9}\ \mathrm{C^2N^{-1}m^{-2}},\quad \varepsilon_{33} = 0.059\ 295 \times 10^{-9}\ \mathrm{C^2N^{-1}m^{-2}},$$

$$\rho = 1800\ \mathrm{kgm^{-3}}$$

The thickness of the plate is taken to be h=1 mm.

The first mode of Lamb wave propagation for two different kinds of material is depicted in figure 12.2. It can be observed from the graph that the first mode of Lamb waves has an increasing trend.

Figure 12.3 shows the Lamb wave dispersion curve for the first mode in the functionally graded orthotropic piezoelectric material described by Data 1. The functionally grading parameter α takes two values, corresponding to cases (i) without functionally grading, i.e. $\alpha = 0$ and (ii) a piezoelectric medium with functionally grading, i.e. $\alpha = 0.5$.

Figure 12.4 shows the Lamb wave dispersion curve for the material corresponding to Data 2. Here, the first and second modes of the dispersion relation are

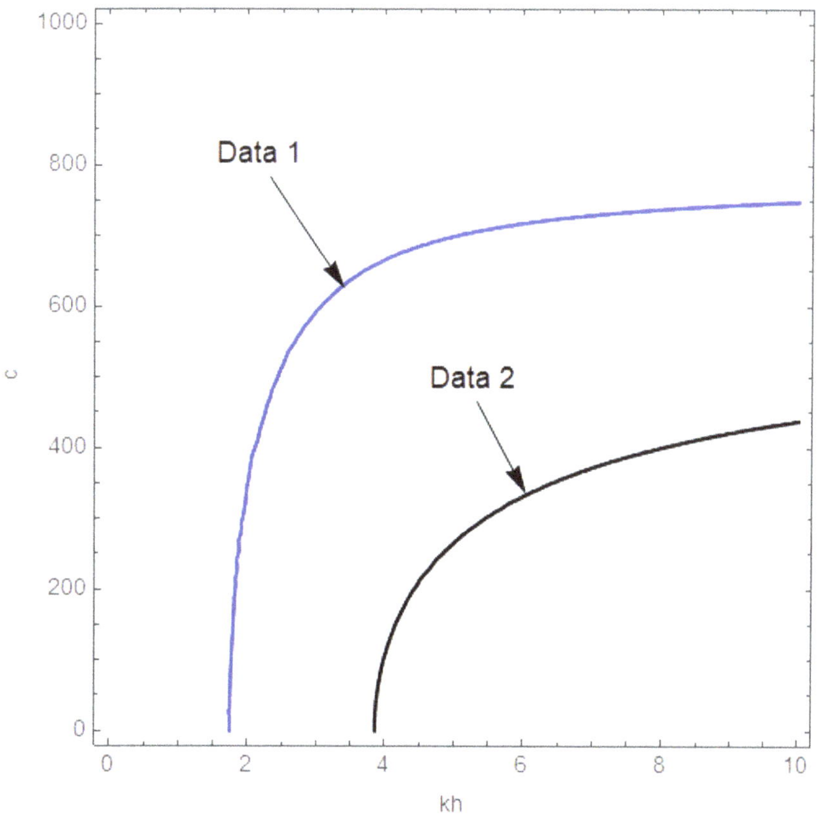

Figure 12.2. The dispersion curves of Lamb waves in FGOP plates described by Data 1 and Data 2.

presented. The first mode of the phase velocity of the Lamb wave increases with increasing wave numbers. In the case of the second mode, the phase velocity of the Lamb wave decays with increasing wave numbers. In addition, the effect of the functionally grading parameter α is prominent for both modes.

12.6 Conclusions

This chapter investigated Lamb wave propagation in FGOP plates. The expression of the dispersion relation for electrically short case was derived using the power series technique. The effect of the functionally graded parameter of the medium on the wave velocity was analyzed and presented graphically for two different types of orthotropic piezoelectric material. The critical points of this chapter are as follows:

i. The power series technique was employed to solve Lamb wave propagation in the FGOP plate.
ii. The functionally grading parameter significantly affects the Lamb wave's dispersion curve.
iii. The curve representing the first mode of the Lamb wave follows an increasing trend, while the second mode has a decreasing trend.

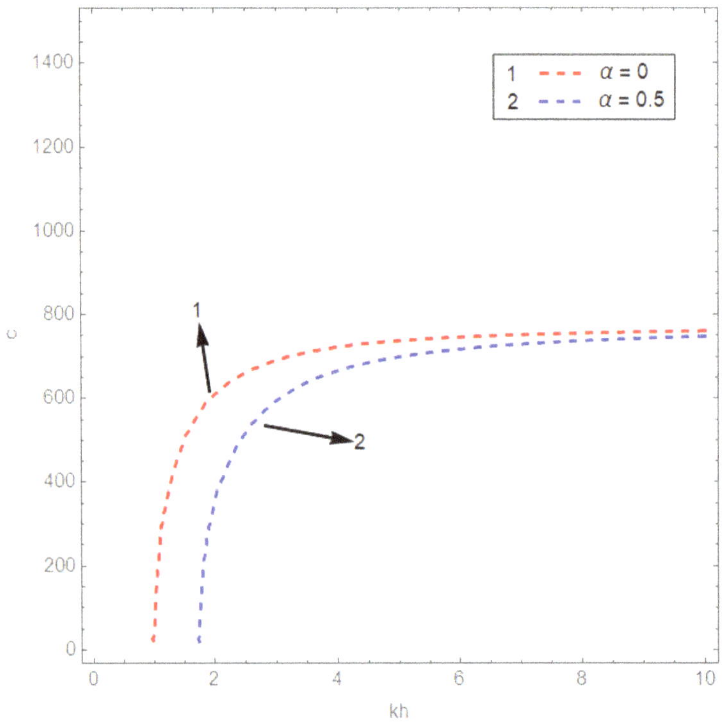

Figure 12.3. The effect of functionally grading parameter on the first mode of Lamb wave in an FGOP plate described by Data 1.

iv. The desired solutions can be improved as required by incorporating more terms (i.e. higher values of n) into the series solution expression.
v. These findings will be helpful in the development and production of nano/micro electromechanical structures.

Appendix A

$$s_2 = \frac{kh(2c_{55}^1 + 2e_{15}^1 + c_{13}^1 + e_{13}^1) + (kh)^2(\rho^0 c^2 - c_{11}^0) - c_{55}^1}{2c_{55}^0}$$

$$t_2 = \frac{\begin{array}{c} e_{33}^0(\varepsilon_{33}^1 - e_{33}^1 + kh(e_{15}^0 + e_{13}^0 + e_{13}^1) + (kh)^2(e_{15}^0 - \varepsilon_{11}^0)) \\ - c_{55}^0(c_{33}^1 + e_{33}^1 + kh(c_{13}^0 + c_{55}^0 + c_{13}^1) + (kh)^2(\rho^0 c^2 - a_{55}^0 - e_{15}^0)) \end{array}}{2(e_{33}^0)^2 + 2c_{33}^0\varepsilon_{33}^0}$$

$$r_2 = \frac{\begin{array}{c} c_{33}^0(\varepsilon_{33}^1 - e_{33}^1 + kh(e_{15}^0 + e_{13}^0 + e_{13}^1) + (kh)^2(e_{15}^0 - \varepsilon_{11}^0)) \\ + e_{33}^0(c_{33}^1 + e_{33}^1 + kh(c_{13}^0 + c_{55}^0 + c_{13}^1) + (kh)^2(\rho^0 c^2 - a_{55}^0 - e_{15}^0)) \end{array}}{2(e_{33}^0)^2 + 2c_{33}^0\varepsilon_{33}^0}$$

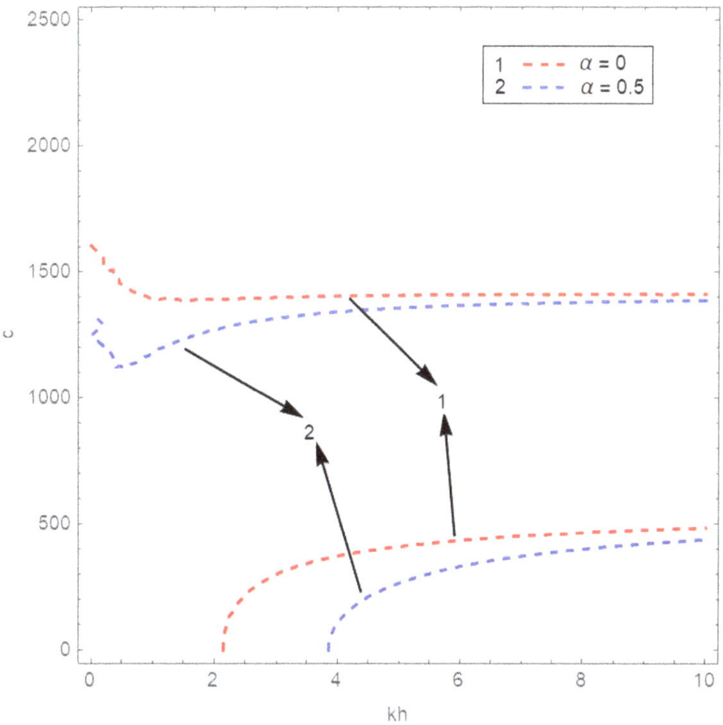

Figure 12.4. The effect of functionally grading parameter on the first and second modes of Lamb waves in an FGOP plate described by Data 2.

$$M_{11}^s = ikc_{13}, \quad M_{14}^s = \frac{c_{33}}{h}, \quad M_{16}^s = \frac{e_{33}}{h}, \quad M_{21}^s = ikc_{13}, \quad M_{22}^s = ikc_{13}(1 + s_2),$$
$$M_{24}^s = \frac{c_{33}}{h}(1 + 2t_2), \quad M_{26}^s = \frac{e_{33}}{h}(1 + 2r_2),$$
$$M_{32}^s = \frac{c_{55}}{h}, \quad M_{33}^s = ikc_{55}, \quad M_{35}^s = ike_{15}, \quad M_{42}^s = \frac{c_{55}}{h}(1 + 2s_2), \quad M_{43}^s = ikc_{55},$$
$$M_{44}^s = ikc_{55}(1 + t_2), \quad M_{45}^s = ike_{15}, \quad M_{46}^s = ike_{15}(1 + r_2)$$
$$M_{55}^s = 1, \quad M_{65}^s = 1, \quad M_{66}^s = (1 + r_2)$$

All the remaining M_{ij}^s which are not mentioned are equal to zero.

Acknowledgments

The authors convey their sincere thanks to NIT Andhra Pradesh, India for providing Institute Fellowships to Mr Arindam Nath and Ms Arpita Maji and supporting them with its best research facility.

References

[1] Shindo Y, Narita F and Ozawa E 1999 Impact response of a finite crack in an orthotropic piezoelectric ceramic *Acta Mech.* **137** 99–107

[2] Ghayesh M H and Farajpour A 2019 A review on the mechanics of functionally graded nanoscale and microscale structures *Int. J. Eng. Sci.* **137** 8–36
[3] Li Y S and Pan E 2015 Static bending and free vibration of a functionally graded piezoelectric microplate based on the modified couple-stress theory *Int. J. Eng. Sci.* **97** 40–59
[4] Komijani M, Reddy J N and Eslami M R 2014 Nonlinear analysis of microstructure-dependent functionally graded piezoelectric material actuators *J. Mech. Phys. Solids* **63** 214–27
[5] Nath A and Dhua S 2023 Effect of an impulsive source on shear wave propagation in a piezo-electro-magnetic structure mounted over a heterogeneous isotropic substrate carrying a viscoelastic thin film *Int. Conf. on Wave Mechanics and Vibrations* (Cham: Springer) 1145–57
[6] Liu G R, Dai K Y, Han X and Ohyoshi T 2003 Dispersion of waves and characteristic wave surfaces in functionally graded piezoelectric plates *J. Sound Vib.* **268** 131–47
[7] Guha S and Singh A K 2022 Transference of SH waves in a piezoelectric fiber-reinforced composite layered structure employing perfectly matched layer and infinite element techniques coupled with finite elements *Finite Elem. Anal. Des.* **209** 103814
[8] Qian Z, Jin F, Wang Z and Kishimoto K 2007 Transverse surface waves on a piezoelectric material carrying a functionally graded layer of finite thickness *Int. J. Eng. Sci.* **45** 455–66
[9] Chattopadhyay A, Gupta S, Sahu S A and Dhua S 2012 Torsional surface waves in heterogeneous anisotropic half-space under initial stress *Arch. Appl. Mech.* **83** 357–66
[10] Singh A K, Kumar S and Chattopadhyay A 2015 Love-type wave propagation in a piezoelectric structure with irregularity *Int. J. Eng. Sci.* **89** 35–60
[11] Cao X, Jin F, Wang Z and Lu T 2009 Bleustein–Gulyaev waves in a functionally graded piezoelectric material layered structure *Sci. China Ser. G – Phys. Mech. Astron.* **52** 613–25
[12] Vlasie V and Rousseau M 2004 Guided modes in a plane elastic layer with gradually continuous acoustic properties *NDT E Int.* **37** 633–44
[13] Cao X, Jin F and Wang Z 2008 Theoretical investigation on horizontally shear waves in a functionally gradient piezoelectric material plate *Adv. Mater. Res.* **33–37** 707–12
[14] Bin W, Jiangong Y and Cunfu H 2008 Wave propagation in non-homogeneous magneto-electro-elastic plates *J. Sound Vib.* **317** 250–64
[15] Pal M K and Singh A K 2021 Analysis of reflection and transmission phenomenon at distinct bonding interfaces in a rotating pre-stressed functionally graded piezoelectric-orthotropic structure *Appl. Math. Comput.* **409** 126398
[16] Cao X, Jin F, Jeon I and Lu T J 2009 Propagation of Love waves in a functionally graded piezoelectric material (FGPM) layered composite system *Int. J. Solids Struct.* **46** 4123–32
[17] Pal M K, Singh A K and Kumari R 2020 Reflection of plane waves on the stress-free and rigid boundary surfaces of pre-stressed piezoelectric-orthotropic substrate: a comparative approach *Mech. Adv. Mater. Struct.* **29** 816–27

IOP Publishing

Functionally Graded Structures
Modelling and computation of static and dynamical problems
Snehashish Chakraverty, Subrat Kumar Jena and Ömer Civalek

Chapter 13

Analysis of Lamb wave propagation in a generalized thermoelastic functionally graded plate

Sudarshan Dhua, Arpita Maji and Arindam Nath

This chapter addresses the dynamic analysis of Lamb wave propagation in a functionally graded plate using the dual-phase-lag (DPL) thermoelastic model. The plate surface is assumed to be isothermal and stress free. Legendre orthogonal polynomial series expansion is used to solve the problem analytically, and the secular equation is obtained. The characteristics of the thermal modes are illustrated by comparing the dispersion curves of an inhomogeneous thermoelastic plate with those of an equivalent pure elastic plate. This study will be helpful in the automotive and aerospace fields, in military industries, and in structural design.

13.1 Introduction

Functionally graded materials (FGMs) are advanced composite materials consisting of two or more constituent phases, each of which exhibits a continuous variation in material properties from one surface to another. Japanese materials scientists came up with the idea of FGMs in 1984 [1]. A typical FGM is constructed using a material that is a combination of a ceramic and a metal. The applications of FGMs are described in a report by Jha *et al* [2]. They can be utilized as the heat insulation material in fields that use extremely high temperatures and high temperature gradients due to the material's gradually changing properties, which relieve thermal stress, minimize stress concentration, and so on. Liu and Tuni [3] discussed surface waves in functionally gradient piezoelectric plates. In addition, Cao *et al* [4] investigated Lamb wave propagation in a plate made from a functionally graded piezoelectric–piezomagnetic material.

Thermoelasticity is the study of dynamical systems whose interactions with their environments involve not only mechanical and external work but also the exchange of heat. Thermoelasticity was first described by Biot [5], who derived dilatation from the thermodynamics of irreversible processes and coupled it with elastic

doi:10.1088/978-0-7503-5301-4ch13

deformation. Beginning with the previous theory, the theory of thermoelasticity has become well established over time [6]. However, the thermal signal transmission speed was projected to be unlimited based on the diffusion-type heat theory used in this investigation. A hyperbolic equation of heat conduction with a relaxation time was derived by Lord and Shulman [7], which guarantees a finite speed for thermal signals in the generalized theory of thermoelasticity. Another generalized theory of thermoelasticity was proposed by Green and Lindsay [8]; it used the concept of two relaxation times. Chandrasekhariah [9] provides a thorough review of the earlier studies of generalized thermoelasticity in his work. Some notable work in thermo-elastic medium has been carried out by the authors of [10–18].

This article investigates Lamb wave propagation in a functionally graded thermoelastic plate using the DPL model. The Legendre polynomial approach has been used to solve the governing equations of this problem. The dispersion curves and displacement distributions are illustrated. In this chapter, stress-free, isothermal boundary conditions are assumed.

13.2 Fundamental equation

The basic governing equations of the generalized orthotropic DPL thermoelastic model in the absence of body forces and sources are as follows:

1. The equation of motion,

$$\tau_{ij,\,j} = \rho\frac{\partial^2 u_i}{\partial t^2}, \tag{13.1}$$

2. The strain–displacement relation,

$$e_{ij} = \frac{1}{2}(u_{i,\,j} + u_{j,\,i}), \tag{13.2}$$

3. and the Fourier law for the DPL model in thermoelasticity, which becomes

$$\left(1 + \tau_T\frac{\partial}{\partial t}\right)k_{ij}T_{,ii} = \left(1 + \tau_q\frac{\partial}{\partial t} + \frac{\tau_q^2}{2}\frac{\partial^2}{\partial t^2}\right)\left(\rho C_e\dot{T} + T_0\beta_{ij}\dot{e}_{ij}\right), \tag{13.3}$$

where e_{ij} is the component of the strain tensor. The component of the displacement vector T_0 is the reference temperature and T the temperature above the reference temperature. ρ is the mass density and τ_T, τ_q are the phase lags of the heat flux and temperature gradient respectively. The component of thermal conductivity is denoted by k_{ij}. C_e is the specific heat under constant strain. β_{ij} are the thermal moduli. The comma notation is used for spatial derivatives.

13.3 The formulation of the problem and its solution

Consider an orthotropic thermoelastic plate that is infinite in the horizontal direction and has a thickness of h. Plane strain acts parallel to $x - z$ in the plane, whereas the boundary $z = 0$ is considered to be stress free. Lamb waves propagate in the direction of x on the axis (figure 13.1).

Figure 13.1. A schematic representation of the problem.

The displacement components are considered to be

$$u = u(x, z, t), \quad v = 0, \quad w = w(x, z, t). \tag{13.4}$$

In this chapter, the body force and external heat sources are taken to be zero for wave propagation. So, based on the DPL generalized thermoelastic theory, the dynamic equations for the thermoelastic plate are

$$T_{11,1} + T_{12,2} + T_{13,3} = \rho\frac{\partial^2 u}{\partial t^2} \tag{13.5}$$

$$T_{12,1} + T_{22,2} + T_{23,3} = \rho\frac{\partial^2 v}{\partial t^2} \tag{13.6}$$

$$T_{13,1} + T_{32,2} + T_{33,3} = \rho\frac{\partial^2 w}{\partial t^2} \tag{13.7}$$

$$\begin{aligned} &k_{11}\left(1 + \tau_T\frac{\partial}{\partial t}\right)\frac{\partial^2 T}{\partial x^2} + k_{22}\left(1 + \tau_T\frac{\partial}{\partial t}\right)\frac{\partial^2 T}{\partial y^2} + k_{33}\left(1 + \tau_T\frac{\partial}{\partial t}\right)\frac{\partial^2 T}{\partial z^2} \\ &= \left(1 + T_q\frac{\partial}{\partial t} + \frac{T_q^2}{2}\frac{\partial^2}{\partial t^2}\right)\left(\rho c_e\dot{T} + T_0\beta_1\frac{\partial\dot{u}}{\partial x} + T_0\beta_2\frac{\partial}{\partial y} + T_0\beta_3\frac{\partial\dot{w}}{\partial z}\right). \end{aligned} \tag{13.8}$$

The relationship between strain and displacement can be written as

$$\begin{aligned} &e_{11} = \frac{\partial u}{\partial x}, \quad e_{22} = \frac{\partial v}{\partial y}, \quad e_{33} = \frac{\partial w}{\partial z}, \quad e_{23} = \frac{1}{2}\left(\frac{\partial v}{\partial z} + \frac{\partial w}{\partial y}\right), \\ &e_{13} = \frac{1}{2}\left(\frac{\partial u}{\partial z} + \frac{\partial w}{\partial x}\right), \quad e_{12} = \frac{1}{2}\left(\frac{\partial u}{\partial y} + \frac{\partial v}{\partial x}\right), \end{aligned} \tag{13.9}$$

which uses the rectangular window function, which is dependent on the material's constants:

$$\pi(z) = \begin{Bmatrix} 1, & 0 \leqslant z \leqslant h \\ 0, & \text{elsewhere} \end{Bmatrix}. \tag{13.10}$$

For the functionally graded orthotopic plate, constitutive equations can be written in the form

$$\begin{aligned}
T_{11} &= C_{11}e_{11} + C_{12}e_{22} + C_{13}e_{33} - \beta_1 T \\
T_{22} &= C_{12}e_{11} + C_{22}e_{22} + C_{23}e_{33} - \beta_2 T \\
T_{33} &= [C_{13}e_{11} + C_{23}e_{22} + C_{33}e_{33} - \beta_3 T]\pi(z) \\
T_{13} &= 2C_{44}e_{13}\pi(z) \\
T_{23} &= 2C_{44}e_{23}\pi(z), \\
T_{12} &= 2C_{66}e_{12}.
\end{aligned} \tag{13.11}$$

The elastic constants of the medium are a function of z because the properties of the material vary in the direction of the thickness, thus, the material constants can be written as

$$\begin{aligned}
&C_{ij}(z) = C_{ij}^{(l)}\left(\frac{z}{h}\right)^l, \quad \rho(z) = \rho^{(l)}\left(\frac{z}{h}\right)^l, \quad \beta_i = \beta_i^{(l)}\left(\frac{z}{h}\right)^l, \\
&K_i = K_i^{(l)}\left(\frac{z}{h}\right)^l \quad l = 0,1,2, \ldots, L.
\end{aligned} \tag{13.12}$$

We assume that the solution of the displacement component and temperature change can be written in the form

$$u(x, z, t) = \exp(ikx - i\omega t)U(z) \tag{13.13}$$

$$v(x, z, t) = \exp(ikx - i\omega t)V(z) \tag{13.14}$$

$$w(x, z, t) = \exp(ikx - i\omega t)W(z) \tag{13.15}$$

$$T(x, z, t) = \exp(ikx - i\omega t)X(z). \tag{13.16}$$

$U(z)$, $V(z)$, $W(z)$ represent the amplitude of vibration in the x-, y-, and z-directions, respectively, and $X(z)$ is the amplitude of temperature change. k is the wave number and ω is the angular frequency.

Substituting equations (13.9)–(13.16) into equations (13.5)–(13.8), the governing differential equations can be derived in terms of temperature change and displacement components as follows:

$$\begin{aligned}
&\left(\frac{z}{h}\right)^l [(C_{44}^l U'' + lC_{44}^l z^{-1}U' + ikC_{44}^l W' + ikWlC_{44}^l z^{-1})\pi(z) \\
&\quad + ikC_{13}^l \frac{\partial W}{\partial z} - k^2 C_{11}^l U - ik\beta_1^l X] \\
&+ C_{44}^l\left(\frac{z}{h}\right)^l (\delta(z - 0) - \delta(z - h))(U' + ikW) = -\rho^l\left(\omega^2\left(\frac{z}{h}\right)^l\right)U
\end{aligned} \tag{13.17}$$

$$\begin{aligned}
&\left(\frac{z}{h}\right)^l [(C_{44}^l V'' + lz^{-1}C_{44}^l V')\pi(z) - k^2 C_{66}^l V] + (\delta(z - 0) - \delta(z - h))\left(\frac{z}{h}\right)^l C_{44}^l \\
&V' = -\rho^l\left(\frac{z}{h}\right)^l \omega^2 V
\end{aligned} \tag{13.18}$$

$$
\begin{aligned}
&\left(\frac{z}{h}\right)^{l} ikC_{44}^{l}U'\pi(z) - C_{44}^{l}k^{2}W\pi(z) + lz^{-1}C_{13}^{l}ikU\pi(z) \\
&\quad + C_{13}^{l}\left(\frac{z}{h}\right)^{l} ik\pi(z)U' + lz^{-1}C_{33}^{l}W'\pi(z) \\
&\quad + C_{33}^{l}\left(\frac{z}{h}\right)^{l} W''\pi(z) - lz^{-1}\beta_{3}^{l}\pi(z)X - \beta_{3}^{l}\pi(z)X' \\
&\quad + \left(ikUC_{13}^{l}\left(\frac{z}{h}\right)^{l} + C_{33}^{l}\left(\frac{z}{h}\right)^{l} W' - \beta_{3}^{l}\left(\frac{z}{h}\right)^{l} X\right)(\delta(z-0) - \delta(z-h)) \\
&\quad = -\rho^{l}\left(\frac{z}{h}\right)^{l}\omega^{2}W
\end{aligned}
\tag{13.19}
$$

$$
\left(\frac{z}{h}\right)^{l}(k_{33}^{l}X'' - k^{2}k_{11}^{l}X) + \left(\frac{z}{h}\right)^{l} T_{1}T_{0}(i\omega\beta_{3}W' - k\omega\beta_{1}U) = -T_{1}\frac{z^{l}}{h^{l}}\rho C_{e}(i\omega)X. \tag{13.20}
$$

The propagation of a shear wave can be represented by equation (13.18), which is independent of the other three equations, i.e. (13.17), (13.19), and (13.20). Here, equations (13.17) and (13.19) represent the propagation of a Lamb-type wave. In this chapter, we discuss the solution for a coupled thermoelastic Lamb-type wave.

The solution of the thermoelastic Lamb wave can be expanded into a Legendre orthogonal polynomial series solution as follows:

$$
\begin{aligned}
&U(z) = \sum_{m=0}^{\infty} P_{m}^{1}Q_{m}(z),\ W(z) = \sum_{m=0}^{\infty} P_{m}^{2}Q_{m}(z) \text{ and} \\
&\qquad X(z) = z(z-h)\sum_{m=0}^{\infty} P_{m}^{3}Q_{m}(z),
\end{aligned}
\tag{13.21}
$$

where P_{m}^{i} ($i = 1, 2, 3$) are the expansion coefficients. The coefficient of thermal expansion is a measurement that can be used to indicate how the size of an object shifts when the temperature of the object varies. $Q_{m}(z)$ can be defined as

$$
Q_{m}(z) = \sqrt{\frac{2m+1}{h}}\,P_{m}\left(\frac{2z-h}{h}\right). \tag{13.22}
$$

In equation (13.22), P_{m} is the mth order Legendre polynomial. Here, the order m runs from zero to ∞. For this problem, the value of m is considered to be finite, i.e. $m = M$, and higher-order terms become negligible.

Substituting equation (13.21) into equations (13.17), (13.19), and (13.20), multiplying both sides by Q_{j}, and integrating both sides over z from zero to h, the equations can be written in the following forms:

$$
S_{11}^{j,m}P_{m}^{1} + S_{12}^{j,m}P_{m}^{2} + S_{13}^{j,m}P_{m}^{3} = 0 \tag{13.23}
$$

$$S_{21}^{j,m}P_m^1 + S_{22}^{j,m}P_m^2 + S_{23}^{j,m}P_m^3 = 0 \tag{13.24}$$

$$S_{31}^{j,m}P_m^1 + S_{32}^{j,m}P_m^2 + S_{33}^{j,m}P_m^3 = 0, \tag{13.25}$$

where the values of $S_{kl}^{j,m}$ $(k, l = 1,2,3)$ are the elements of a non-symmetric matrix. They can be obtained from equations (13.17), (13.18), and (13.19) and are presented in appendix A. Equations (13.23)–(13.25) can be written in matrix form, i.e.

$$\begin{bmatrix} S_{11}^{j,m} & S_{12}^{j,m} & S_{13}^{j,m} \\ S_{21}^{j,m} & S_{22}^{j,m} & S_{23}^{j,m} \\ S_{31}^{j,m} & S_{32}^{j,m} & S_{33}^{j,m} \end{bmatrix} \begin{Bmatrix} P_m^1 \\ P_m^2 \\ P_m^3 \end{Bmatrix} = 0. \tag{13.26}$$

For nonzero solutions of equation (13.26), the determinant of the coefficient matrix must be equal to zero, which yields

$$f(k, \omega) = \begin{vmatrix} S_{11}^{j,m} & S_{12}^{j,m} & S_{13}^{j,m} \\ S_{21}^{j,m} & S_{22}^{j,m} & S_{23}^{j,m} \\ S_{31}^{j,m} & S_{32}^{j,m} & S_{33}^{j,m} \end{vmatrix} = 0. \tag{13.27}$$

Equation (13.27) gives the dispersion curves for Lamb wave propagation in a functionally graded orthotropic plate under the DPL generalized thermoelastic model.

13.4 Numerical results

For the numerical computation, we have considered the following data:

$$C_{11} = 16.28 \times 10^{10}\ \text{N m}^{-2},\quad C_{13} = 5.08 \times 10^{10}\ \text{N m}^{-2},$$
$$C_{33} = 6.27 \times 10^{10}\ \text{N m}^{-2},\quad C_{55} = 7.70 \times 10^{10}\ \text{N m}^{-2},$$
$$\beta_1 = 5.75 \times 10^{6}\ \text{N m}^{-2}\ \text{deg}^{-1},\quad \beta_3 = 5.17 \times 10^{6}\ \text{N m}^{-2}\ \text{deg}^{-1},$$
$$K_1 = 124\ \text{W m}^{-1}\ \text{deg}^{-1},\quad K_3 = 124\ \text{W m}^{-1}\ \text{deg}^{-1},$$
$$\rho = 7.14 \times 10^{3}\ \text{kg m}^{-1},\quad C_e = 390\ \text{J kg}^{-1}\ \text{deg}^{-1},\quad \tau_q = 2 \times 10^{-7}\ \text{s},$$
$$\tau_q = 1.5 \times 10^{-7}\ \text{s},\quad T_0 = 296\ \text{deg}$$

Based on the previous formulations, computer programs have been written to draw the dispersion curve for M=1. Figure 13.2 represents the dispersion curve ω vs kh.

13.5 Conclusions

This chapter presented Lamb wave propagation in an FG orthotropic thermoelastic plate using the DPL model. The problem has been solved using the orthogonal polynomial method. The dispersion curve ω vs kh has an increasing trend. The dispersion curve can be made more accurate by considering higher values of M.

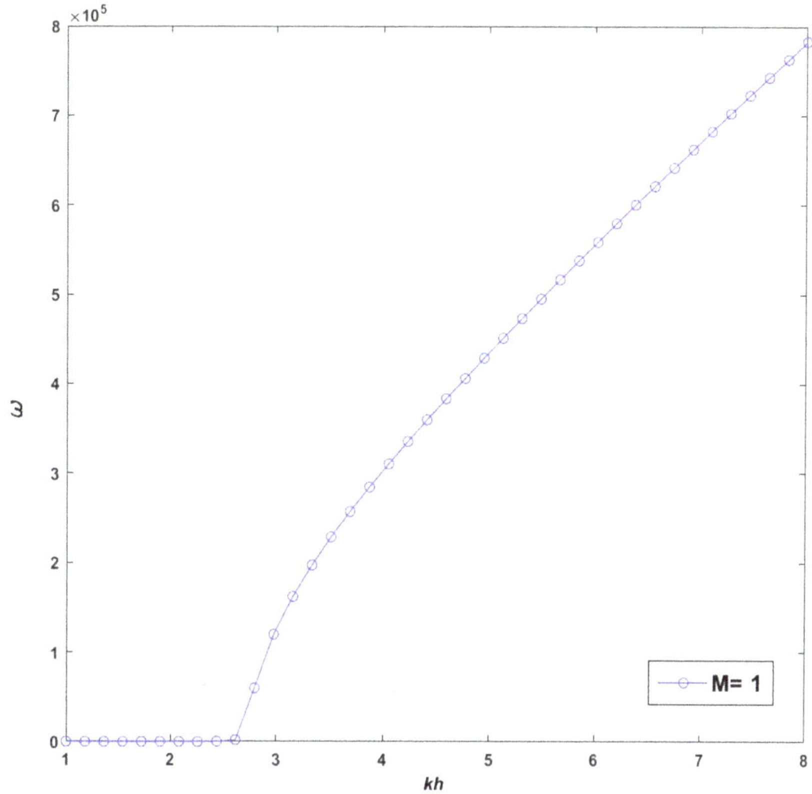

Figure 13.2. Dispersion curve of the FGM thermoelastic plate for M=1.

Appendix

$$S_{11}^{j,m} = h^{-l}\left[\left(C_{44}^{l}\int_0^h Q_j(z)z^l\frac{\partial^2 Q_m}{\partial z^2}dz\right) + \left(lC_{44}^{l}\int_0^h Q_j(z)z^{l-1}\frac{\partial Q_m}{\partial z}dz\right)\right.$$
$$\left.-\left(k^2C_{11}^{l}\int_0^h z^l Q_m(z)Q_j(z)dz\right) + \left(C_{44}^{l}\int_0^h Q_j z^l\frac{\partial Q_m}{\partial z}\frac{\delta(z-0)-\delta(z-h)}{\delta z}dz\right)\right] + \omega^2 M_1^j$$

$$S_{12}^{j,m}$$
$$= ikh^{-l}$$
$$\left[\left(\left(C_{13}^{l} + C_{44}^{l}\right)\int_0^h z^l Q_j\frac{\partial Q_m}{\partial z}dz\right) + \left(lC_{44}^{l}\int_0^h z^{l-1}Q_j(z)Q_m(z)dz\right)\right.$$
$$\left.+\left(C_{44}^{l}\int_0^h z^l Q_j\frac{\partial(H(z)-H(z-h))}{\partial z}Q_m(z)dz\right)\right]$$

$$S_{13}^{j,m} = ikh^{-l}\beta_1^l\int_0^h z^{l+1}(z-h)Q_m(z)Q_j(z)dz,\quad M_1^j = -h^{-l}\rho^l\int_0^h z^l Q_j(z)Q_m(z)dz$$

$$S_{21}^{j,m} = ikh^{-l}\left[\left(\left(C_{44}^{l} + C_{13}^{l}\right)\int_0^h \frac{\partial Q_m}{\partial z} Q_j z^l dz\right) + \left(lC_{13}^{l}\int_0^h z^{l-1} Q_m Q_j dz\right) + \left(C_{13}^{l}\int_0^h Q_j z^l Q_m \frac{\partial (H(z) - H(z-h))}{\partial z} dz\right)\right]$$

$$S_{22}^{j,m} = h^{-l}\left[\left(C_{33}^{l}\int_0^h z^l \frac{\partial^2 Q_m}{\partial z^2} Q_j dz\right) + \left(lC_{33}^{l}\int_0^h z^{l-1}\frac{\partial Q_m}{\partial z} Q_j dz\right) - \left(k^2 C_{44}^{l}\int_0^h z^l Q_j Q_m dz\right) - \left(C_{33}^{l}\int_0^h z^l \frac{\partial Q_m}{\partial z}\frac{\partial (H(z) - H(z-h))}{\partial z} dz\right)\right] + \omega^2 M_1^j$$

$$S_{23}^{j,m} = -h^{-l}\beta_3^l\left[\left(\int_0^h Q_j(m) z^l (z-h) Q_m(z) dz\right) + \left(\int_0^h z^{l+1}(z-h) Q_m(z)\frac{\partial (H(z) - H(z-h))}{\partial z} dz\right) + \cdots\right.$$

$$\begin{aligned}
S_{31}^{j,m} = -T_1 T_0\left(\frac{h^{-l}}{\omega}\right)\Bigg[&\left(-C_{44}^{l} k\int_0^h Q_j z^l \frac{\partial^2 Q_m}{\partial z^2}\left(\sum_{l=0}^{L}\frac{z^l}{h^l}\beta_3^l\right) dz\right)\\
&- \left(lkC_{13}^{l}\int_0^h Q_j \frac{\partial Q_m}{\partial z} z^{l-1}\left(\sum_{l=0}^{L}\frac{z^l}{h^l}\beta_3^l\right) dz\right) - \left(C_{13}^{l} k\int_0^h Q_j \frac{\partial^2 Q_m}{\partial z^2} z^{l-1}\left(\sum_{l=0}^{L}\frac{z^l}{h^l}\beta_3^l\right) dz\right)\\
&- \left(kC_{13}^{l}\int_0^h Q_j z^l \frac{\partial Q_m}{\partial z}\left(\sum_{l=0}^{L}\frac{z^l}{h^l}\beta_3^l\right)\left(\frac{\partial (H(z) - H(z-h))}{\partial z}\right) dz\right) + \left(klh^{-l} C_{44}^{l}\int_0^h Q_j z^{l-1}\frac{\partial Q_m}{\partial z}\left(\sum_{l=0}^{L}\frac{z^l}{h^l}\beta_3^l\right) dz\right)\\
&+ \left(l^2 C_{13}^{l} h^{-l} k\int_0^h Q_j Q_m z^{l-2}\left(\sum_{l=0}^{L}\frac{z^l}{h^l}\beta_3^l\right) dz\right) + \left(lkC_{13}^{l}\int_0^h z^{l-1} Q_j \frac{\partial Q_m}{\partial z}\left(\sum_{l=0}^{L}\frac{z^l}{h^l}\beta_3^l\right) dz\right)\\
&- \left(kC_{13}^{l} h^{-l}\int_0^h Q_m Q_j z^l \left(\frac{\partial (H(z) - H(z-h))}{\partial z}\right) dz\right)\Bigg]\\
&+ \left(T_0 T_1 \frac{k}{\omega} h^{-l} C_{44}^{l}\int_0^h z^l \frac{\partial^2 Q_m}{\partial z^2} Q_j\left(\sum_{l=0}^{L}\frac{z^l}{h^l}\beta_1^l\right)\left(\frac{\rho^0 + \sum_{l=1}^{L}\frac{lz^{l-1}}{h^l}\rho^l}{\sum_{l=0}^{L}\frac{z^l}{h^l}\rho^l}\right) dz\right)\\
&+ \left(h^{-l} T_1 T_0 \frac{k}{\omega} lC_{44}^{l}\int_0^h z^{l-1}\frac{\partial Q_m}{\partial z} Q_j\left(\sum_{l=0}^{L}\frac{z^l}{h^l}\beta_1^l\right)\right)\left(\frac{\rho^0 + \sum_{l=1}^{L}\frac{lz^{l-1}}{h^l}\rho^l}{\sum_{l=0}^{L}\frac{z^l}{h^l}\rho^l}\right) dz\\
&+ \left(T_1 T_0 \frac{k}{\omega} h^{-l} C_{44}^{l}\int_0^h z^l Q_j \frac{\partial Q_m}{\partial z}\sum_{l=0}^{L}\frac{z^l}{h^l}\beta_1^l\right)\left(\frac{\rho^0 + \sum_{l=1}^{L}\frac{lz^{l-1}}{h^l}\rho^l}{\sum_{l=0}^{L}\frac{z^l}{h^l}\rho^l}\right)\left(\frac{\partial (H(z) - H(z-h))}{\partial z}\right) dz
\end{aligned}$$

$$
\begin{aligned}
&S_{32}^{j,m} \\
&= -T_0T_1\left(i\frac{h^{-l}}{\omega}\right) \\
&\left[-\left(C_{44}^l k^2\int_0^h Q_j z^l\frac{\partial Q_m}{\partial z}\left(\sum_{l=0}^{L}\frac{z^l}{h^l}\beta_3^l\right)dz\right)+\left(lC_{33}^l\int_0^h Q_j\frac{\partial^2 Q_m}{\partial z^2}z^{l-1}\left(\sum_{l=0}^{L}\frac{z^l}{h^l}\beta_3^l\right)dz\right)\right. \\
&+\left(C_{33}^l\int_0^h Q_j z^l\frac{\partial^3 Q_m}{\partial z^3}\left(\sum_{l=0}^{L}\frac{z^l}{h^l}\beta_3^l\right)dz\right) \\
&+\left(C_{33}^l\int_0^h Q_j z^l\frac{\partial^2 Q_m}{\partial z^2}\left(\sum_{l=0}^{L}\frac{z^l}{h^l}\beta_3^l\right)\left(\frac{\partial(H(z)-H(z-h))}{\partial z}\right)dz\right) \\
&+\left(C_{44}^l k^2 l h^{-l}\int_0^h Q_j Q_m z^{l-1}\left(\sum_{l=0}^{L}\frac{z^l}{h^l}\beta_3^l\right)dz\right)-\left(l^2C_{33}^l\int_0^h z^{l-2}Q_j\frac{\partial Q_m}{\partial z}\left(\sum_{l=0}^{L}\frac{z^l}{h^l}\beta_3^l\right)dz\right) \\
&\left.-\left(lC_{33}^l\int_0^h z^{l-1}Q_j\frac{\partial^2 Q_m}{\partial z^2}\left(\sum_{l=0}^{L}\frac{z^l}{h^l}\beta_3^l\right)dz\right)+\left(C_{33}^l h^{-l}\int_0^h Q_j z^l\frac{\partial Q_m}{\partial z}\left(\frac{\partial(H(z)-H(z-h))}{\partial z}\right)dz\right)\right] \\
&+\left(ikC_{44}^l T_1T_0\frac{k}{\omega}\int_0^h z^l\frac{\partial Q_m}{\partial z}Q_j\left(\sum_{l=0}^{L}\frac{z^l}{h^l}\beta_1^l\right)\left(\frac{\rho^0+\sum_{l=1}^{L}\frac{lz^{l-1}}{h^l}\rho^l}{\sum_{l=0}^{L}\frac{z^l}{h^l}\rho^l}\right)dz\right) \\
&+\left(T_0T_1\frac{ik^2 l}{\omega}C_{44}^l\int_0^h Q_jQ_m z^{l-1}\left(\sum_{l=0}^{L}\frac{z^l}{h^l}\beta_1^l\right)\left(\frac{\rho^0+\sum_{l=1}^{L}\frac{lz^{l-1}}{h^l}\rho^l}{\sum_{l=0}^{L}\frac{z^l}{h^l}\rho^l}\right)dz\right) \\
&+\left(ikT_0T_1\frac{k}{\omega}h^{-l}C_{44}^l\int_0^h Q_j z^l Q_m\left(\sum_{l=0}^{L}\frac{z^l}{h^l}\beta_1^l\right)\left(\frac{\rho^0+\sum_{l=1}^{L}\frac{lz^{l-1}}{h^l}\rho^l}{\sum_{l=0}^{L}\frac{z^l}{h^l}\rho^l}\right)\right. \\
&\left.\left(\frac{\partial(H(z)-H(z-h))}{\partial z}\right)dz\right)
\end{aligned}
$$

$$S_{33}^{j,m} = -T_1T_0\left(\frac{h^{-l}}{\omega}\right)\left[\left(-l\beta_3^l\int_0^h z^lQ_jQ_m(z-h)\left(\sum_{l=0}^{L}\frac{z^l}{h^l}\beta_3^l\right)dz\right)\right.$$

$$-\left(\beta_3^l\int_0^h Q_jz^{l+1}\left(z-h\right)\frac{\partial^2 Q_m}{\partial z^2}\left(\sum_{l=0}^{L}\frac{z^l}{h^l}\beta_3^l\right)dz\right)-\left(\beta_3^l\int_0^h Q_jz^{l+1}(z-h)\right.$$

$$\left.\frac{\partial Q_m}{\partial z}\left(\sum_{l=0}^{h}\frac{z^l}{h^l}\beta_3^l\right)\left(\frac{\partial(H(z)-H(z-h))}{\partial z}\right)dz\right)+\left(l^2\beta_3^l\int_0^h z^{l-1}(z-h)Q_jQ_m\left(\sum_{l=0}^{L}\frac{z^l}{h^l}\beta_3^l\right)dz\right)$$

$$+\left(l\beta_3^l\int_0^h z^{l+1}Q_j(z-h)\frac{\partial Q_m}{\partial z}\left(\sum_{l=0}^{L}\frac{z^l}{h^l}\beta_3^l\right)dz\right)$$

$$-\left(\beta_3^lh^{-l}\int_0^h Q_jz^{l+1}(z-h)Q_m\left(\frac{\partial(H(z)-H(z-h))}{\partial z}\right)dz\right)$$

$$+\left(k_{33}^lh^{-l}\int_0^h Q_jz^{l+1}(z-h)\frac{\partial^2 Q_m}{\partial z^2}\left(\sum_{l=0}^{L}\frac{z^l}{h^l}\rho^l\right)dz\right)-(k^2k_{11}^lh^{-l}\int_0^h Q_jz^{l+1}(z-h)Q_m$$

$$\left.\left(\sum_{l=0}^{L}\frac{z^l}{h^l}\rho^l\right)dz\right)+i\omega M_2^j$$

$$M_2^j = h^{-l}T_1C_e^l\int_0^h Q_jQ_mz^{l+1}(z-h)\left(\sum_{l=0}^{L}\frac{z^l}{h^l}\rho^l\right)^2dz\Bigg).$$

References

[1] Koizumi M 1997 FGM activities in Japan *Composites* B **28** 1–4

[2] Jha D K, Kant T and Singh R K 2013 A critical review of recent research on functionally graded plates *Compos. Struct.* **96** 833–49

[3] Liu G R and Tani J 1994 Surface waves in functionally gradient piezoelectric plates *J. Vib. Acoust.* **116** 440–8

[4] Cao X, Shi J and Jin F 2012 Lamb wave propagation in the functionally graded piezoelectric–piezomagnetic material plate *Acta Mech.* **223** 1081–91

[5] Biot M A 2005 Theory of propagation of elastic waves in a fluid-saturated porous solid. II. Higher frequency range *J. Acoust. Soc. Am.* **28** 179

[6] Nowacki W 1975 *Dynamic Problems of Thermoelasticity* (Dordrecht: Springer) https://link.springer.com/book/9789028600454

[7] Lord H and Shulman Y 1967 A generalized dynamical theory of thermoelasticity *J. Mech. Phys. Solids* **15** 299–309

[8] Green A E and Lindsay K A 1972 Thermoelasticity *J. Elast.* **2** 1–7

[9] Chandrasekharalah D S 1986 Thermoelasticity with second sound: a review *Appl. Mech. Rev.* **39** 355–76

[10] Warren W E and Chen P J 1973 Wave propagation in the two temperature theory of thermoelasticity *Acta Mech.* **16** 21–33

[11] Abbas I A, Abdalla A E N N, Alzahrani F S and Spagnuolo M 2016 Wave propagation in a generalized thermoelastic plate using eigenvalue approach *J. Therm. Stresses* **39** 1367–77

[12] Yu J, Wu B and He C 2011 Guided thermoelastic wave propagation in layered plates without energy dissipation *Acta Mech. Solida Sin.* **24** 135–43

[13] Hosseini S M 2009 Coupled thermoelasticity and second sound in finite length functionally graded thick hollow cylinders (without energy dissipation) *Mater. Des.* **30** 2011–23

[14] Verma K L and Hasebe N 2001 Dispersion of thermoelastic waves in a plate with and without energy dissipation *Int. J. Thermophys.* **22** 957–78

[15] Verma K L and Hasebe N 2001 Wave propagation in plates of general anisotropic media in generalized thermoelasticity *Int. J. Eng. Sci.* **39** 1739–63

[16] Verma K L 2002 On the propagation of waves in layered anisotropic media in generalized thermoelasticity *Int. J. Eng. Sci.* **40** 2077–96

[17] Verma K L and Hasebe N 2002 Wave propagation in transversely isotropic plates in generalized thermoelasticity *Arch. Appl. Mech.* **72** 470–82

[18] Jiangong Y and Tonglong X 2010 Generalized thermoelastic waves in spherical curved plates without energy dissipation *Acta Mech.* **212** 39–50

www.ingramcontent.com/pod-product-compliance
Ingram Content Group UK Ltd.
Pitfield, Milton Keynes, MK11 3LW, UK
UKHW051902100926
8556IPUK00005B/731

9 780750 352994